Six Sigma Case Studies with Minitab®

Six Sigma Case Studies with Minitab®

Kishore K. Pochampally
Surendra M. Gupta

CRC Press is an imprint of the
Taylor & Francis Group, an **informa** business

CRC Press
Taylor & Francis Group
6000 Broken Sound Parkway NW, Suite 300
Boca Raton, FL 33487-2742

Printed on acid-free paper
Version Date: 20140526

International Standard Book Number-13: 978-1-4822-0557-2 (Hardback)

Library of Congress Cataloging-in-Publication Data

Pochampally, Kishore K.
Six Sigma case studies with Minitab / authors, Kishore K. Pochampally, Surendra M. Gupta.
pages cm
Includes bibliographical references and index.
ISBN 978-1-4822-0557-2 (hardback)
1. Six sigma (Quality control standard)--Case studies. 2. Quality control--Data processing. 3. Minitab. I. Gupta, Surendra M. II. Title.

TS156.17.S59P63 2014
658.5'62--dc23 2013047013

Visit the Taylor & Francis Web site at
http://www.taylorandfrancis.com

and the CRC Press Web site at
http://www.crcpress.com

Dedicated to Our Families

Seema Dasgupta, Buddy Pochampally, Sharada Pochampally,

Narasimha Rao Pochampally, Sharmila Dasgupta, and Ram Dasgupta

—Kishore K. Pochampally

Sharda Gupta, Monica Gupta, and Neil Gupta

—Surendra M. Gupta

Contents

Section II Six Sigma Case Studies

Preface

Fierce competition among companies is motivating them to continuously improve customer satisfaction. Similarly, due to the numerous options available to customers today, there has been an exponential rise in customers' expectations from companies to reduce or eliminate defects, such as late delivery of a pizza, production of an unsafe car, a surgical error, long waiting time at a theme park ride, and so on. Reduction of defects invariably means improvement of process quality, and hence, increase in customer satisfaction level.

Six Sigma is one of the most widely used methodologies to improve the quality of an existing process at a company. The methodology was developed and used by Motorola in 1986, and followed by General Electric and other companies in both manufacturing and service industries. Six Sigma uses a systematic five-phase approach called DMAIC (define-measure-analyze-improve-control) to improve a process—(i) *define*: the problem faced by the process is defined in this phase; (ii) *measure*: in this phase, the current performance of the process is measured; (iii) *analyze*: this phase analyzes the process to identify the root causes of the problem; (iv) *improve*: in this phase, recommendations are made to minimize or eliminate the root causes of the problem and then those recommendations are implemented to improve the process; and (v) *control*: this phase ensures that the improved process is controlled so that the process does not slide back to the previous problem.

Minitab® is the world's leading statistical software used for quality improvement. It is user-friendly, has hundreds of sample datasets, and can perform very basic to very advanced statistical analyses of both fractional and integer data.

This book illustrates the application of Minitab® for Six Sigma projects, using case studies in a variety of sectors, including health care, manufacturing, airline, and fast food. Detailed steps and screenshots are provided to explain how to use a number of quality analysis and improvement tools in Minitab®.

Acknowledgments

We would like to thank William Gillett (Southern New Hampshire University), Paul LeBlanc (Southern New Hampshire University), Patricia Lynott (Southern New Hampshire University), Noah Japhet (Northeastern University), Cindy Renee Carelli (Taylor & Francis), Kate Gallo (Taylor & Francis), and Iris Fahrer (Taylor & Francis) for their help and support during this project.

Kishore K. Pochampally, PhD

Surendra M. Gupta, PhD

About the Authors

Kishore K. Pochampally, PhD, is an associate professor of quantitative studies, operations and project management at Southern New Hampshire University (SNHU) in Manchester (NH). His prior academic experience is as a post-doctoral fellow at Massachusetts Institute of Technology (MIT) in Cambridge (MA). He earned a PhD in industrial engineering from Northeastern University in Boston. He teaches Six Sigma courses at SNHU and conducts Lean Six Sigma training at service organizations. He is a Six Sigma Black Belt (American Society for Quality) and a Project Management Professional (PMP®).

Surendra M. Gupta, PhD, PE, is a professor of mechanical and industrial engineering and the director of the Laboratory for Responsible Manufacturing, Northeastern University. He earned his BE in electronics engineering from Birla Institute of Technology and Science, his MBA from Bryant University, and his MSIE and PhD in industrial engineering from Purdue University. He is a registered professional engineer in the state of Massachusetts. Dr. Gupta's research interests are in the areas of production/manufacturing systems and operations research. He is mostly interested in environmentally conscious manufacturing, reverse and closed-loop supply chains, disassembly modeling, and remanufacturing. He has authored or coauthored more than 450 technical papers published in books, journals, and international conference proceedings. His publications have been cited by thousands of researchers all over the world in journals, proceedings, books, and dissertations. He has traveled to all seven continents—Africa, Antarctica, Asia, Australia, Europe, North America, and South America—and presented his work at international conferences on six continents. Dr. Gupta has taught more than 100 courses in such areas as operations research, inventory theory, queueing theory, engineering economy, supply chain management, and production planning and control. He has received many recognitions, including the Outstanding Research Award and Outstanding Industrial Engineering Professor Award (in recognition of teaching excellence) from Northeastern University and the National Outstanding Doctoral Dissertation Advisor Award.

Section I

Background

1

Introduction to Six Sigma Quality

Six Sigma was developed and used by Motorola in 1986, followed by General Electric and other companies in both manufacturing and service industries. Essentially, Six Sigma is a methodology that is used to improve an existing process (manufacturing/service). This chapter illustrates a number of terms used in this methodology and also gives an introduction to the DMAIC (define-measure-analyze-improve-control) approach used to improve a process.

Section 1.1 gives definitions of some important terms with examples. Section 1.2 introduces the DMAIC approach of Six Sigma with an example. It also gives the difference between DMAIC and another less popular approach called the DMADV (define-measure-analyze-design-verify) approach. Finally, Section 1.3 gives an outline of this book.

1.1 Definitions

Normal Distribution: If the histogram for a population of data looks like a symmetrical bell curve, it is likely that the data are normally distributed. Consider the sample of 150 valve diameters shown in Table 1.1. The histogram for the data is shown in Figure 1.1. It is evident from Figure 1.1 that the data are most likely normally distributed. Notice that the histogram in Figure 1.1 shows the process mean (5.998) and the process standard deviation (0.01985) as well.

Process: A process is a set of tasks that convert inputs to outputs. For example, the process that manufactures the Toyota Camry (output) on an assembly line uses inputs such as capital, workforce, machines, facilities, and so on. If the output is a tangible product (e.g., a book by CRC Press), it is called a manufacturing process. If the output is a service (e.g., delivery of a book by Amazon.com), it is called a service process.

Process Mean: It is the average of the characteristic values of a population (e.g., an entire batch) of products produced by a process. For example, the process mean for the process that bottles 2-liter soda bottles may be 1.999 liters. The closer the process mean is to the process target, the better the process is.

TABLE 1.1
Valve Diameters

5.99	5.99	5.99	6.04	6.02	6.01	6.00	5.99	6.01	5.99
6.00	6.00	5.98	6.01	5.98	6.01	5.97	6.02	6.00	5.97
6.01	6.01	5.99	6.00	5.99	6.01	5.95	6.00	5.98	6.04
6.00	6.00	5.99	6.01	5.99	5.98	5.99	5.99	5.98	6.01
5.99	6.01	6.02	5.96	6.00	5.95	5.99	5.99	6.00	6.00
6.01	5.96	5.98	5.98	6.03	6.01	5.97	5.98	6.01	6.03
6.02	5.99	5.99	6.01	5.97	5.98	6.03	6.01	6.00	6.00
6.00	5.98	5.95	5.97	5.98	6.00	6.03	5.97	5.99	5.99
5.99	6.00	5.98	6.00	6.00	5.99	6.00	6.00	5.97	6.01
5.97	5.98	6.05	6.00	5.99	6.01	5.99	6.03	6.02	5.99
5.98	6.01	6.02	6.01	6.01	6.03	5.99	6.02	5.97	6.05
5.97	5.98	6.00	5.97	5.99	6.00	6.03	5.98	6.02	6.01
6.00	6.03	6.00	6.03	5.99	6.01	6.03	6.01	5.99	5.99
6.03	6.05	5.99	6.00	5.98	5.97	6.00	6.00	6.01	6.01
6.00	6.01	6.02	6.00	6.00	5.95	6.00	6.00	6.01	6.00

Process Standard Deviation: It measures the average deviation of the characteristic values of a population of products from the process mean. For example, the process standard deviation for the process that bottles 2-liter soda bottles may be 0.002 liters. The lower the process standard deviation, the better (i.e., more consistent) the process is.

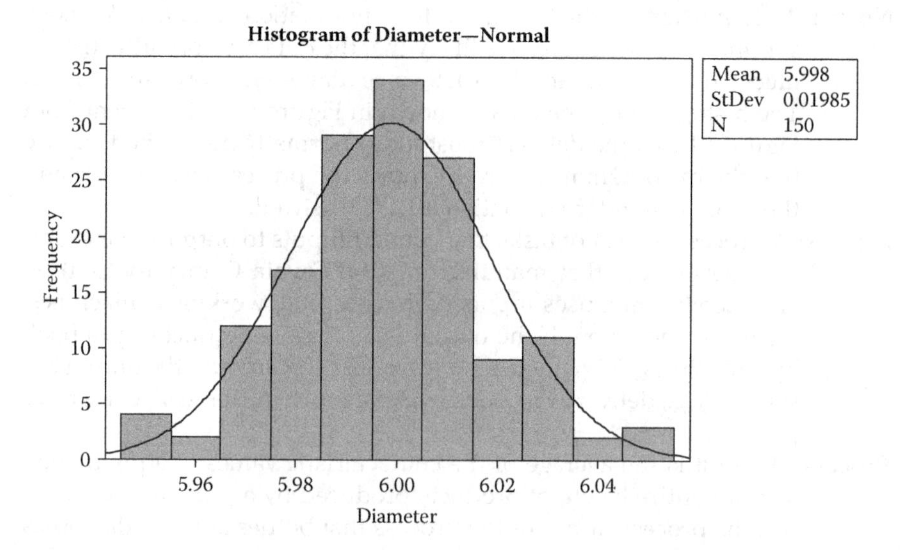

FIGURE 1.1
Histogram.

Process Target: It is the best possible value for a product characteristic. For example, the process target for a process that bottles 2-liter soda bottles is 2 liters.

Process Tolerance: It is the deviation of the product characteristic value from the process target that the customer is willing to tolerate. For example, the process tolerance for the process that bottles 2-liter soda bottles may be ±0.003 liter. This means, the lower specification limit (LSL) is 1.997 liters (i.e., 2 – 0.003) and the upper specification limit (LSL) is 2.003 liters (i.e., 2 + 0.003). In other words, the customer in this example is satisfied if the volume is between 1.997 liters and 2.003 liters.

Quality: Quality is the extent to which customers are satisfied. If most of the customers are satisfied with a product, we can say that quality is high for the product as well as for the process that produces that product.

Voice of the Customer (VOC): This term is used to define the customer's wants or needs. It includes process target and process tolerance.

Voice of the Process (VOP): This term is used to define the process performance. It includes process mean and process standard deviation.

1.1.1 Defects Per Million Opportunities (DPMO)

This is the measure of defects out of every million products produced by a process. For example, if 10 customers, out of 100 customers served on a given day, receive the pizza late, the DPMO of the pizza delivery process is calculated as follows:

$$\text{DPMO} = \frac{\text{Number of Defects}}{\text{Population Size}} * 1,000,000 = \frac{10}{100} * 1,000,000 = 100,000$$

1.1.1.1 DPMO Example 1

Consider the sample of 150 values in Table 1.1. Assume that the lower specification limit (LSL) is 5.95 and the upper specification limit (USL) is 6.05. Figure 1.2 shows the LSL and USL as well as the histogram. Calculate the DPMO of the valve manufacturing process.

1.1.1.1.1 Solution

Because this is a sample instead of a population, we use the normal curve (instead of the data themselves) in Figure 1.2 to calculate the DPMO of the valve manufacturing process. We also assume here that the population is normally distributed.

Clearly, in Figure 1.2, the proportion of defects is the sum of the area on the left side of 5.95 and the area on the right side of 6.05. Using

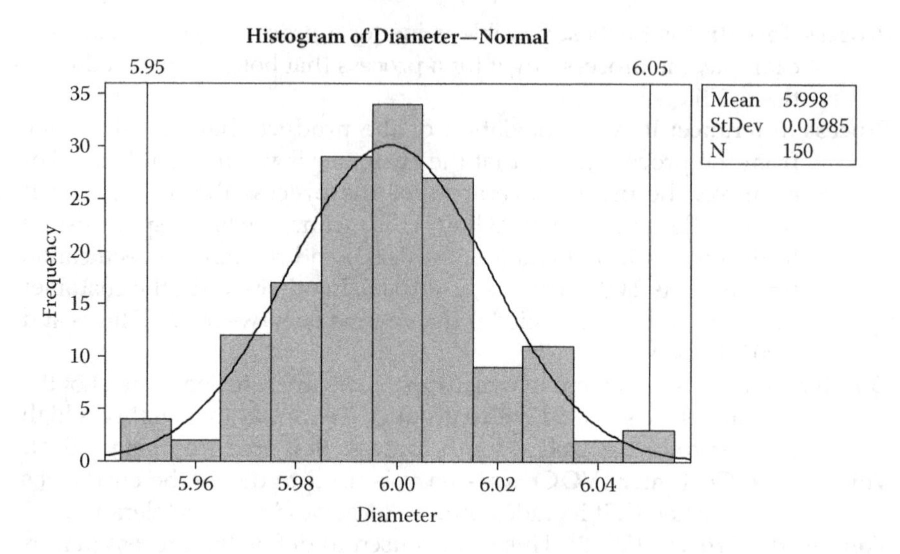

FIGURE 1.2
Histogram with specification limits.

Minitab®, the area on the left side of 5.95 can be calculated as 0.0078001 (see Figure 1.3).

Similarly, Minitab® can be used to calculate the area on the left side of 6.05 as shown in Figure 1.4. Notice in Figure 1.4, that the area on the left side of 6.05 is 0.995599. This means that the area on the right side of 6.05 is 0.004401 (i.e., 1 – 0.995599). Hence, the proportion of defectives

$$= \text{Area on the left side of } 5.95 + \text{Area on the right side of } 6.05$$
$$= 0.0078001 + 0.004401$$
$$= 0.012201$$

Hence, DPMO = 0.012201 * 1,000,000 = 12,201.

See Table 1.2 for the sigma level of this process. The higher the sigma level, the lower the DPMO value is, and hence, the better the process is. Notice that, for a Six Sigma process, the DPMO should be at most 3.4. It is impossible to

Cumulative Distribution Function

```
Normal with mean = 5.998 and standard deviation = 0.01985

   x  P( X <= x )
5.95    0.0078001
```

FIGURE 1.3
Area on the left side of LSL for DPMO Example 1.

Cumulative Distribution Function

```
Normal with mean = 5.998 and standard deviation = 0.01985

   x   P( X <= x )
6.05       0.995599
```

FIGURE 1.4
Area on the left side of USL for DPMO Example 2.

achieve a DPMO lower than or equal to 3.4 for most processes. However, the approach of Six Sigma is much more important than the numbers in Table 1.2. (Reasoning for the numbering of sigma levels (1–6) and proofs for the DPMO values of the respective sigma levels, is beyond the scope of this book.)

Based on Table 1.2 and the DPMO value (12,201) for this process, the sigma level of this process is between 3 and 4.

1.1.1.2 DPMO Example 2

Bearingo, Inc. manufactures stainless steel ball bearings 25 mm in diameter. A tolerance of 6 mm above or below 25 mm is acceptable to customers. The mean and standard deviation of the manufacturing process (assuming normal distribution) are calculated to be 28 mm and 2 mm, respectively. Calculate the defects per million opportunities (DPMO).

1.1.1.2.1 Solution

Process target = 25 mm (given)

Process tolerance = ± 6 mm (given)

Hence, LSL = 25 – 6 = 19 mm, and USL = 25 + 6 = 31 mm

Process mean = 28 mm (given)

Process standard deviation = 2 mm (given)

See Figure 1.5 for the process data distribution.

TABLE 1.2
Sigma Levels and DPMOs

Sigma Level	DPMO
1	690,000
2	308,000
3	66,800
4	6,210
5	233
6	3.4

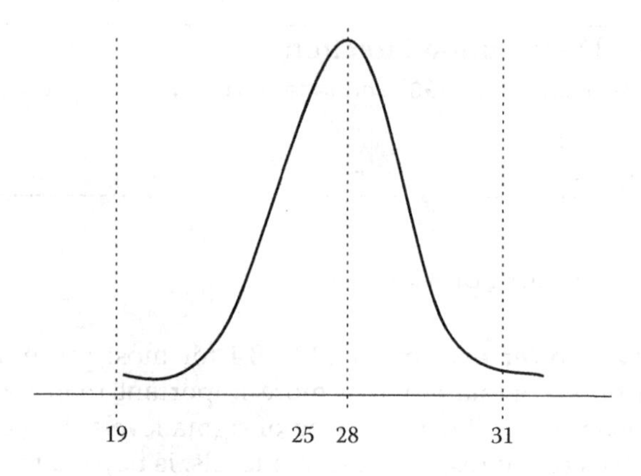

FIGURE 1.5
Process data distribution for DPMO Example 2.

It is evident from Figure 1.5 that the proportion of defects is the sum of the area on the left side of 19 and the area on the right side of 31. Using Minitab®, the area on the left side of 19 can be calculated as 0.0000034 (see Figure 1.6).

Similarly, Minitab® can be used to calculate the area on the left side of 31 as shown in Figure 1.7. Notice in Figure 1.7 that the area on the left side of 31 is 0.933193. This means that the area on the right side of 31 is 0.066807 (i.e., 1 – 0.933193). Hence, the proportion of defectives

= Area on the left side of 19 + Area on the right side of 31
= 0.0000034 + 0.066807
= 0.0668104

Hence, DPMO = 0.0668104 * 1,000,000 = 66,810.4

Based on Table 1.2 and the DPMO value (66,810.4) for this process, the sigma level of this process is very close to 3.

Cumulative Distribution Function

```
Normal with mean = 28 and standard deviation = 2

  x   P( X <= x )
 19     0.0000034
```

FIGURE 1.6
Area on the left side of LSL for DPMO Example 2.

Cumulative Distribution Function

```
Normal with mean = 28 and standard deviation = 2

  x  P( X <= x )
 31     0.933193
```

FIGURE 1.7
Area on the left side of USL for DPMO Example 2.

1.1.1.3 DPMO Example 3

Suppose that the pizza restaurant in your town delivers pizza in 1,750 seconds on average, with a standard deviation of 90 seconds. The target delivery time is 30 minutes or less. Calculate the DPMO for this delivery process. What is the sigma level of this process? Assume normal distribution for the delivery times.

1.1.1.3.1 Solution

Process target = 30 minutes = 1,800 seconds (given)

LSL = None (given)

USL = Process target = 30 minutes = 1,800 seconds (given)

Process mean = 1,750 seconds (given)

Process standard deviation = 90 seconds (given)

See Figure 1.8 for the process data distribution.

It is evident from Figure 1.8 that the proportion of defects is the area on the right side of 1,800. Minitab® can be used to calculate the area on the left side of 1,800 as shown in Figure 1.9.

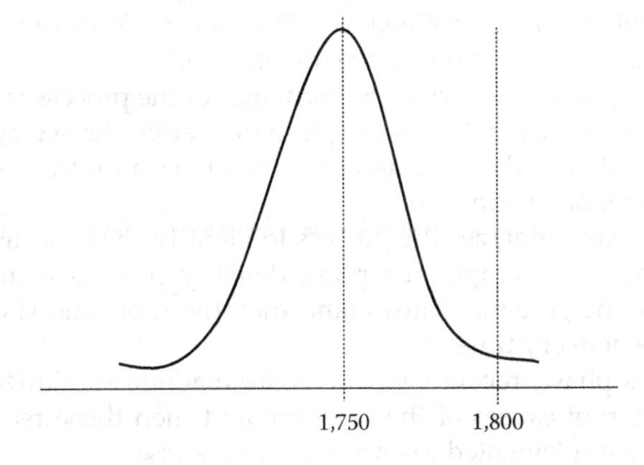

FIGURE 1.8
Process data distribution for DPMO Example 3.

Cumulative Distribution Function

```
Normal with mean = 1750 and standard deviation = 90

   x  P( X <= x )
1800     0.710743
```

FIGURE 1.9
Area on the left side of USL for DPMO Example 3.

Notice in Figure 1.9 that the area on the left side of 1,800 is 0.710743. This means that the area on the right side of 1,800 is 0.289257 (i.e., 1 – 0.710743). Hence, the proportion of defectives

= Area on the right side of 1,800

= 0.289257

Hence, DPMO = 0.289257 * 1,000,000 = 289,257.

Based on Table 1.2 and the DPMO value (289,257) for this process, the sigma level of the pizza delivery process is between 2 and 3.

1.2 DMAIC Approach

Six Sigma uses a systematic approach called DMAIC (define-measure-analyze-improve-control) to improve a given process. Each of the five phases of this approach is illustrated in Figure 1.10. See Pyzdek and Keller (2009) for an excellent introduction to the DMAIC approach.

Define: The problem is defined in this phase. For example, if the customers of a pizza delivery process are complaining about late delivery, then the problem may be defined as, "The average delivery time is long, and many of the customers are dissatisfied."

Measure: In this phase, the current performance of the process is measured. For example, after taking a sample of deliveries, the average delivery time is calculated as 41 minutes, which is 11 minutes over the process target of 30 minutes.

Analyze: This phase analyzes the process to identify the root causes of the problem. For example, the pizza delivery process is analyzed to identify the potential causes (and then the root causes) of the long average delivery time.

Improve: In this phase, recommendations are made to minimize or eliminate the root causes of the problem, and then those recommendations are implemented to improve the process.

Control: This phase ensures that the improved process is controlled so that the process does not slide back to the previous problem.

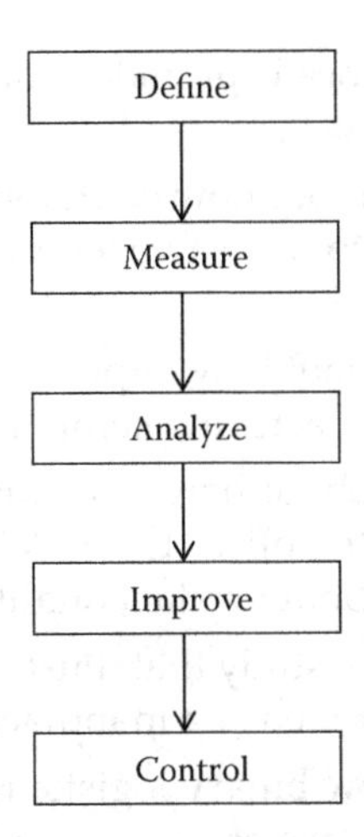

FIGURE 1.10
DMAIC approach.

Although the DMAIC approach is typically used to improve an existing process, there is a relatively less popular approach called DMADV (define-measure-analyze-design-verify) that is used to design a new process or product. In DMADV, *define* means the purpose of the new process or product, *measure* means collecting voice of the customer (VOC) data from potential customers, *analyze* means analyzing the different ways to satisfy the VOC for the new process/product and selecting the optimal way, *design* means designing the new process/product with the optimal way, and *verify* means verifying whether the VOC is actually satisfied with the new process/product. See Gitlow, Levine, and Popovich (2006) for an excellent introduction to the DMADV approach.

1.3 Book Outline

The following is the outline of the rest of this book:

Chapter 2 gives a brief definition/description of each of the following quality analysis and improvement tools/techniques used in the case studies presented in this book: confidence interval estimation, hypothesis testing, chi-square analysis, process capability analysis, binary logistic regression, item analysis, cluster analysis, mixture design and analysis of experiments, multivariate analysis, Pareto chart, cause-and-effect diagram, gage repeatability and reproducibility analysis, Taguchi design and analysis of experiments, factorial design and analysis of experiments, and statistical control charts. In addition, a useful reference is provided for the interested reader for each tool/technique.

Chapter 3 is a case study about how confidence intervals can be used in a Six Sigma project to assess variation in fat content at a fast-food restaurant.

Chapter 4 is a case study about how hypothesis testing can be used for quality control at a manufacturing company.

Chapter 5 is a case study about how chi-square analysis can be used in a Six Sigma project to collect the VOC data and then to check whether a claim made about product quality is true.

Chapter 6 is a Six Sigma case study that illustrates how process capability analysis can be performed at a manufacturing company.

Chapter 7 demonstrates how binary logistic regression can be used to predict customer satisfaction at a restaurant.

Chapter 8 is a Six Sigma case study that illustrates how item analysis and cluster analysis can be used to gather the VOC data from employees at a service firm.

Chapter 9 shows how to use the DMADV approach and mixture design and analysis of experiments to optimize pollution level and temperature of fuels.

Chapter 10 is a Six Sigma case study that demonstrates how to use multivariate analysis to reduce patient waiting time at a medical center.

Chapter 11 shows how to use a Pareto chart and fishbone diagram to minimize recyclable waste disposal in a town.

Chapter 12 illustrates how to perform gage repeatability and reproducibility analysis at a medical equipment manufacturer.

Chapter 13 is a Six Sigma case study that demonstrates how to perform Taguchi design and analysis of experiments to improve customer satisfaction at an airline company.

Chapter 14 is a Six Sigma case study that illustrates the use of factorial design and analysis of experiments to optimize a chemical process.

Finally, Chapter 15 is a Six Sigma case study that demonstrates how to perform chi-square analysis to verify source association with parts purchased and products produced at a manufacturing company.

The book also has an Appendix with the contents of the Minitab® worksheets that are referred to in most of the chapters.

Every Six Sigma project must implement the DMAIC approach to improve an existing process or the DMADV approach to design a new process/product. Although the case studies in this book (Chapters 3–15) do follow either of these five-phase approaches, not all of the five phases are explained in depth in order to allow sufficient room for significant emphasis on the Minitab® tools used in these case studies. Also, it is important to note that, although these case studies are fictitious, the problems addressed in them are often faced in the real world in a variety of industries.

References

Gitlow, Howard S., Levine, David M., and Popovich, Edward A. 2006. *Design for Six Sigma for Green Belts and Champions: Applications for Service Operations—Foundations, Tools, DMADV, Cases, and Certification*. Upper Saddle River, NJ: PH Professional Business.

Pyzdek, T. and Keller, P. 2009. *The Six Sigma Handbook*. New York: McGraw-Hill Professional.

2

Quality Analysis and Improvement Tools/Techniques Used in This Book

This chapter gives a brief definition/description of each of the quality analysis and improvement tools/techniques used in the case studies presented in this book. Because this is not a statistics textbook, in-depth coverage of each of these tools or techniques is beyond the scope of the book. However, a useful reference is provided for the interested reader for each tool and technique.

2.1 Confidence Interval Estimation

Confidence interval estimation is a technique to estimate a population parameter (such as population proportion) using sample data. The estimate is calculated for a given confidence level and is expressed as an interval. The higher the confidence level is, the less precise the interval estimate. See Montgomery and Runger (2011) for an excellent introduction to confidence interval estimation.

An application of this technique to Six Sigma is illustrated using Minitab® in Chapter 3.

2.2 Hypothesis Testing

Hypothesis testing is a technique to test whether there is enough statistical evidence to reject a claim. Typically, the claim is expressed as the "null hypothesis," and an "alternative hypothesis" is considered to verify which of these two hypotheses is true. These two hypotheses are mutually exclusive (if one is true, the other one is not) and collectively exhaustive (no other hypothesis is possible). Hypothesis testing is explained in detail in Montgomery and Runger (2011).

An application of this technique to Six Sigma is illustrated using Minitab® in Chapter 4.

2.3 Chi-Square Analysis

Chi-square analysis is a type of hypothesis testing where a sample statistic (called chi-square value) used in the test is assumed to follow a chi-square distribution. This technique is explained in detail in Black (2011).

Applications of this technique to Six Sigma are illustrated using Minitab® in Chapters 4 and 15.

2.4 Process Capability Analysis

If *USL* is the upper specification limit for a process, *LSL* is the lower specification limit for a process, μ is the process mean, and σ is the process standard deviation, the following process capability ratios can measure process performance:

First-generation process capability ratio,

$$C_p = \frac{USL - LSL}{6\sigma}$$

Second-generation process capability ratio with respect to *LSL*,

$$C_{pl} = \frac{\mu - LSL}{3\sigma}$$

Second-generation process capability ratio with respect to *USL*,

$$C_{pu} = \frac{USL - \mu}{3\sigma}$$

Second-generation process capability ratio,

$$C_{pk} = \text{MINIMUM of } \{C_{pl}, C_{pu}\}$$

The higher the C_p and C_{pk} values are, the better the process is. See Ryan (2011) for further explanation of these ratios.

An application of the process capability ratios to Six Sigma is illustrated using Minitab® in Chapter 6.

2.5 Binary Logistic Regression

Binary logistic regression is a technique used to predict the outcome of a binary categorical variable with exactly two possible outcomes (e.g., Yes or No for whether a product is defective). This technique is explained in detail in Black (2011).

An application of this technique to Six Sigma is illustrated using Minitab® in Chapter 7.

2.6 Item Analysis

Item analysis is used to check whether there is a correlation among categorical responses to multiple questions in a customer survey. See Boslaugh (2012) for an excellent introduction to item analysis. An application of this technique to Six Sigma is illustrated using Minitab® in Chapter 8.

2.7 Cluster Analysis

Cluster analysis helps group customers into various clusters, using coordinate systems and Euclidean distances. See Boslaugh (2012) for a thorough introduction to cluster analysis.

An application of this technique to Six Sigma is illustrated using Minitab® in Chapter 8.

2.8 Mixture Design and Analysis of Experiments

Mixture design and analysis of experiments is a technique used to optimize the proportion of each of the components of a mixture such as a fuel mixture or a juice blend. This technique is explained in detail in Perry and Bacon (2006).

An application of this technique to Six Sigma is illustrated using Minitab® in Chapter 9.

2.9 Multivariate Analysis

Multivariate analysis is a technique that analyzes three or more variables at a time. The dependent variables are typically called "response variables" and the independent variables are typically called "factors." This technique is explained in detail in Ryan (2011).

An application of this technique to Six Sigma is illustrated using Minitab® in Chapter 10.

2.10 Pareto Chart

The Pareto chart is a tool that can prioritize problems using their frequencies. Essentially, it is used to identify what few problems are causing the largest negative impact. This tool is explained in detail in Ryan (2011).

An application of this technique to Six Sigma is illustrated using Minitab® in Chapter 11.

2.11 Cause-and-Effect Diagram

The cause-and-effect diagram is a tool that can help identify root causes of a problem ("effect"). It can also be used to divide numerous causes of a problem into several categories for easier analysis. This tool is explained in detail in Ryan (2011).

An application of this technique to Six Sigma is illustrated using Minitab® in Chapter 11.

2.12 Gage Repeatability and Reproducibility Analysis

Gage repeatability and reproducibility (R&R) analysis is a technique that can verify whether the measurement or inspection system used to collect the data (e.g., number of defectives in a sample) is efficient. Repeatability measures the consistency of an inspector with himself or herself. Reproducibility measures the consistency of an inspector with other inspectors. This technique is explained in detail in Gitlow and Levine (2004).

An application of this technique to Six Sigma is illustrated using Minitab® in Chapter 12.

2.13 Taguchi Design and Analysis of Experiments

Taguchi design and analysis of experiments is a technique that uses signal-to-noise ratios to check whether the variable for improvement depends on a certain set of controllable factors amid a set of uncontrollable ("noise") factors. This technique is explained in detail in Ryan (2011).

An application of this technique to Six Sigma is illustrated using Minitab® in Chapter 13.

2.14 Factorial Design and Analysis of Experiments

Factorial design and analysis of experiments is a technique that can statistically verify whether the output variable depends on a particular input variable. It also helps check if there is interaction among the input variables. This technique is explained in detail in Montgomery and Runger (2011).

An application of this technique to Six Sigma is illustrated using Minitab® in Chapter 14.

2.15 Statistical Control Charts

It is important to verify whether a process is stable and predictable before measuring the performance of the process. Statistical control charts help in such verification. A variety of statistical control charts is explained in detail in Montgomery and Runger (2011).

Applications of some of the statistical control charts to Six Sigma are illustrated using Minitab® in Chapters 3, 6, and 12.

2.16 Normality Test

The normality test is a tool to verify whether the collected data are normally distributed. This verification is important because many of the tools (e.g., statistical control charts; see Section 2.15) assume that the collected data are normally distributed. See Wedgwood (2006) for an excellent explanation of the normality test.

Applications of this tool to Six Sigma are illustrated using Minitab® in Chapters 3, 4, and 6.

2.17 Analysis of Variance (ANOVA)

Analysis of variance (ANOVA) is a technique that is often used in conjunction with factorial design and analysis of experiments (see Section 2.14) to compare multiple population means while minimizing the error of rejecting a null hypothesis (see Section 2.2) when the null hypothesis is true. This technique is explained in detail in Black (2011).

An application of this technique to Six Sigma is illustrated using Minitab® in Chapter 14.

References

Black, Ken. 2011. *Business Statistics: For Contemporary Decision Making*. Somerset, NJ: John Wiley & Sons. 7th ed.

Boslaugh, Sarah. 2012. *Statistics in a Nutshell*. Sebastopol, CA: O'Reilly Media, Inc. 2nd ed.

Gitlow, Howard S. and Levine, David M. 2004. *Six Sigma for Green Belts and Champions: Foundations, DMAIC, Tools, Cases, and Certification*. Upper Saddle River, NJ: PH Professional Business.

Montgomery, Douglas C. and Runger, George C. 2011. *Applied Statistics and Probability for Engineers*. Somerset, NJ: John Wiley & Sons. 5th ed.

Perry, Randy C. and Bacon, David W. 2006. *Commercializing Great Products with Design for Six Sigma*. Upper Saddle River, NJ: Prentice Hall.

Ryan, Thomas P. 2011. *Statistical Methods for Quality Improvement*. Somerset, NJ: John Wiley & Sons. 3rd ed.

Wedgwood, I. 2006. *Lean Sigma: A Practitioner's Guide*. Upper Saddle River, NJ: Prentice Hall.

Section II

Six Sigma Case Studies

3

Confidence Intervals to Assess Variation in Fat Content at a Fast-Food Restaurant

Hamburrgerr, Inc. is a fast-food restaurant serving hamburgers among a few other items. The restaurant claims that the average fat content in the hamburgers is 15 grams.

Section 3.1 gives a brief description of the define phase. Section 3.2 illustrates the measure phase with detailed instructions for using Minitab®. The analyze phase is briefly discussed in Section 3.3. Section 3.4 illustrates the improve phase with detailed instructions for using Minitab®. Finally, the control phase is briefly discussed in Section 3.5.

3.1 Define Phase

A few of the customers complained to the operations manager recently that the fat content in the hamburgers appeared to be higher than the restaurant's claim of 15 grams. The operations manager wishes to use 95% confidence intervals to verify whether the restaurant's claim of 15 grams (average) of fat in the hamburgers is correct. He also wishes to verify his assumption that the standard deviation of the fat content is less than 1 gram.

3.2 Measure Phase

The operations manager randomly selects 20 hamburgers and measures the fat content (in grams) in each of them as follows: 15.5, 12.3, 15.4, 16.5, 15.9, 17.1, 16.9, 14.3, 19.1, 18.2, 18.5, 16.3, 20.0, 19.0, 15.6, 13.5, 14.0, 16.5, 19.0, and 18.6.

Open the CHAPTER_3_BEFORE.MTW worksheet that has the above data (the worksheet is available at the publisher's website; the data from the worksheet are also provided in the Appendix). Before constructing a confidence interval for the above data, it is important to check whether the data are in statistical control. Because each number is for one hamburger, the appropriate set of control charts is I-MR (individual and moving range) charts. Figure 3.1 shows how to select I-MR charts in Minitab®. Doing so will open

	C1
	Fat
1	15.5
2	12.3
3	15.4
4	16.5
5	15.9
6	17.1
7	16.9
8	14.3
9	19.1
10	18.2
11	18.5
12	16.3
13	20.0
14	19.0
15	15.6
16	13.5
17	14.0
18	16.5

FIGURE 3.1
Selection of "I-MR Chart" before improvement.

the dialog box shown in Figure 3.2. Select "Fat" for "Variable", and click on "OK." The I-MR charts shown in Figure 3.3 will be the result. As is evident, the data are in statistical process control.

Now that the data are in statistical control, the operations manager wishes to construct the 95% confidence interval for the mean of fat content in

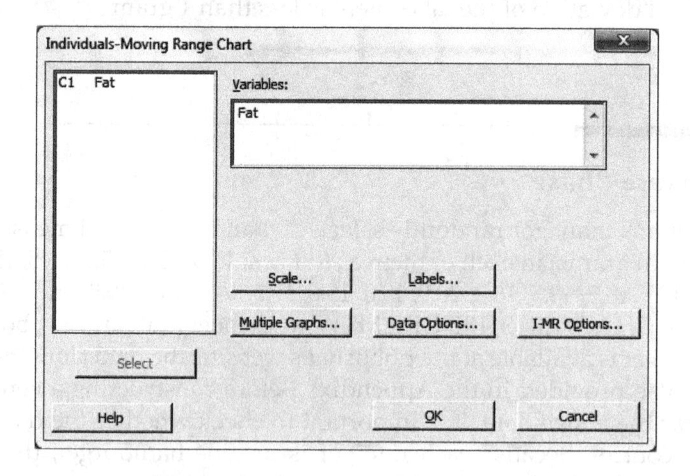

FIGURE 3.2
Variable of "Fat" for I-MR chart before improvement.

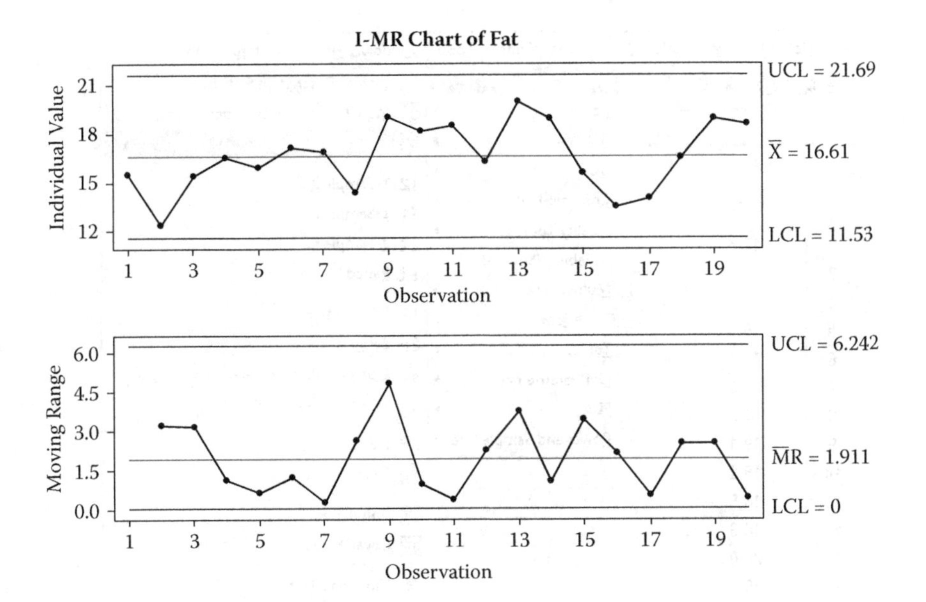

FIGURE 3.3
I-MR charts before improvement.

population of the hamburgers. He also wishes to check whether the data are normally distributed before relying on the confidence interval. The "Graphical Summary" option in Minitab® produces both the normality test result as well as the confidence interval. Figure 3.4 shows to how to select the graphical summary. Doing so will open the dialog box shown in Figure 3.5. Select "Fat" for "Variable" and click on "OK". The graphical summary shown in Figure 3.6 will be the result. Because the *P*-value (0.624) of the Anderson–Darling normality test is greater than 0.05, the data are normally distributed. Notice that the population mean is estimated to be between 15.625 grams and 17.595 grams. The entire confidence interval (15.625, 17.595) is greater than the claim of 15 grams, therefore the operations manager considers this a serious issue.

The operations manager then proceeds to verify his assumption that the standard deviation of the fat content is less than 1 gram. Figure 3.7 shows how to select "1 Variance" in Minitab®. Doing so will open the dialog box shown in Figure 3.8. Select "Samples in columns" for "Data" and "Fat" for "columns". Click on "Options" and it opens the dialog box shown in Figure 3.9. Select "less than" for "Alternative" because the operations manager is interested in the standard deviation being less than 1 gram. Click on "OK" and it takes you back to the dialog box shown in Figure 3.8. Click on "OK" and the output shown in Figure 3.10 is the result. Because the data are normally distributed, we must look at the upper bound (2.89 grams) of the standard deviation given by the chi-square method. This means that 95% of

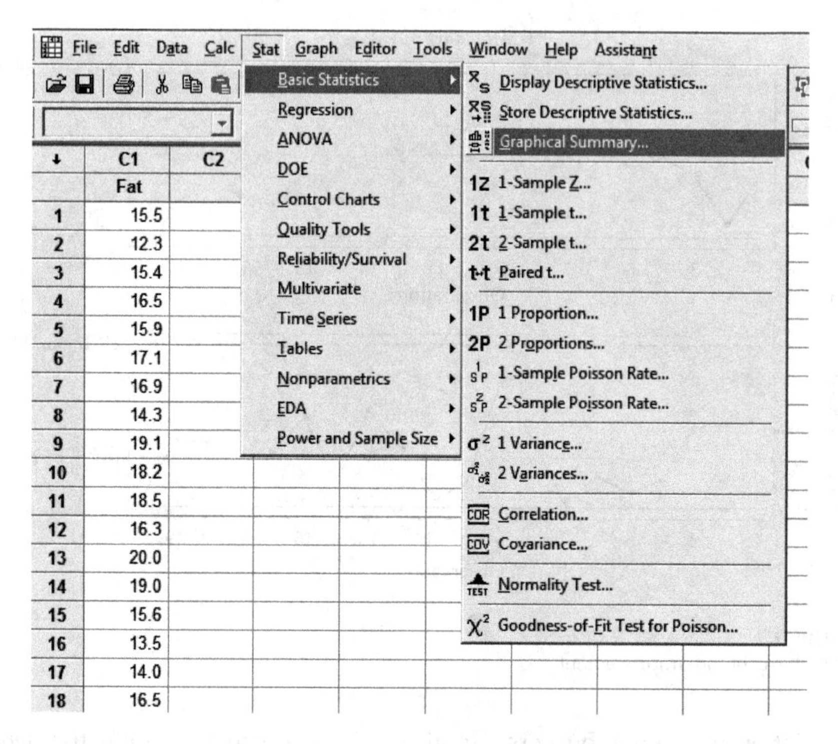

FIGURE 3.4
Selection of "Graphical Summary" before improvement.

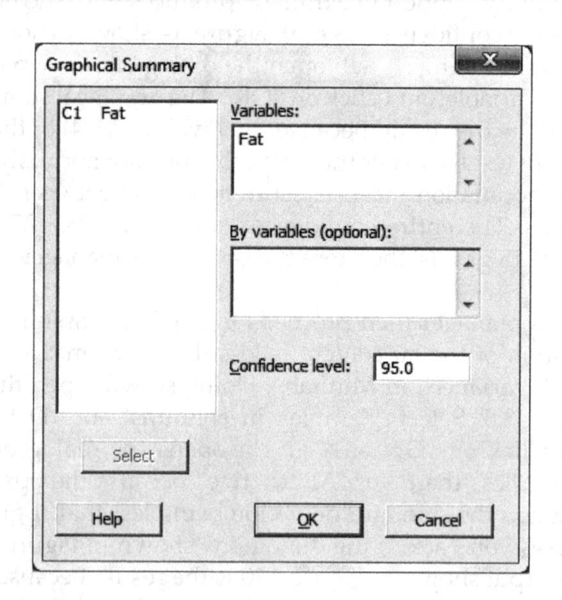

FIGURE 3.5
Variable of "Fat" for graphical summary before improvement.

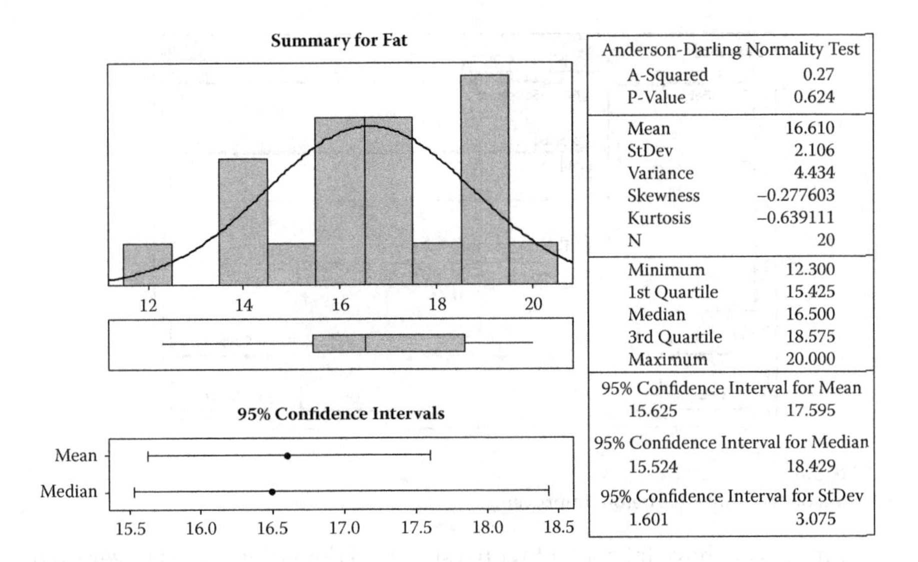

FIGURE 3.6
Graphical summary before improvement.

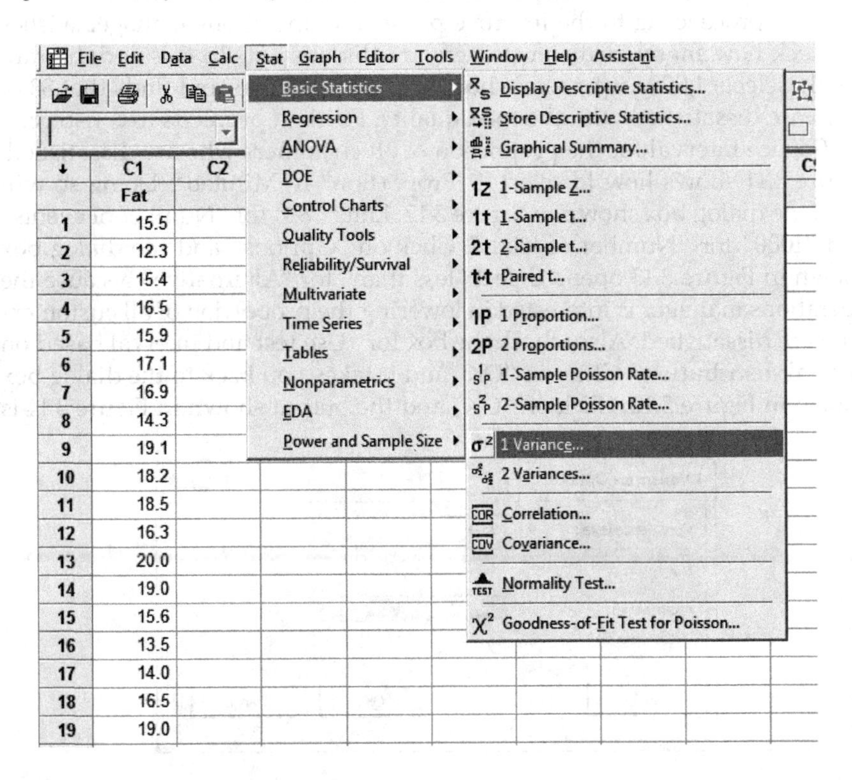

FIGURE 3.7
Selection of "1 Variance" before improvement.

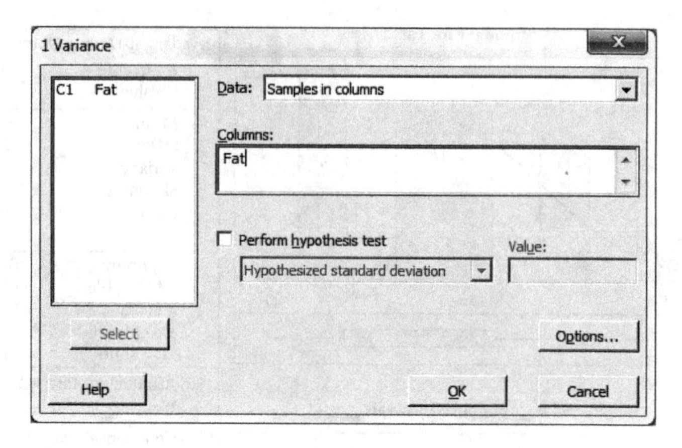

FIGURE 3.8
Selection of "Fat" variable before improvement.

the hamburgers have fat content with a standard deviation that is lower than 2.89 grams. Inasmuch as 2.89 grams is much greater than the assumption of 1 gram, the operations manager considers this also a serious issue.

Before proceeding to the improve phase, the operations manager wishes to check how many of the customers are dissatisfied. To this end, he randomly selects 1,000 customers who bought hamburgers and finds that 83 of them are dissatisfied with the food quality. He then proceeds to construct a confidence interval for the proportion of all customers who are dissatisfied. Figure 3.11 shows how to select "1 Proportion" in Minitab®. Doing so will open the dialog box shown in Figure 3.12. Enter "83" for "Number of events" and "1000" for "Number of trials". Click on "Options" and the dialog box shown in Figure 3.13 opens. Select "less than" for "Alternative" because the operations manager is interested in lowering the proportion of all customers who are dissatisfied. Also, check the box for "Use test and interval based on normal distribution". Click on "OK" and it takes you back to the dialog box shown in Figure 3.12. Click on "OK" and the output shown in Figure 3.14 is

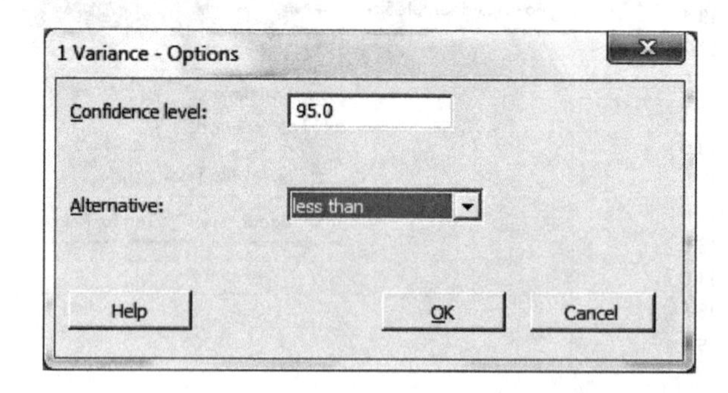

FIGURE 3.9
Selection of "less than" option for estimating standard deviation before improvement.

Test and CI for One Variance: Fat

```
Method
The chi-square method is only for the normal distribution.
The Bonett method is for any continuous distribution.

Statistics
Variable    N  StDev  Variance
Fat        20   2.11      4.43


95% One-Sided Confidence Intervals

                          Upper
                          Bound
                            for   Upper Bound
Variable  Method          StDev  for Variance
Fat       Chi-Square       2.89          8.33
          Bonett           2.81          7.92
```

FIGURE 3.10
One-sided 95% confidence interval for standard deviation before improvement.

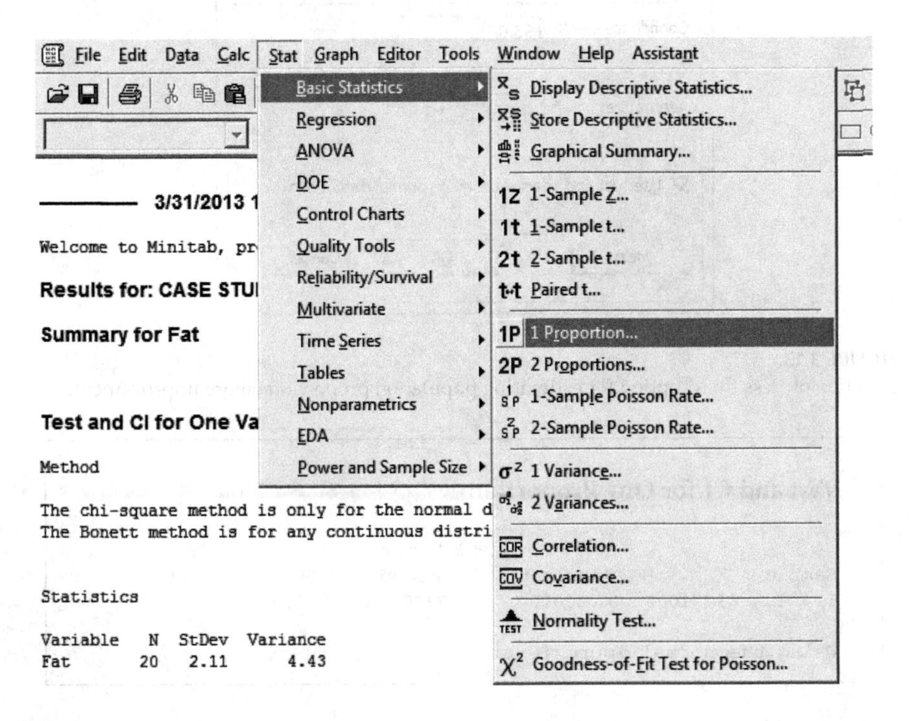

FIGURE 3.11
Selection of "1 Proportion" before improvement.

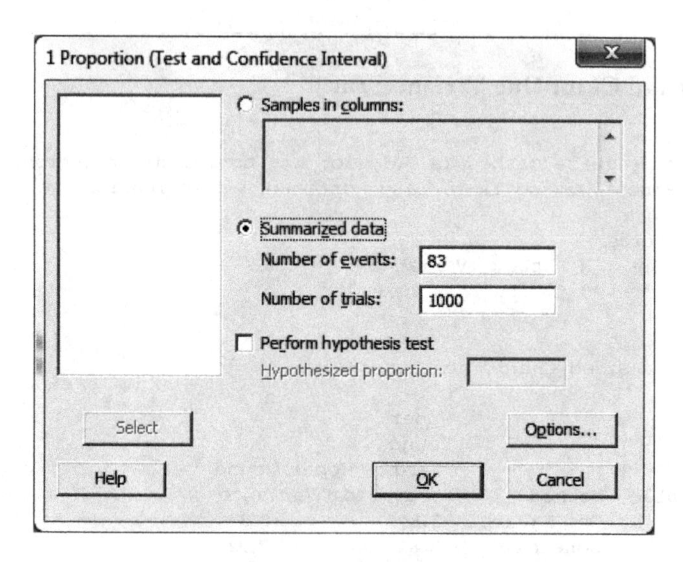

FIGURE 3.12
Entry of "Number of events" and "trials" before improvement.

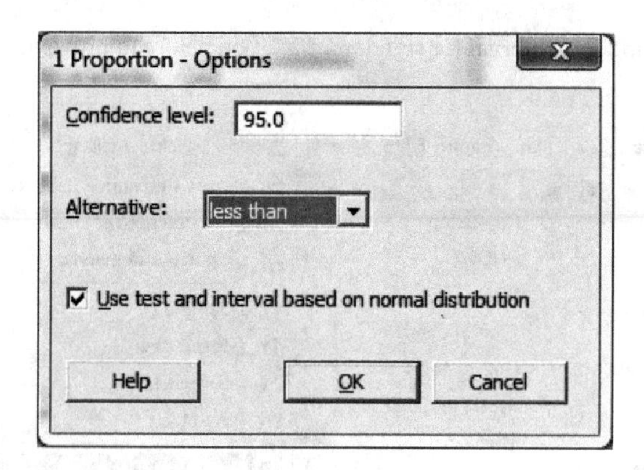

FIGURE 3.13
Selection of "less than" option for estimating population proportion before improvement.

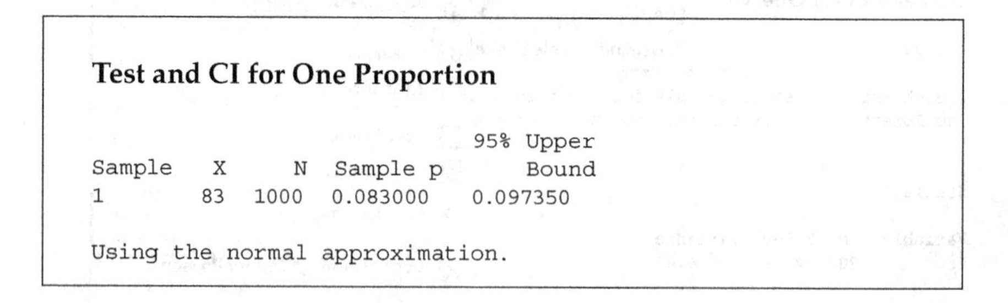

Test and CI for One Proportion

```
                                95% Upper
Sample   X     N  Sample p      Bound
1       83  1000  0.083000   0.097350

Using the normal approximation.
```

FIGURE 3.14
One-sided 95% confidence interval for population proportion before improvement.

the result. It is clear that that there is a 95% probability that the proportion of all customers who are dissatisfied is less than 0.097 (or 9.7%). The operations manager wishes to lower this upper bound estimate.

3.3 Analyze Phase

The operations manager analyzes the process and discovers that the fat content and its variation are highly affected by the amount of oil used by the employees on the three different grills used to make hamburgers.

3.4 Improve Phase

The operations manager standardizes the process so that the use of oil on the three different grills is controlled. An automatic machine is bought and installed to dispense oil on a grill each time a hamburger is made.

In order to check whether the process has really improved, the operations manager randomly selects 20 hamburgers and measures the fat content (in grams) in each of them as follows: 14.9, 15.0, 15.4, 15.3, 15.2, 15.1, 14.9, 14.8, 15.6, 14.5, 15.3, 15.8, 15.0, 15.0, 14.3, 15.3, 15.2, 14.7, 15.1, and 14.7.

Open the CHAPTER_3_AFTER.MTW worksheet that has the above data (the worksheet is available at the publisher's website; the data from the worksheet are also provided in the Appendix). Before constructing a confidence interval for the above data, it is important to check whether the data are in statistical control. Because each number is for one hamburger, the appropriate set of control charts is I-MR (individual moving range) charts. The I-MR charts for the above data are shown in Figure 3.15. As is evident, the data are in statistical process control.

Now that the data are in statistical control, the operations manager wishes to construct the 95% confidence interval for the mean of fat content in the population of hamburgers. He also wishes to check whether the data are normally distributed before relying on the confidence interval. Figure 3.16 shows the graphical summary. Because the *P*-value (0.925) of the Anderson–Darling normality test is greater than 0.05, the data are normally distributed. Notice that the population mean is estimated to be between 14.887 grams and 15.223 grams. Inasmuch as the claim of 15 grams is within the interval (14.887, 15.223), it is clear that the process has improved with respect to the mean fat content.

The operations manager then proceeds to verify his assumption that the standard deviation of the fat content is less than 1 gram. As is clear from the

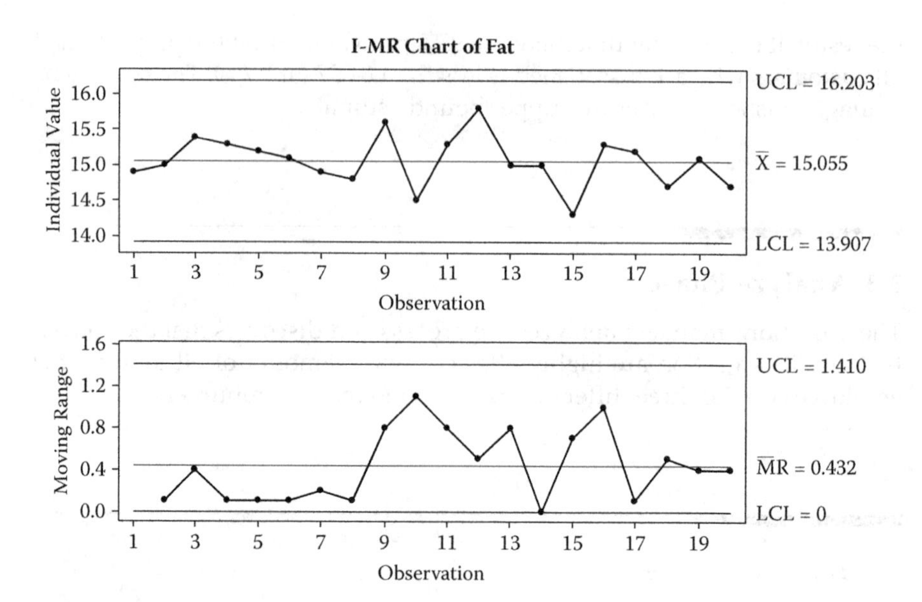

FIGURE 3.15
I-MR charts after improvement.

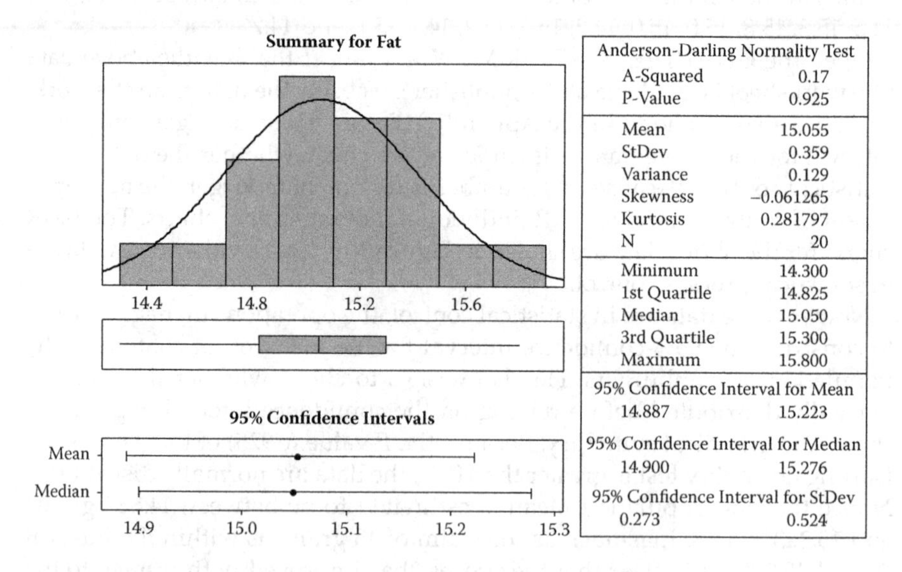

FIGURE 3.16
Graphical summary after improvement.

```
Test and CI for One Variance: Fat

Method

The chi-square method is only for the normal distribution.
The Bonett method is for any continuous distribution.


Statistics

Variable   N   StDev   Variance
Fat       20   0.359      0.129


95% One-Sided Confidence Intervals

                                Upper Bound   Upper Bound
Variable   Method                for StDev   for Variance
Fat        Chi-Square                0.492          0.242
           Bonett                    0.504          0.254
```

FIGURE 3.17
One-sided 95% confidence interval for standard deviation after improvement.

output shown in Figure 3.17, the upper bound is 0.492 gram for the standard deviation (given by the chi-square method). This means that 95% of the hamburgers have fat content with a standard deviation that is lower than 0.492 gram. Because 0.492 gram is less than the assumption of 1 gram, it is evident that the process has improved with respect to the standard deviation as well.

One can also simulate confidence intervals using Minitab®. The operations manager wishes to perform the simulation for 10 normally distributed samples with a sample size of 20 hamburgers. Doing what is shown in Figure 3.18 opens the dialog box shown in Figure 3.19. Select "20" for "Number of rows of data to generate". Enter "C3-C12" for "Store in column(s)". Because, after process improvement, the population mean is approximately 15 grams and the population standard deviation is approximately 0.5 grams, enter "15" for "mean" and "0.5" for "Standard deviation". Click on "OK" and the data shown in Figure 3.20 are the result.

Before proceeding to simulation, it is a good idea to check whether the data generated are in statistical control. The sample size is greater than 10, therefore the appropriate control charts are Xbar-S charts (sample means and sample standard deviations). Figure 3.21 shows how to select "Xbar-S". Doing so will open the dialog box shown in Figure 3.22. Select "Observations for a subgroup are in one row of columns" from the drop-down menu, and select "C3-C12" in the empty box below the menu. Click on "OK" and the Xbar-S charts shown in Figure 3.23 are the result. It is clear that the data are in statistical control.

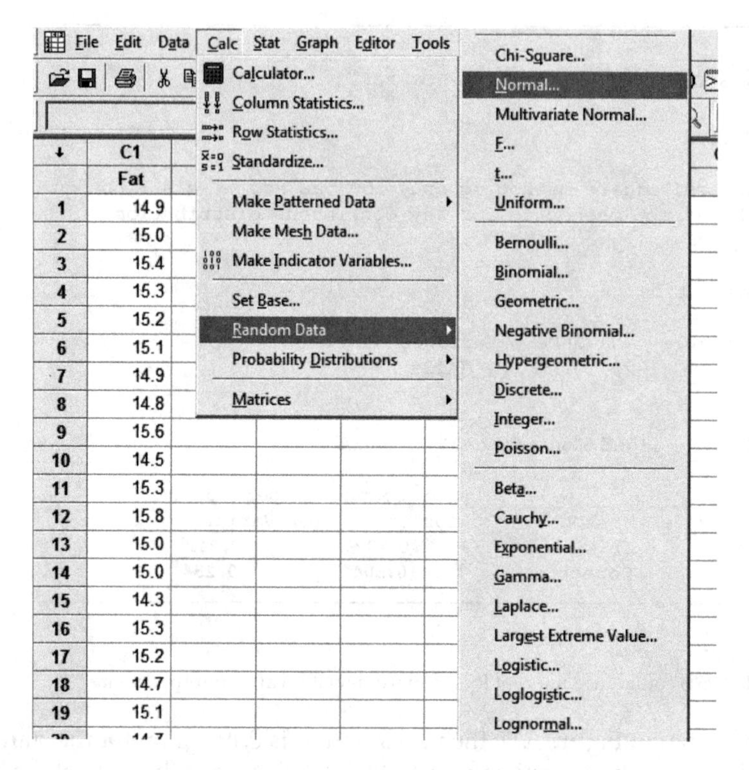

FIGURE 3.18
Selection of "Normal" to generate normally distributed data.

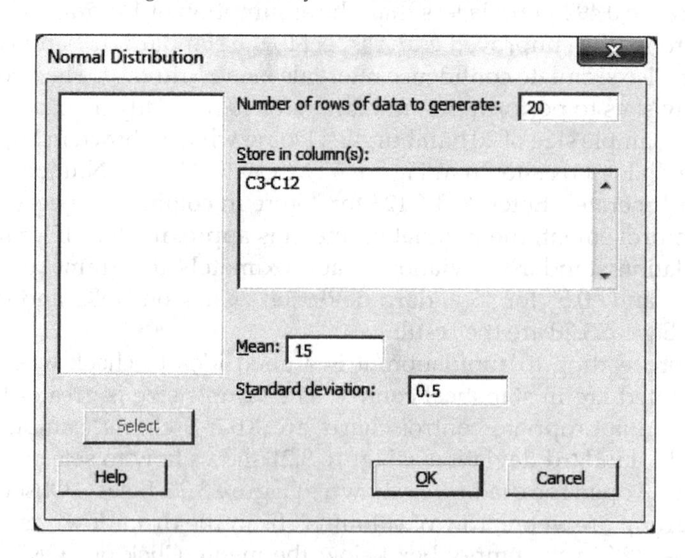

FIGURE 3.19
Mean and standard deviation for generation of random samples.

Stat Graph Editor Tools Window Help Assistant

C3	C4	C5	C6	C7	C8	C9	C10	C11	C12
14.5005	15.0615	15.3180	15.0453	15.6137	15.2391	15.1003	14.4816	15.3697	15.5550
15.8619	15.0352	15.4866	15.4720	14.7087	15.0590	15.7989	14.9403	14.5047	14.8268
16.0734	15.5290	14.9963	14.4413	15.1784	14.5252	15.6091	14.7555	15.1575	15.2456
15.0604	14.0656	15.9527	14.7279	15.4551	15.2834	14.6864	14.7153	15.6713	14.9993
15.0238	14.9675	14.4500	14.9712	15.9254	14.9253	14.8719	15.3898	14.6208	15.9183
15.5331	14.9850	14.8421	15.1611	15.1169	14.4918	14.6385	15.0241	15.2196	15.5929
15.0069	14.9272	14.9978	14.2482	14.5279	14.4722	14.5577	14.9531	15.5514	16.3353
15.6438	15.3550	15.5513	15.0008	14.7496	14.3828	15.8827	13.4386	15.5228	15.2980
14.8808	15.2983	14.9399	14.6704	15.5762	14.7267	14.9465	15.3734	15.1594	15.2588
14.7897	15.5399	14.7708	15.3025	15.0310	14.4046	14.8088	14.6081	15.3152	15.1458
14.5837	15.7337	15.0453	15.1426	14.7760	14.9770	14.6273	14.3773	15.9340	14.2630
14.5624	13.9011	14.5378	15.4423	15.4841	15.7054	14.7800	15.6339	15.2047	15.1385
14.9980	15.5109	14.8571	15.4504	14.8212	15.6376	15.4373	15.1999	14.5589	15.0355
15.4724	15.6305	16.1623	14.8468	16.2342	14.2581	13.8463	14.7384	15.0293	15.0925
14.2872	15.8478	15.1231	15.6011	14.8971	15.5123	14.2790	15.4921	14.7830	15.5205
14.5219	16.2434	14.5973	15.1377	15.5711	15.7814	14.4165	15.1727	15.5683	15.7207
16.1166	14.7879	15.8140	14.3520	15.5345	14.7504	15.5359	14.8842	15.8614	15.5887
14.7562	14.9886	14.9110	15.0878	14.9427	15.3621	15.8416	15.3940	15.4287	14.7872
15.2872	14.7014	14.5969	15.7314	16.6402	15.3543	14.5002	15.0465	15.1714	15.0964

FIGURE 3.20
Ten random samples with sample size of 20.

For simulating the confidence intervals for the 10 samples generated, select "Interval Plot" as shown in Figure 3.24. Doing so opens the dialog box shown in Figure 3.25. Select "Simple" under "Multiple Ys" and click on "OK". That will open the dialog box shown in Figure 3.26. Select "C3-C12" columns for "Graph variables" and click on "OK". The intervals plot (95% confidence intervals) for the 10 samples shown in Figure 3.27 are created. As shown in Figure 3.28, right-click anywhere on the interval plot and select "Reference Lines". This opens the dialog box shown in Figure 3.29. Enter "15" for "Show reference lines at Y values" and click on "OK". This adds the reference line of 15, as shown in Figure 3.30. Because most of the confidence intervals contain the population mean of 15 grams, this simulation confirms that the population mean estimate is indeed 15 grams. If the simulation is performed for a large number of samples, 95% of the confidence intervals are expected to contain the population mean of 15 grams.

The operations manager now wishes to check how many of the customers are dissatisfied after the process has improved. To this end, he randomly selects 1,200 customers who bought hamburgers and finds that 25 of them are dissatisfied with the food quality. He then proceeds to construct a confidence interval for the proportion of all customers who are dissatisfied. In the dialog box shown in Figure 3.31, enter "25" for "Number of events" and

Stat Graph Editor Tools Window Help Assistant

Basic Statistics, Regression, ANOVA, DOE, Control Charts, Quality Tools, Reliability/Survival, Multivariate, Time Series, Tables, Nonparametrics, EDA, Power and Sample Size

Box-Cox Transformation..., Variables Charts for Subgroups, Variables Charts for Individuals, Attributes Charts, Time-Weighted Charts, Multivariate Charts, Rare Event Charts

Xbar-R..., Xbar-S..., I-MR-R/S (Between/Within)..., Xbar..., R..., S..., Zone...

C3	C4	C5	C6	C7	C8	C9	C10	C11	C12
14.7897	15.5399	14.7708	15.3025	15.0310	14.4046	[illegible]	[illegible]	[illegible]	15.1458
14.5837	15.7337	15.0453	15.1426	14.7760	14.9770	14.6273	14.3773	15.9340	14.2630
14.5624	13.9011	14.5378	15.4423	15.4841	15.7054	14.7800	15.6339	15.2047	15.1385
14.9980	15.5109	14.8571	15.4504	14.8212	15.6376	15.4373	15.1999	14.5589	15.0355
15.4724	15.6305	16.1623	14.8468	16.2342	14.2581	13.8463	14.7384	15.0293	15.0925
14.2872	15.8478	15.1231	15.6011	14.8971	15.5123	14.2790	15.4921	14.7830	15.5205
14.5219	16.2434	14.5973	15.1377	15.5711	15.7814	14.4165	15.1727	15.5683	15.7207
16.1166	14.7879	15.8140	14.3520	15.5345	14.7504	15.5359	14.8842	15.8614	15.5887
14.7562	14.9886	14.9110	15.0878	14.9427	15.3621	15.8416	15.3940	15.4287	14.7872
15.2872	14.7014	14.5969	15.7314	16.6402	15.3543	14.5002	15.0465	15.1714	15.0964

FIGURE 3.21
Selection of "Xbar-S".

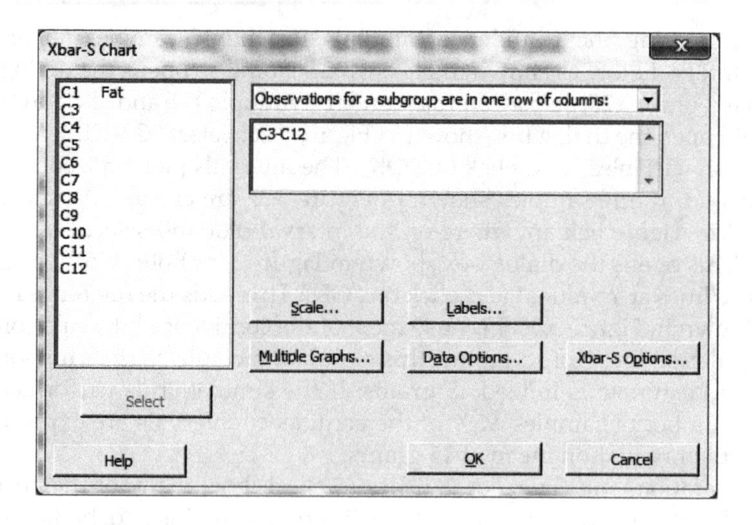

FIGURE 3.22
Selection of variables for Xbar-S charts.

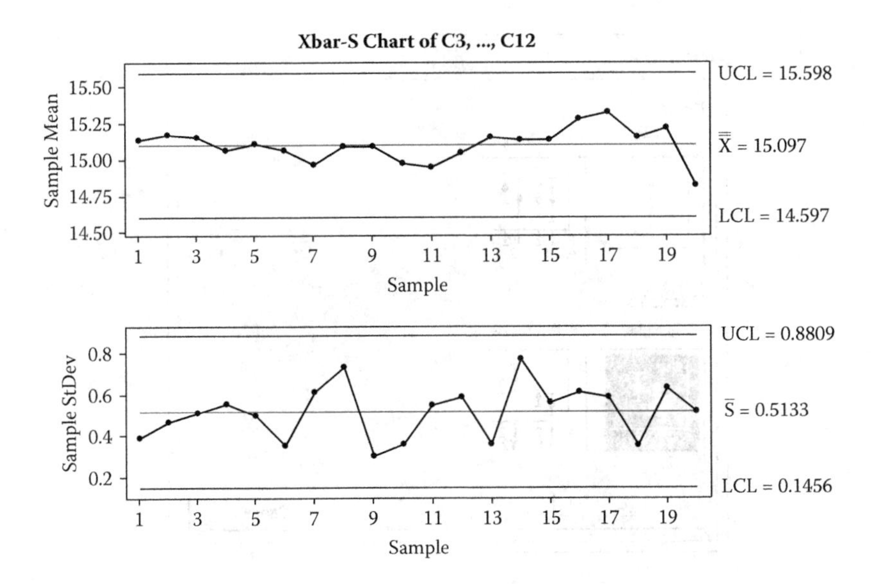

FIGURE 3.23
Xbar-S charts.

Graph Editor Tools Window Help Assistant

- Scatterplot...
- Matrix Plot...
- Marginal Plot...
- Histogram...
- Dotplot...
- Stem-and-Leaf...
- Probability Plot...
- Empirical CDF...
- Probability Distribution Plot...
- Boxplot...
- Interval Plot...
- Individual Value Plot...
- Line Plot...
- Bar Chart...
- Pie Chart...
- Time Series Plot...
- Area Graph...
- Contour Plot...
- 3D Scatterplot...
- 3D Surface Plot...

				C7	C8	C9	C10	C11	C12
			53	15.6137	15.2391	15.1003	14.4816	15.3697	15.5550
			20	14.7087	15.0590	15.7989	14.9403	14.5047	14.8268
			13	15.1784	14.5252	15.6091	14.7555	15.1575	15.2456
			79	15.4551	15.2834	14.6864	14.7153	15.6713	14.9993
			12	15.9254	14.9253	14.8719	15.3898	14.6208	15.9183
			11	15.1169	14.4918	14.6385	15.0241	15.2196	15.5929
			82	14.5279	14.4722	14.5577	14.9531	15.5514	16.3353
			08	14.7496	14.3828	15.8827	13.4386	15.5228	15.2980
			04	15.5762	14.7267	14.9465	15.3734	15.1594	15.2588
			25	15.0310	14.4046	14.8088	14.6081	15.3152	15.1458
			26	14.7760	14.9770	14.6273	14.3773	15.9340	14.2630
			23	15.4841	15.7054	14.7800	15.6339	15.2047	15.1385
			04	14.8212	15.6376	15.4373	15.1999	14.5589	15.0355
			68	16.2342	14.2581	13.8463	14.7384	15.0293	15.0925
			11	14.8971	15.5123	14.2790	15.4921	14.7830	15.5205
			77	15.5711	15.7814	14.4165	15.1727	15.5683	15.7207
			20	15.5345	14.7504	15.5359	14.8842	15.8614	15.5887
			78	14.9427	15.3621	15.8416	15.3940	15.4287	14.7872
15.2872	14.7014	14.5969	15.7314	16.6402	15.3543	14.5002	15.0465	15.1714	15.0964

FIGURE 3.24
Selection of "Interval Plot" for simulation of confidence intervals.

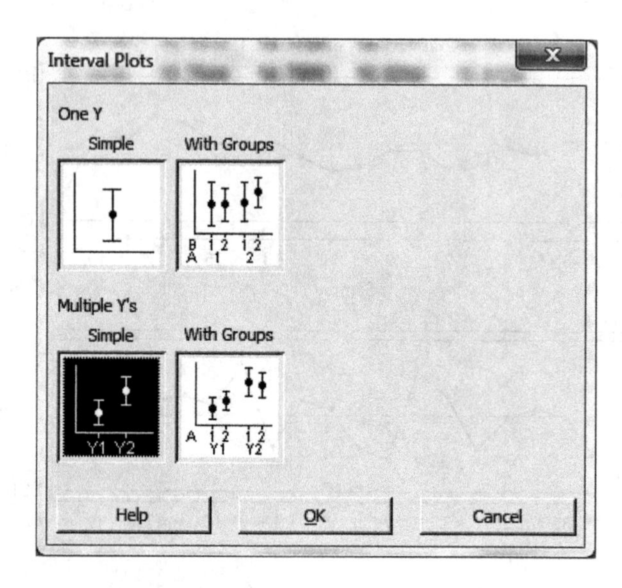

FIGURE 3.25
Selection of "Simple" interval plots under "Multiple Y's".

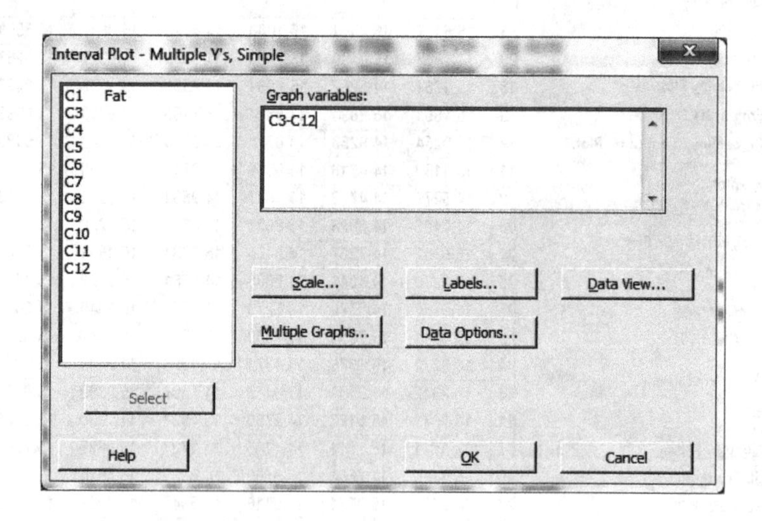

FIGURE 3.26
Selection of variables for simulation.

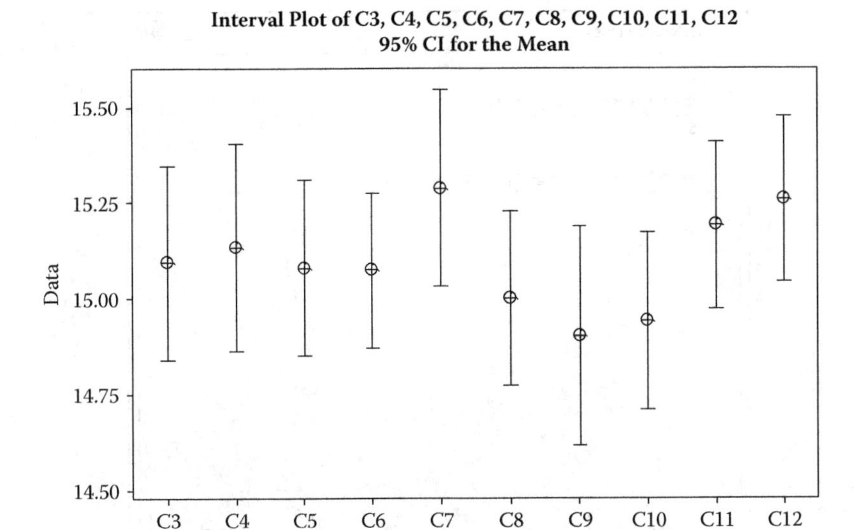

FIGURE 3.27
Interval plot without reference line.

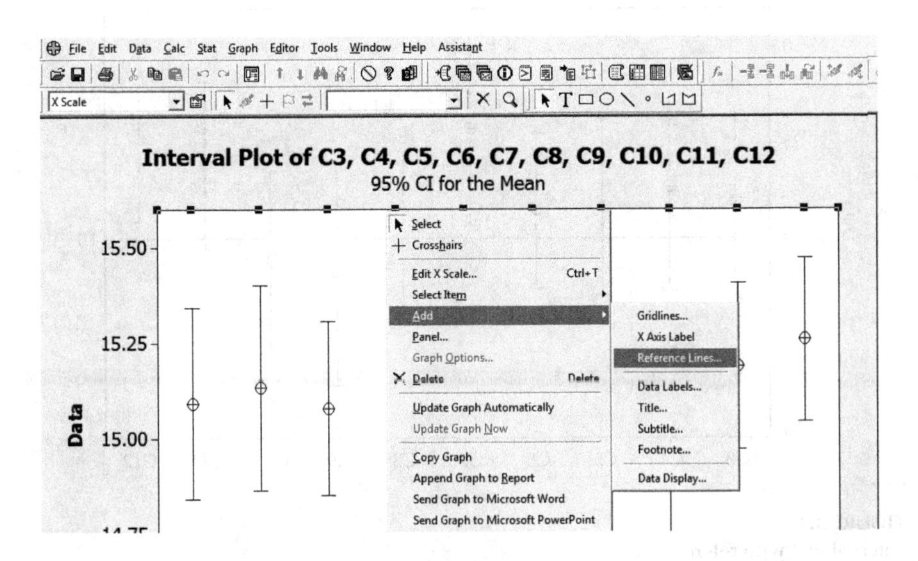

FIGURE 3.28
Process to show reference line.

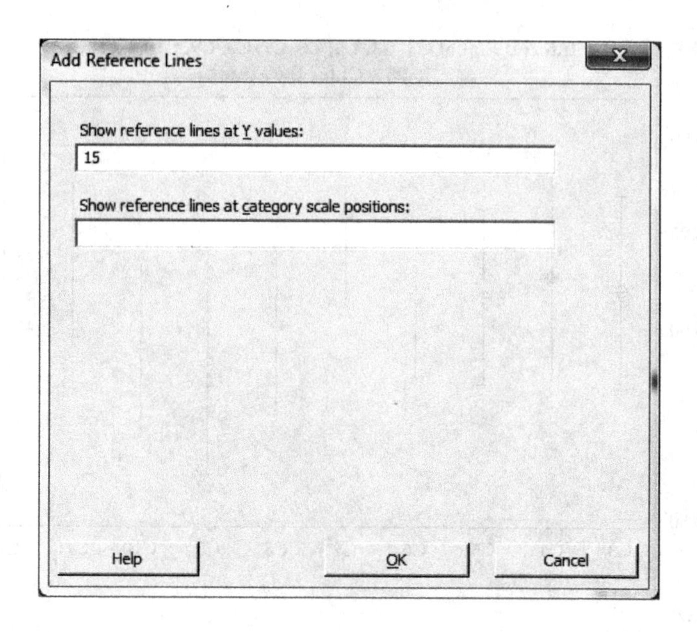

FIGURE 3.29
Entry of reference line.

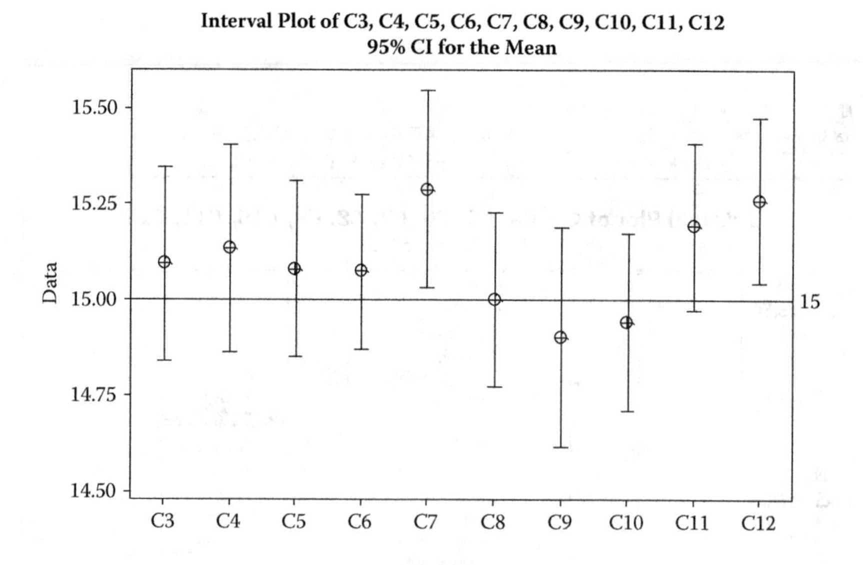

FIGURE 3.30
Interval plot with reference line.

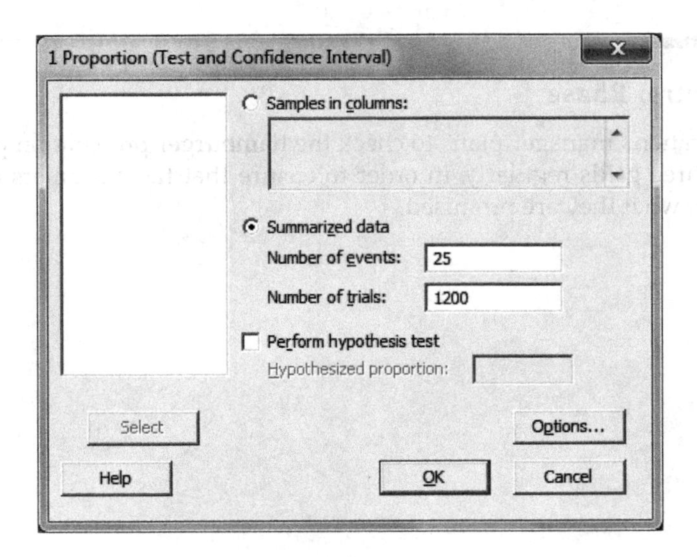

FIGURE 3.31
Entering number of events and trials after improvement.

"1200" for "Number of trials". Click on "Options" and select "less than" for "Alternative" because the operations manager is interested in lowering the proportion of all customers who are dissatisfied. Also, check the box for "Use test and interval based on normal distribution". Click on "OK" and it takes you back to the dialog box shown in Figure 3.31. Click on "OK" and the output shown in Figure 3.32 is the result. It is clear that that there is a 95% probability that the proportion of all customers who are dissatisfied is less than 0.027 (or 2.7%). The operations manager is satisfied that the process has improved.

Test and CI for One Proportion

Sample	X	N	Sample p	95% Upper Bound
1	25	1200	0.020833	0.027615

Using the normal approximation.

FIGURE 3.32
One-sided 95% confidence interval for population proportion after improvement.

3.5 Control Phase

The operations manager plans to check the hamburger preparation process on the three grills regularly, in order to ensure that the customers consistently get what they are promised.

4

Hypothesis Testing for Quality Control at a Manufacturing Company

The process engineer at a manufacturing company decides to perform quality control checks with respect to (1) proportion of defectives for the population of a particular category of products, (2) performance of a particular machine, and (3) comparison of output from three different shifts.

Section 4.1 gives a brief description of the define phase. Section 4.2 illustrates the measure and analyze phases with detailed instructions for using Minitab®. The improve and control phases are briefly described in Section 4.3.

4.1 Define Phase

The process engineer has the following hypothesis tests to perform:

Test 1 is to check whether the population proportion of defectives is less than the target of 5%. In other words,

Null Hypothesis: Population proportion of defectives = 0.05.

Alternative Hypothesis: Population proportion of defectives < 0.05.

Test 2 is to check whether the length of the product from Machine Q is equal to the target of 10 inches. It is suspected to be less than 10 inches. In other words,

Null Hypothesis: Product length from Machine Q = 10.

Alternative Hypothesis: Product length from Machine Q < 10.

Test 3 is to check whether or not the mean diameters of products are all the same across the three shifts. In other words,

Null Hypothesis: Process mean of Shift 1 = process mean of Shift 2 = process mean of Shift 3.

Alternative Hypothesis: Process means are not all equal.

4.2 Measure and Analyze Phases

4.2.1 Test 1

A sample of 200 products is taken, and 3 products have been identified as defective. Figure 4.1 shows how to select "1 Proportion" in Minitab® to test the population proportion of defectives. Doing so will open the dialog box shown in Figure 4.2. Enter "3" for "Number of events" and "200" for "Number of trials". Also, check the box for "Perform hypothesis test" and enter "0.05" for "Hypothesized proportion". Click on "Options" and the dialog box shown in Figure 4.3 opens. Select "less than" from the drop-down menu for "Alternative" and click on "OK". This takes you back to the dialog box shown in Figure 4.2. Click on "OK" and the output shown in Figure 4.4 is the result. Because *P*-value (0.012) is less than 0.05, reject the null hypothesis. In other words, the population proportion of defectives is lower than 5%, which is good news.

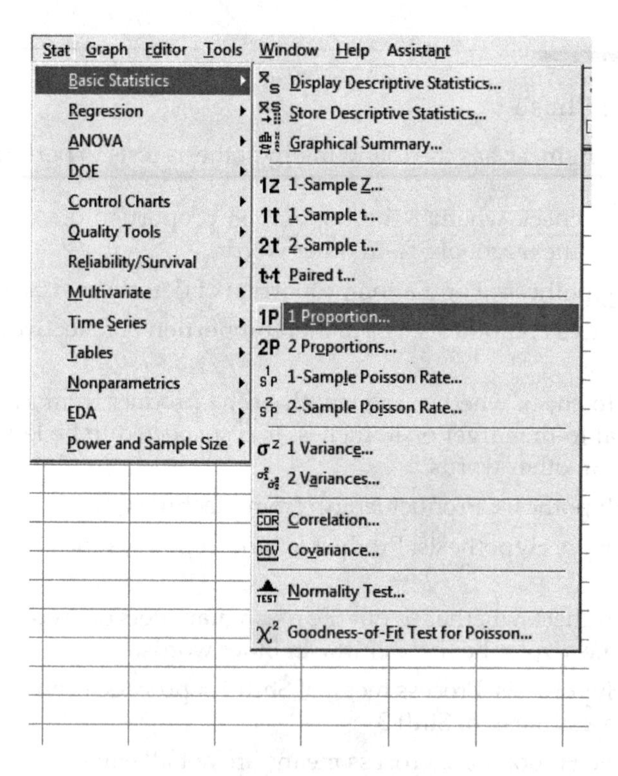

FIGURE 4.1
Selection of "1 Proportion".

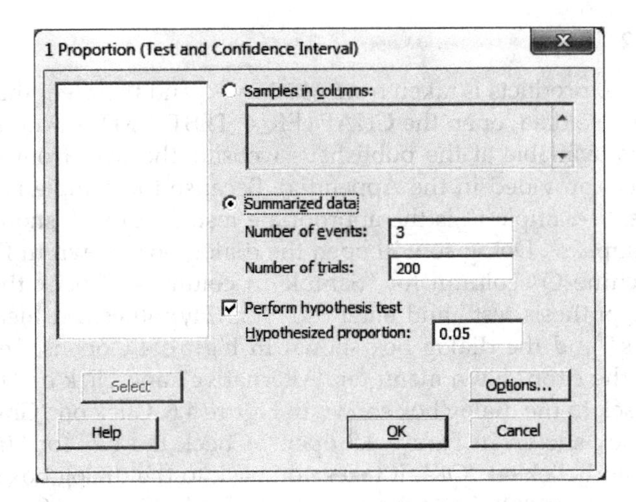

FIGURE 4.2
Entry of summarized data.

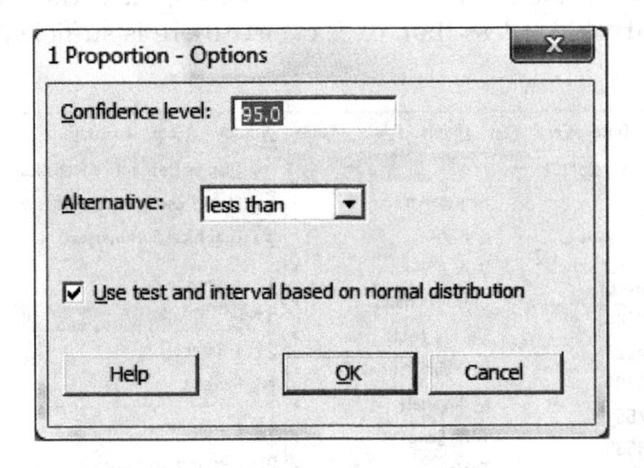

FIGURE 4.3
Selection of "less than" for alternative hypothesis for Test 1.

Test and CI for One Proportion

```
Test of p = 0.05 vs p < 0.05
                                       95% Upper
Sample  X    N  Sample p        Bound  Z-Value  P-Value
1       3  200  0.015000     0.029138    -2.27    0.012

Using the normal approximation.
The normal approximation may be inaccurate for small samples.
```

FIGURE 4.4
Output for Test 1.

4.2.2 Test 2

A sample of 20 products is taken from Machine Q and their lengths are measured. For these data, open the CHAPTER_4_TEST_2.MTW worksheet (the worksheet is available at the publisher's website; the data from the worksheet are also provided in the Appendix). Because the sample is less than 30 products, "1-Sample t" is the appropriate test. Figure 4.5 shows how to select "1-Sample t". Doing so will open the dialog box shown in Figure 4.6. Select "Machine Q" column for "Sample in columns". Check the box for "Perform hypothesis test" and enter "10" for "Hypothesized mean". Click on "Options" and the dialog box shown in Figure 4.7 opens. Select "less than" from the drop-down menu for "Alternative" and click on "OK". This takes you back to the dialog box shown in Figure 4.6. Click on "Graphs" and the dialog box shown in Figure 4.8 opens. Check the box for "Individual value plot" and click on "OK". It takes you back to the dialog box shown in Figure 4.6. Click on "OK" and the outputs shown in Figures 4.9 and 4.10 are the results. It is clear from Figure 4.9 that 95% of the data is less than the H0 (null hypothesis) value of 10. For that same reason, the *P*-value in the output shown in Figure 4.10 is less than 0.05. Hence, there is sufficient evidence to

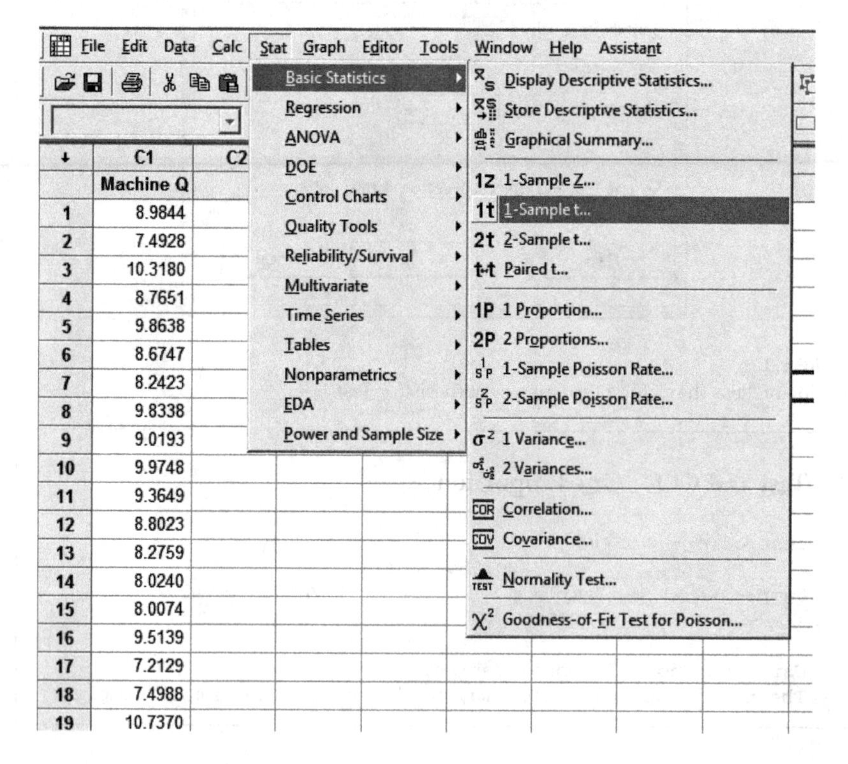

FIGURE 4.5
Selection of "1-Sample t".

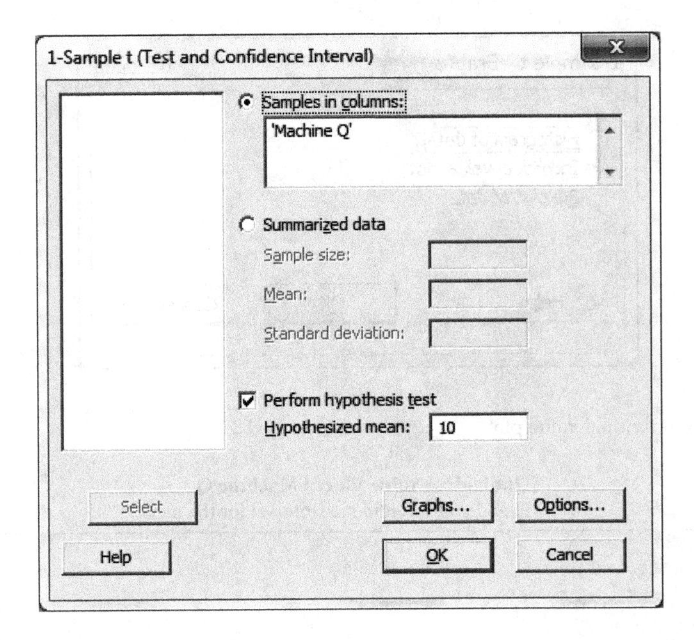

FIGURE 4.6
Selection of variables and entry of hypothesized mean.

reject the null hypothesis. In other words, Machine Q is indeed producing shorter products than desired. Therefore, the process engineer recommends fixing Machine Q.

4.2.3 Test 3

A sample of 20 products is taken from each of the three shifts, and their diameters are measured. For these data, open the CHAPTER_4_TEST_3.MTW worksheet (the worksheet is available at the publisher's website; the data

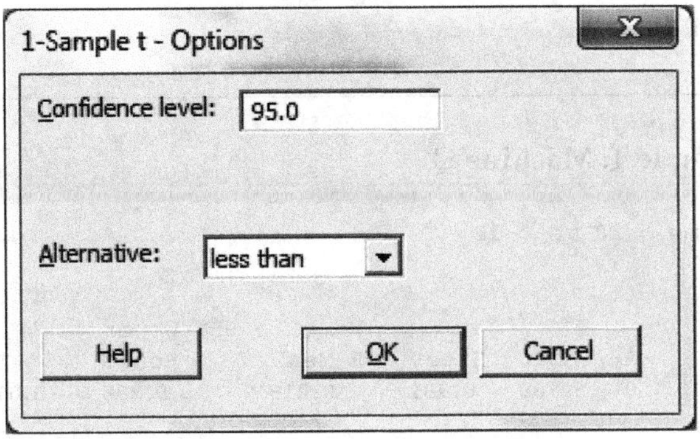

FIGURE 4.7
Selection of "less than" for alternative hypothesis for Test 2.

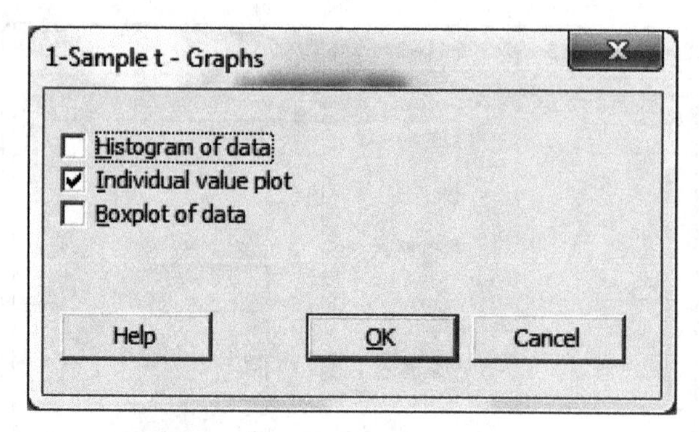

FIGURE 4.8
Selection of individual value plot.

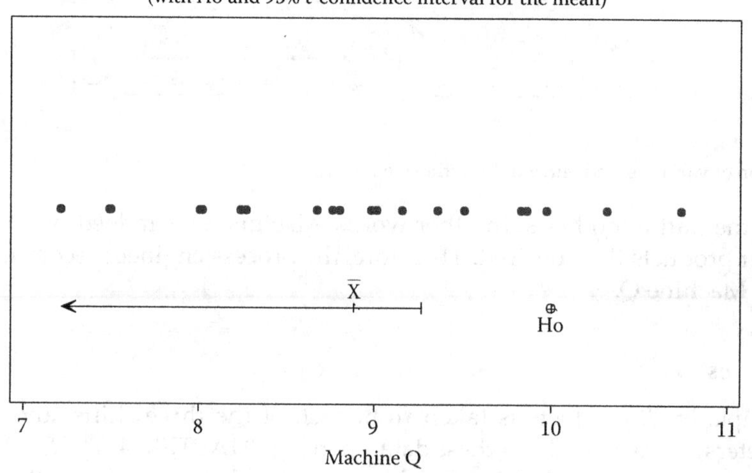

FIGURE 4.9
Individual value plot.

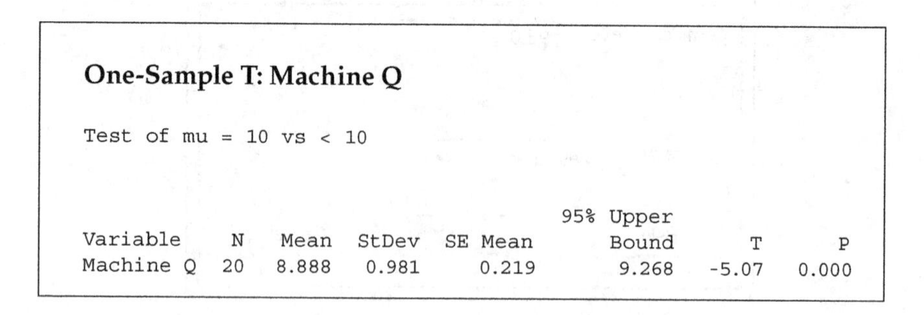

One-Sample T: Machine Q

Test of mu = 10 vs < 10

Variable	N	Mean	StDev	SE Mean	95% Upper Bound	T	P
Machine Q	20	8.888	0.981	0.219	9.268	-5.07	0.000

FIGURE 4.10
Output of 1-sample *t*-test.

from the worksheet are also provided in the Appendix). The process engineer wants to perform an analysis of variance (ANOVA) to check whether the mean diameters of the products from all of these three shifts are the same. However, before performing the ANOVA, it is important to check whether the following conditions for ANOVA are satisfied:

- The data from the three shifts are normally distributed (normality test).
- The data from the three shifts have equal variances (test of equal variances).

For the normality test, select "Probability Plot" as shown in Figure 4.11. Doing so will open the dialog box shown in Figure 4.12. Select "Multiple" and click on "OK". This opens the dialog box shown in Figure 4.13. Select "Shift 1 – Shift 3" columns for "Graph variables" and check the box for "Graph variables form groups". Click on "Multiple graphs" and it opens the dialog box shown in Figure 4.14. Select "On separate graphs" and click on

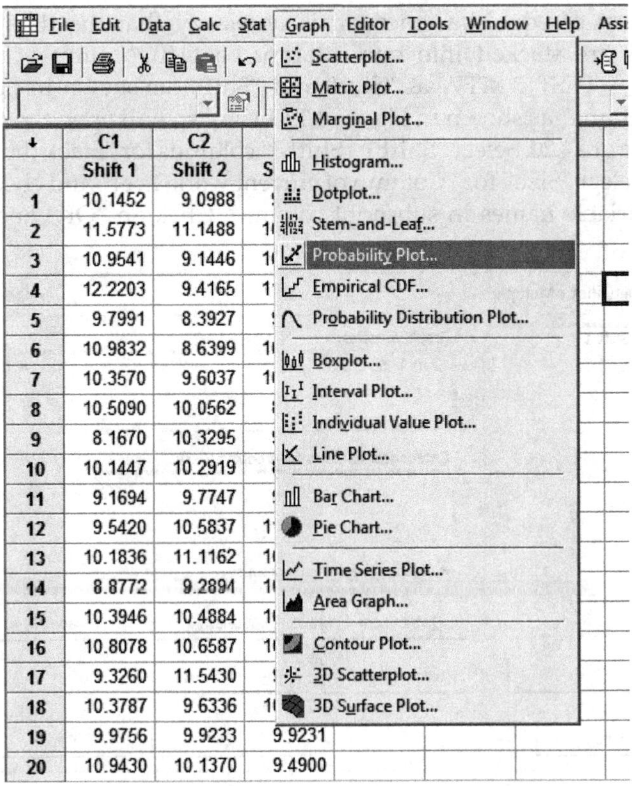

FIGURE 4.11
Selection of probability plot.

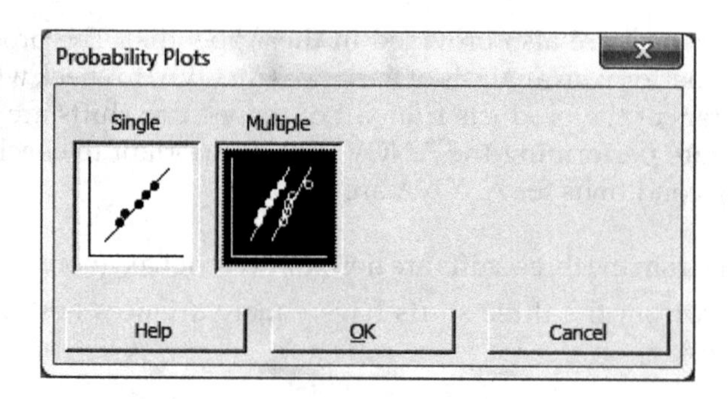

FIGURE 4.12
Selection of "Multiple" option for probability plots.

"OK". This takes you back to the dialog box shown in Figure 4.13. Click on "OK" and the probability plots shown in Figures 4.15–4.17 are the results. Inasmuch as the *P*-values for the three shifts (0.813, 0.996, 0.519, respectively) are greater than 0.05, the data from each of the three shifts are normally distributed.

For the test of equal variances, it is necessary that the data from the three shifts are stacked into one column. Label two empty columns in CHAPTER_4_TEST_3.MTW as "Size" and "Shift", as shown in Figure 4.18. Select "Columns" as shown in Figure 4.19. Doing so will open the dialog box shown in Figure 4.20. Select "Shift 1 – Shift 3" columns for "Stack the following columns", select "Size" for "Column of current worksheet" and check the box for "Use variable names in subscript column". Click on "OK" and the data

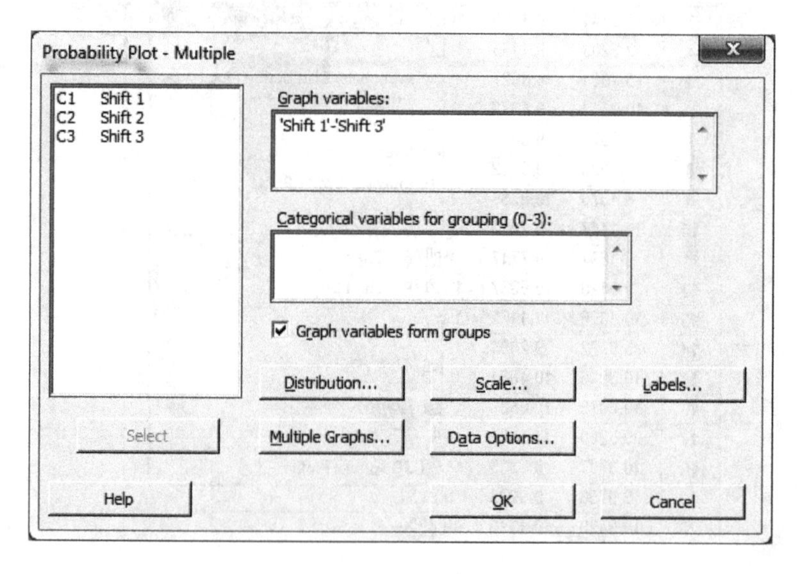

FIGURE 4.13
Selection of variables for probability plots.

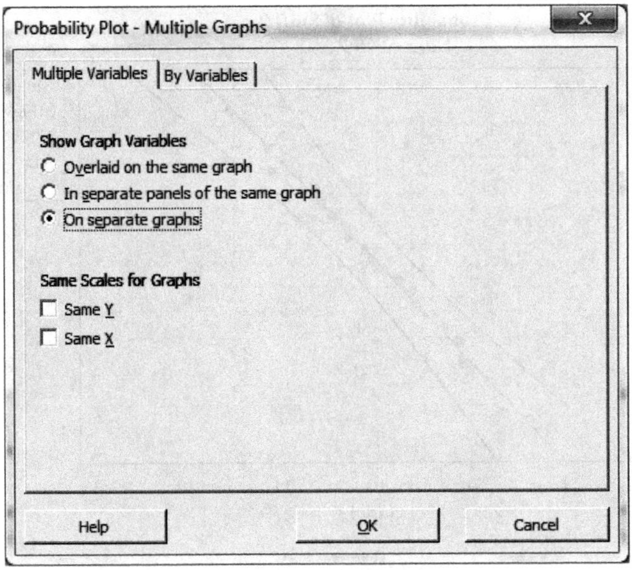

Probability Plot - Multiple Graphs

Multiple Variables | By Variables

Show Graph Variables

- Overlaid on the same graph
- In separate panels of the same graph
- (•) On separate graphs

Same Scales for Graphs

- [ ] Same Y
- [ ] Same X

Help | OK | Cancel

FIGURE 4.14
Selection of separate outputs for probability plots.

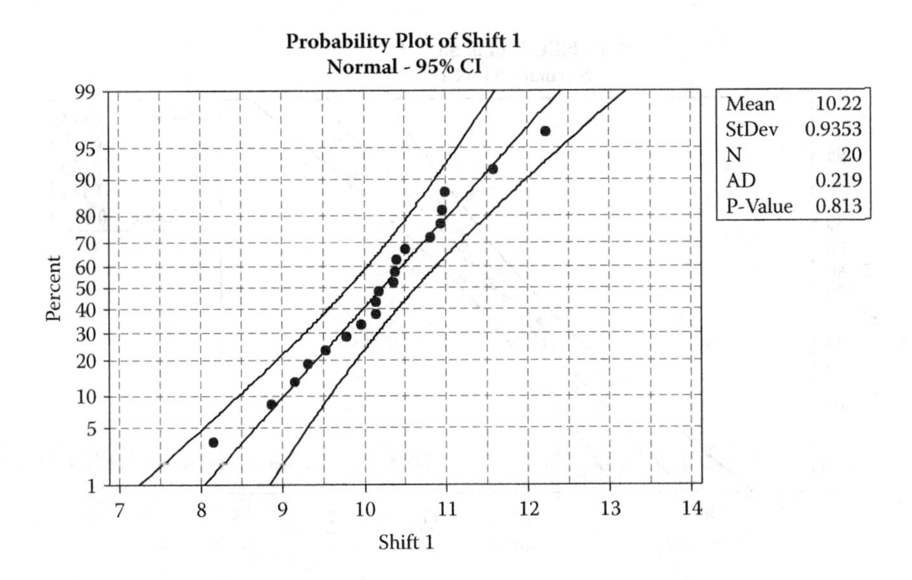

FIGURE 4.15
Probability plot for Shift 1.

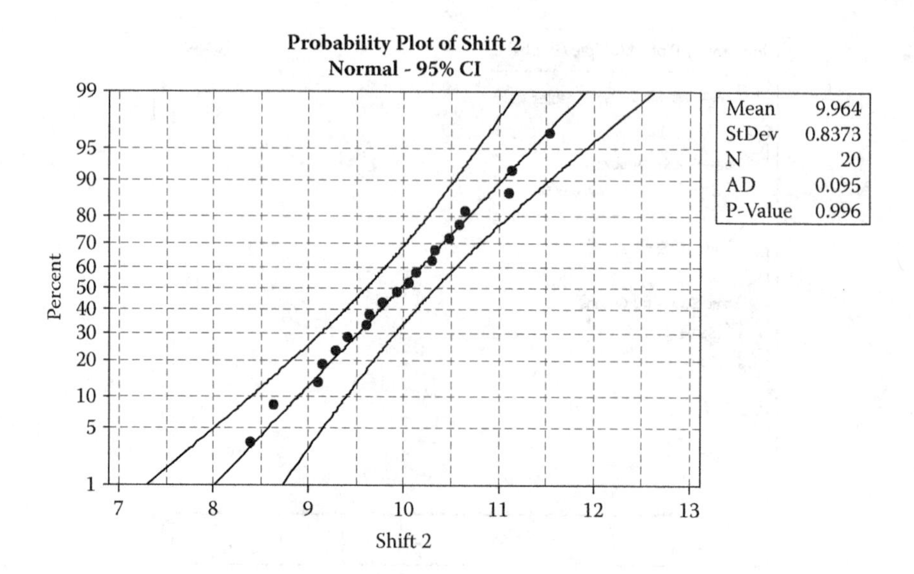

FIGURE 4.16
Probability plot for Shift 2.

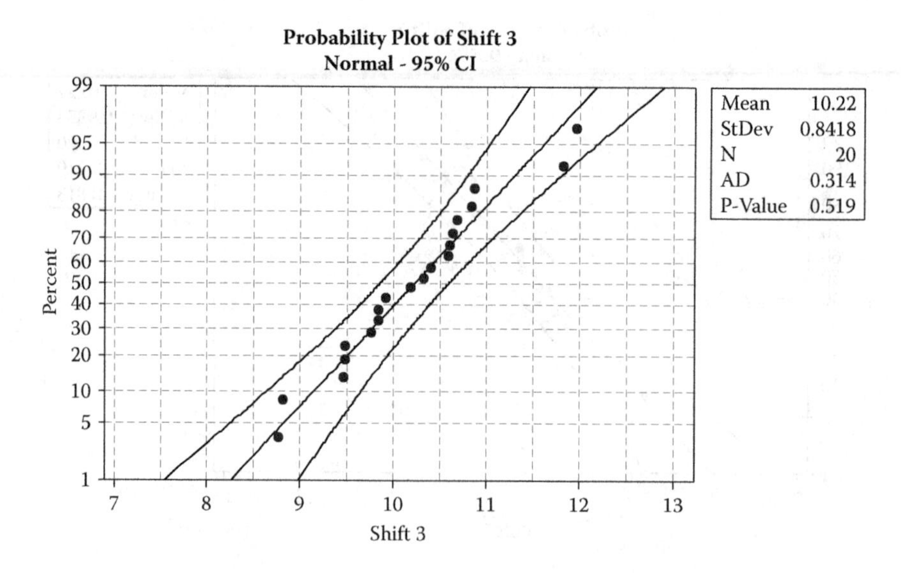

FIGURE 4.17
Probability plot for Shift 3.

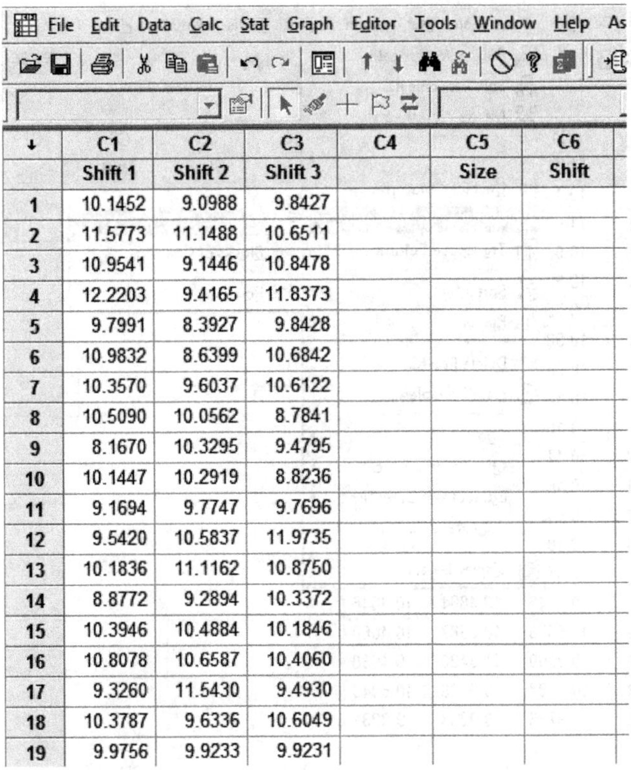

File Edit Data Calc Stat Graph Editor Tools Window Help As

↓	C1	C2	C3	C4	C5	C6
	Shift 1	Shift 2	Shift 3		Size	Shift
1	10.1452	9.0988	9.8427			
2	11.5773	11.1488	10.6511			
3	10.9541	9.1446	10.8478			
4	12.2203	9.4165	11.8373			
5	9.7991	8.3927	9.8428			
6	10.9832	8.6399	10.6842			
7	10.3570	9.6037	10.6122			
8	10.5090	10.0562	8.7841			
9	8.1670	10.3295	9.4795			
10	10.1447	10.2919	8.8236			
11	9.1694	9.7747	9.7696			
12	9.5420	10.5837	11.9735			
13	10.1836	11.1162	10.8750			
14	8.8772	9.2894	10.3372			
15	10.3946	10.4884	10.1846			
16	10.8078	10.6587	10.4060			
17	9.3260	11.5430	9.4930			
18	10.3787	9.6336	10.6049			
19	9.9756	9.9233	9.9231			

FIGURE 4.18
Labeling two empty columns for stacking data.

are stacked in the "Size" column as shown in Figure 4.21. Type in the shift numbers in the "Shift" column as shown in Figure 4.22. Then, select "Test for Equal Variances" as shown in Figure 4.23. This will open the dialog box shown in Figure 4.24. Select "Size" for "Response" and "Shift" for "Factors". Click on "OK" and the outputs shown in Figures 4.25 and 4.26 are the results. As shown in these two figures, the *P*-value (0.862) for Bartlett's test is greater than 0.05. Hence, the data from the three shifts have equal variances.

To perform Test 3, select "One-Way" as shown in Figure 4.27. Doing so will open the dialog box shown in Figure 4.28. Select "Size" for "Response" and "Shift" for "Factor". Click on "OK" and the output shown in Figure 4.29 is the result. Because the *P*-value (0.558) is not less than 0.05, there is insufficient evidence to reject the null hypothesis. Hence, the process means of the three shifts are equal.

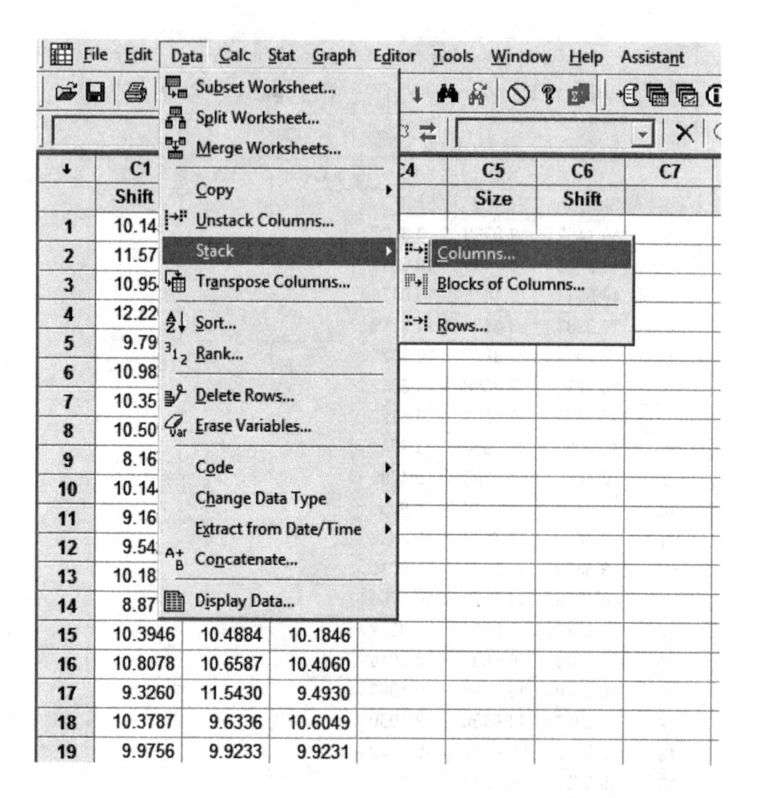

FIGURE 4.19
Selection of columns.

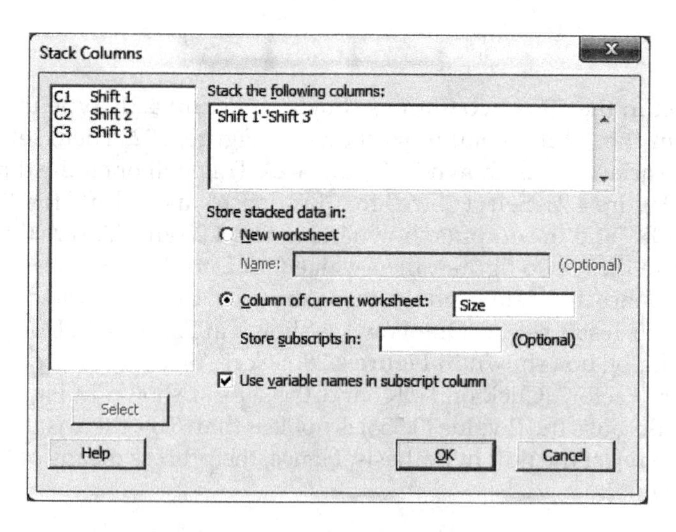

FIGURE 4.20
Selection of variables for stacking.

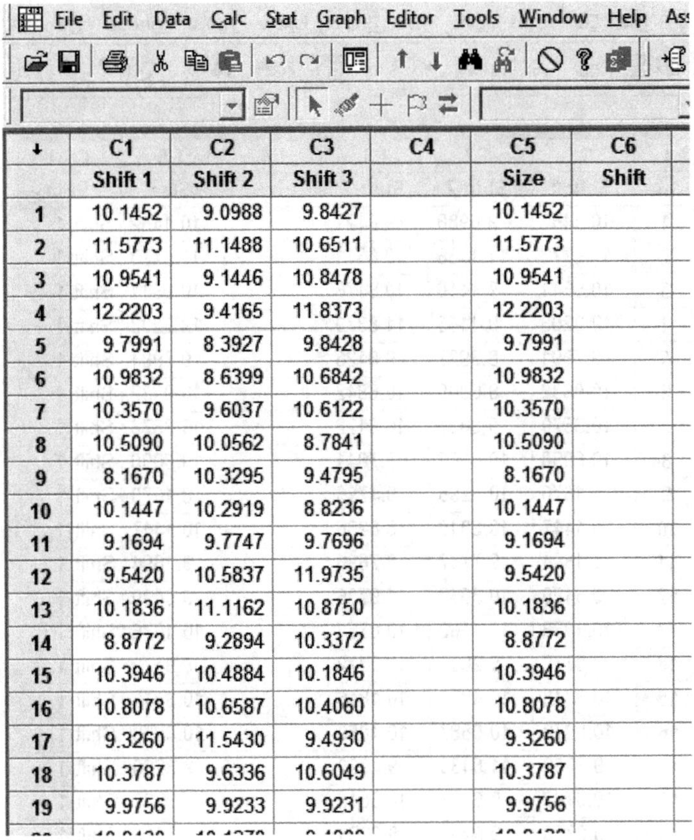

File Edit Data Calc Stat Graph Editor Tools Window Help As

↓	C1	C2	C3	C4	C5	C6
	Shift 1	Shift 2	Shift 3		Size	Shift
1	10.1452	9.0988	9.8427		10.1452	
2	11.5773	11.1488	10.6511		11.5773	
3	10.9541	9.1446	10.8478		10.9541	
4	12.2203	9.4165	11.8373		12.2203	
5	9.7991	8.3927	9.8428		9.7991	
6	10.9832	8.6399	10.6842		10.9832	
7	10.3570	9.6037	10.6122		10.3570	
8	10.5090	10.0562	8.7841		10.5090	
9	8.1670	10.3295	9.4795		8.1670	
10	10.1447	10.2919	8.8236		10.1447	
11	9.1694	9.7747	9.7696		9.1694	
12	9.5420	10.5837	11.9735		9.5420	
13	10.1836	11.1162	10.8750		10.1836	
14	8.8772	9.2894	10.3372		8.8772	
15	10.3946	10.4884	10.1846		10.3946	
16	10.8078	10.6587	10.4060		10.8078	
17	9.3260	11.5430	9.4930		9.3260	
18	10.3787	9.6336	10.6049		10.3787	
19	9.9756	9.9233	9.9231		9.9756	

FIGURE 4.21
Stacked data.

↓	C1	C2	C3	C4	C5	C6-T
	Shift 1	Shift 2	Shift 3		Size	Shift
1	10.1452	9.0988	9.8427		10.1452	Shift 1
2	11.5773	11.1488	10.6511		11.5773	Shift 1
3	10.9541	9.1446	10.8478		10.9541	Shift 1
4	12.2203	9.4165	11.8373		12.2203	Shift 1
5	9.7991	8.3927	9.8428		9.7991	Shift 1
6	10.9832	8.6399	10.6842		10.9832	Shift 1
7	10.3570	9.6037	10.6122		10.3570	Shift 1
8	10.5090	10.0562	8.7841		10.5090	Shift 1
9	8.1670	10.3295	9.4795		8.1670	Shift 1
10	10.1447	10.2919	8.8236		10.1447	Shift 1
11	9.1694	9.7747	9.7696		9.1694	Shift 1
12	9.5420	10.5837	11.9735		9.5420	Shift 1
13	10.1836	11.1162	10.8750		10.1836	Shift 1
14	8.8772	9.2894	10.3372		8.8772	Shift 1
15	10.3946	10.4884	10.1846		10.3946	Shift 1
16	10.8078	10.6587	10.4060		10.8078	Shift 1
17	9.3260	11.5430	9.4930		9.3260	Shift 1
18	10.3787	9.6336	10.6049		10.3787	Shift 1
19	9.9756	9.9233	9.9231		9.9756	Shift 1

FIGURE 4.22
Stacked data with shift numbers.

File Edit Data Calc Stat Graph Editor Tools Window Help Assistant

Stat menu: Basic Statistics ▸ | Regression ▸ | ANOVA ▸ | DOE ▸ | Control Charts ▸ | Quality Tools ▸ | Reliability/Survival ▸ | Multivariate ▸ | Time Series ▸ | Tables ▸ | Nonparametrics ▸ | EDA ▸ | Power and Sample Size ▸

ANOVA submenu: One-Way... | One-Way (Unstacked)... | Two-Way... | Analysis of Means... | Balanced ANOVA... | General Linear Model... | Fully Nested ANOVA... | Balanced MANOVA... | General MANOVA... | Test for Equal Variances... | Interval Plot... | Main Effects Plot... | Interactions Plot...

↓	C1 Shift 1	C2 Shift 2	C3	C4	C5	C6
1	10.1452	9.098				
2	11.5773	11.148				
3	10.9541	9.144				
4	12.2203	9.416				
5	9.7991	8.392				
6	10.9832	8.639				
7	10.3570	9.603				
8	10.5090	10.056				
9	8.1670	10.329				
10	10.1447	10.2919	8.8236		10	
11	9.1694	9.7747	9.7696		9	
12	9.5420	10.5837	11.9735		9	
13	10.1836	11.1162	10.8750		10.1836	Shift 1
14	8.8772	9.2894	10.3372		8.8772	Shift 1
15	10.3946	10.4884	10.1846		10.3946	Shift 1
16	10.8078	10.6587	10.4060		10.8078	Shift 1
17	9.3260	11.5430	9.4930		9.3260	Shift 1
18	10.3787	9.6336	10.6049		10.3787	Shift 1
19	9.9756	9.9233	9.9231		9.9756	Shift 1

FIGURE 4.23
Selection of "Test for Equal Variances".

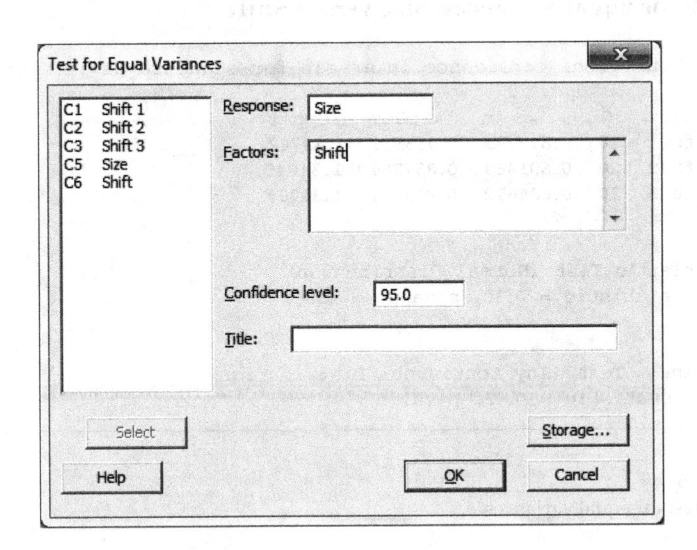

FIGURE 4.24
Selection of variables for test for equal variances.

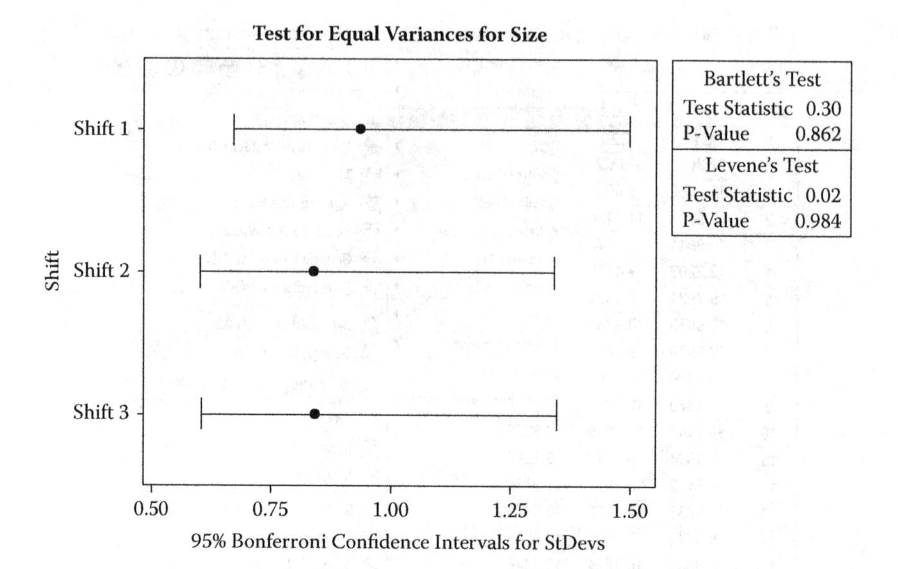

FIGURE 4.25
Confidence intervals for standard deviations of three shifts.

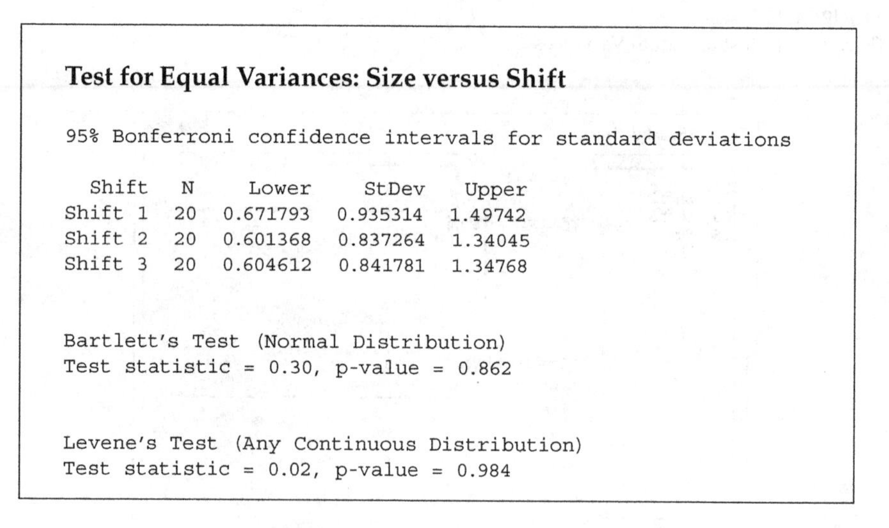

Test for Equal Variances: Size versus Shift

95% Bonferroni confidence intervals for standard deviations

Shift	N	Lower	StDev	Upper
Shift 1	20	0.671793	0.935314	1.49742
Shift 2	20	0.601368	0.837264	1.34045
Shift 3	20	0.604612	0.841781	1.34768

Bartlett's Test (Normal Distribution)
Test statistic = 0.30, p-value = 0.862

Levene's Test (Any Continuous Distribution)
Test statistic = 0.02, p-value = 0.984

FIGURE 4.26
Output of test for equal variances.

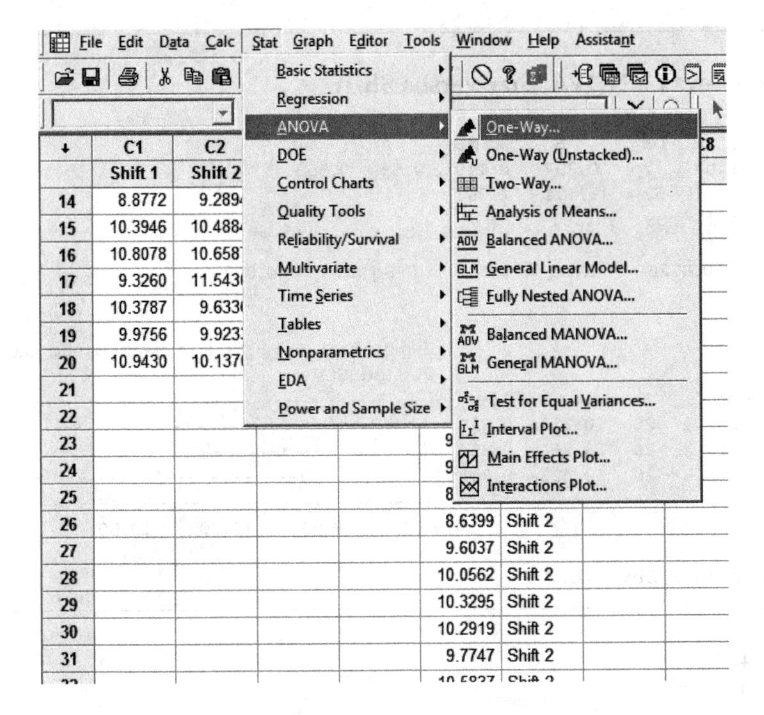

FIGURE 4.27
Selection of "One-Way".

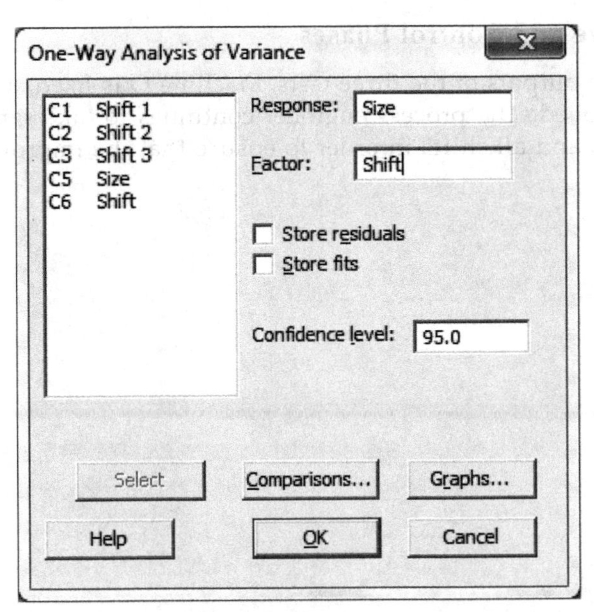

FIGURE 4.28
Selection of variables for ANOVA.

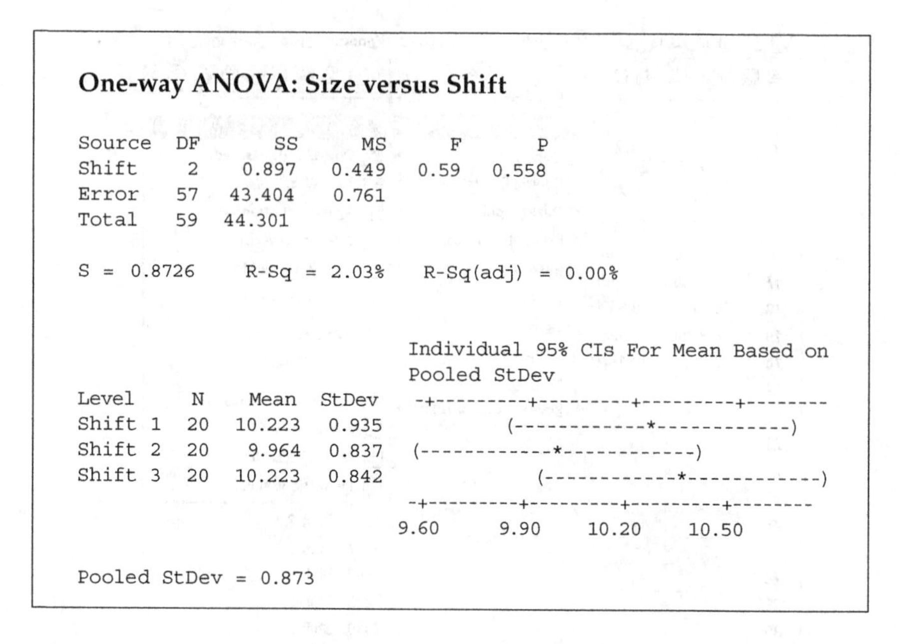

One-way ANOVA: Size versus Shift

Source	DF	SS	MS	F	P
Shift	2	0.897	0.449	0.59	0.558
Error	57	43.404	0.761		
Total	59	44.301			

S = 0.8726   R-Sq = 2.03%   R-Sq(adj) = 0.00%

```
                                   Individual 95% CIs For Mean Based on
                                   Pooled StDev
Level     N    Mean  StDev  -+---------+---------+---------+--------
Shift 1  20  10.223  0.935            (------------*------------)
Shift 2  20   9.964  0.837  (------------*------------)
Shift 3  20  10.223  0.842              (------------*------------)
                            -+---------+---------+---------+--------
                           9.60      9.90     10.20     10.50

Pooled StDev = 0.873
```

FIGURE 4.29
Output of ANOVA.

4.3 Improve and Control Phases

Based on the outputs of the three tests, Machine Q is fixed, and the process is improved. The process engineer continues to take samples from all machines and all shifts in order to ensure that the improved process is in control.

5

Chi-Square Analysis to Verify Quality of Candy Packets

This case study is about how chi-square analysis can be used in a Six Sigma project to collect voice of the customer (VOC) data and then to verify if a claim made about product quality is true.

Colorful Candy, Inc. makes colored chocolate candy and sells them in packets that are claimed to contain 14% yellow candy, 13% red candy, 20% orange candy, 24% blue candy, 16% green candy, and 13% purple candy. The company wants to know whether the customers really care about the variety of colors, and if so, whether the above claim of percentages is correct.

Section 5.1 gives a brief description of the define phase. Section 5.2 explains the measure phase. The analyze phase is illustrated in Section 5.3 with detailed instructions for using Minitab®. Finally, the improve and control phases are briefly discussed in Section 5.4.

5.1 Define Phase

The company wants to ensure that customers are getting what they are expecting regarding the percentages of different colors of candy in each packet they are purchasing. However, before randomly selecting a packet and testing it, the company executives decide to check whether customers really care about the different colors.

5.2 Measure Phase

To gather the VOC data, a number of randomly selected customers in Boston, Cleveland, New York City, San Francisco, and Chicago, are asked to rate on a 1–7 scale, how important the variety of colors is to them, with 7 being "extremely important" and 1 being "not important at all". The collected data are in the CHAPTER_5_1.MTW worksheet (the worksheet is available at the

↓	C1	C2-T	C3
	Importance of Candy Color	City	Observed
1	7	Boston	197
2	7	Cleveland	274
3	7	New York City	642
4	7	San Francisco	210
5	7	Chicago	197
6	6	Boston	257
7	6	Cleveland	405
8	6	New York City	304
9	6	San Francisco	252
10	6	Chicago	203
11	5	Boston	315
12	5	Cleveland	364
13	5	New York City	196
14	5	San Francisco	348
15	5	Chicago	250
16	4	Boston	480
17	4	Cleveland	326
18	4	New York City	263
19	4	San Francisco	486
20	4	Chicago	478
21	3	Boston	98
22	3	Cleveland	82

FIGURE 5.1
Data collected for importance given to candy color.

publisher's website; the data from the worksheet are also provided in the Appendix). A part of the worksheet is shown in Figure 5.1.

5.3 Analyze Phase

A bar chart is plotted for the data collected. Figure 5.2 shows how to select "Bar Chart". Doing so opens the dialog box shown in Figure 5.3. Select "Values from a table" from the drop-down menu, and select the "Stack" option under "One column of values". Click on "OK" and the dialog box shown in Figure 5.4 opens. Select the "Observed" column for "Graph variables" and select "City" and "Importance of Candy Color" for "Categorical variables for grouping". The bar chart shown in Figure 5.5 is the result. Although it is clear from the bar chart that customers in New York City seem to give a lot more importance

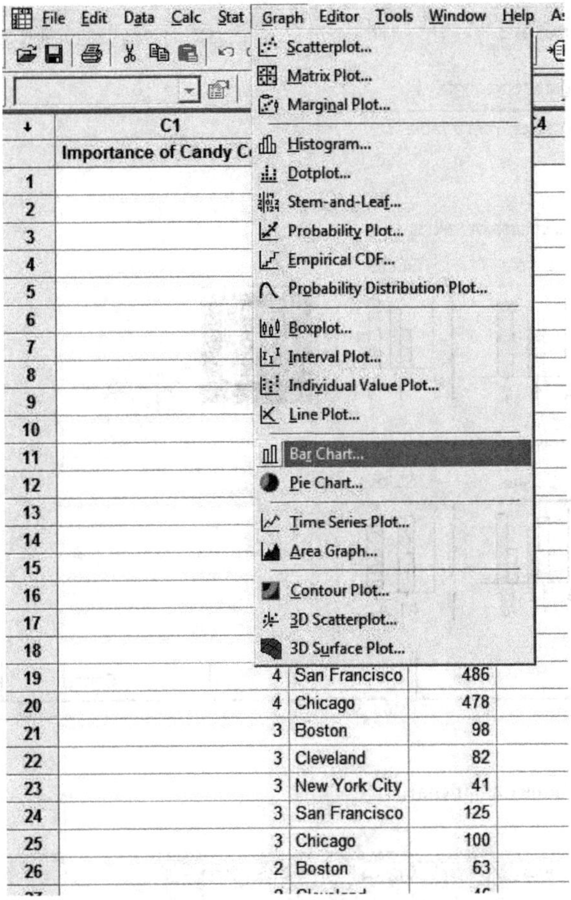

FIGURE 5.2
Selection of "Bar Chart."

to candy color than those in the other cities, the company executives want to perform a chi-square test of homogeneity to check whether the higher importance by customers in New York City is statistically significant. Figure 5.6 shows how to select "Chi-Square". Doing so will open the dialog box shown in Figure 5.7. Select "Importance of Candy Color" for "For rows" and "City" for "For columns". Select "Observed" for "Frequencies in" and check the box for "Counts". Click on "OK" and the output shown in Figure 5.8 is the result. The output shows how many customers were sampled in each of the cities and how many customers gave what importance rating (1–7).

For expected values, click on "Chi-Square" in the dialog box shown in Figure 5.7, and the dialog box shown in Figure 5.9 opens. Check the box for "Expected cell counts" and click on "OK". It takes you back to the dialog box shown in Figure 5.7. Click on "OK" and the output shown in Figure 5.10 is the result.

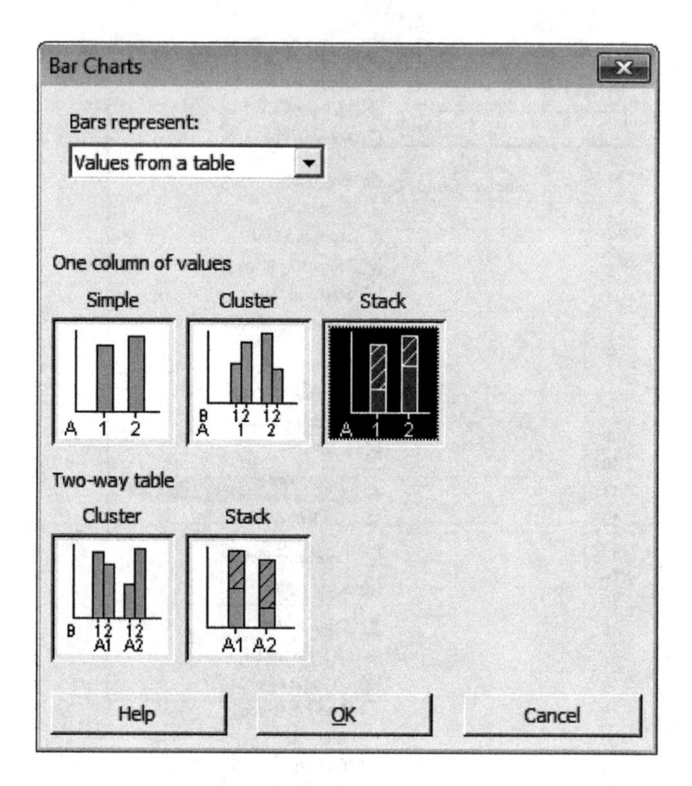

FIGURE 5.3
Selection from options for bar chart.

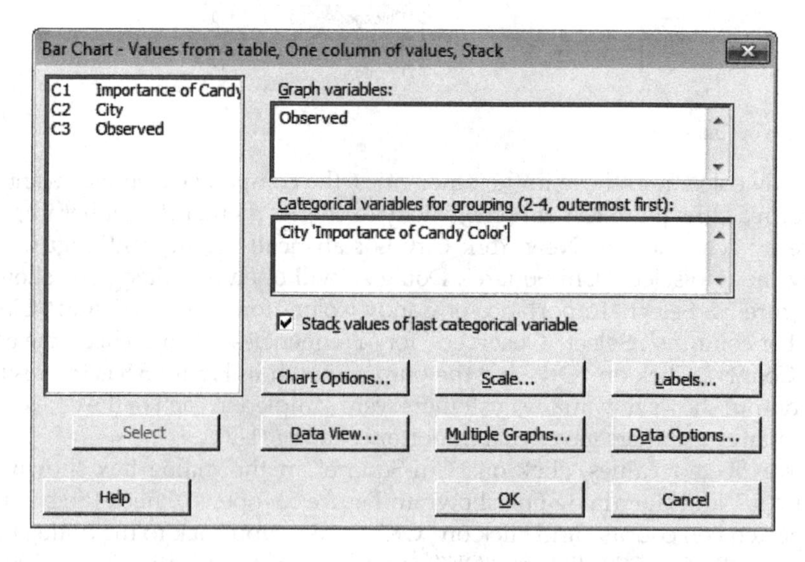

FIGURE 5.4
Selection of variables for bar chart.

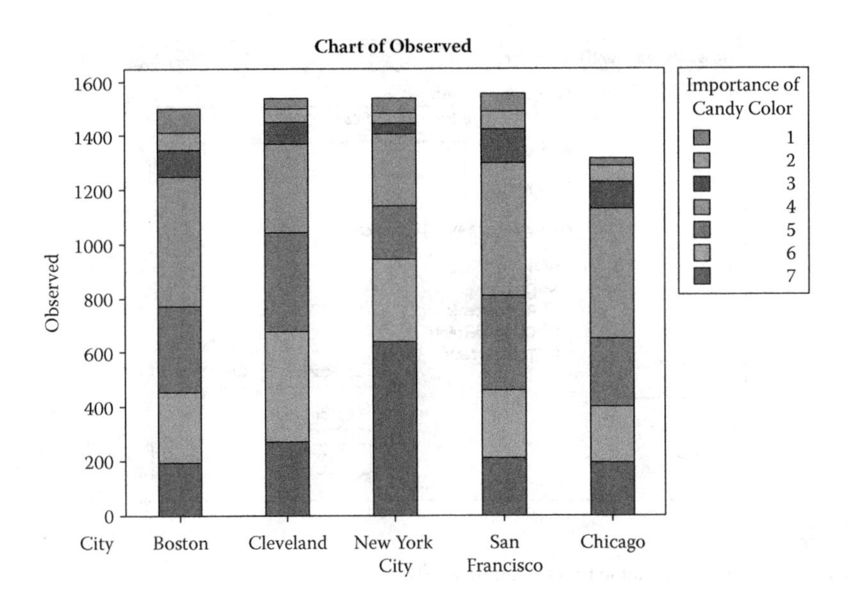

FIGURE 5.5
Bar chart.

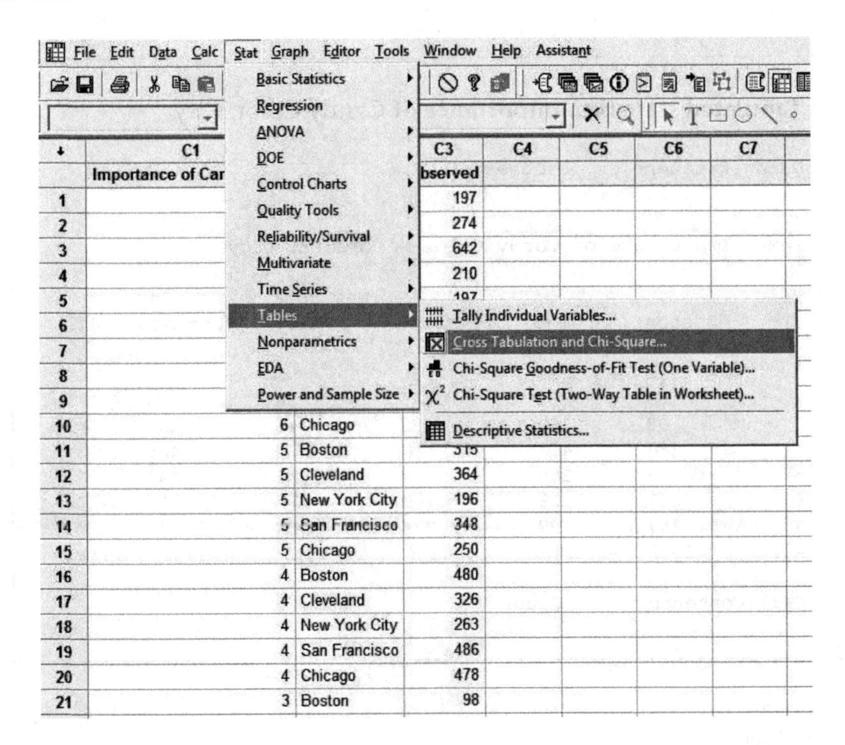

FIGURE 5.6
Selection of chi-square test.

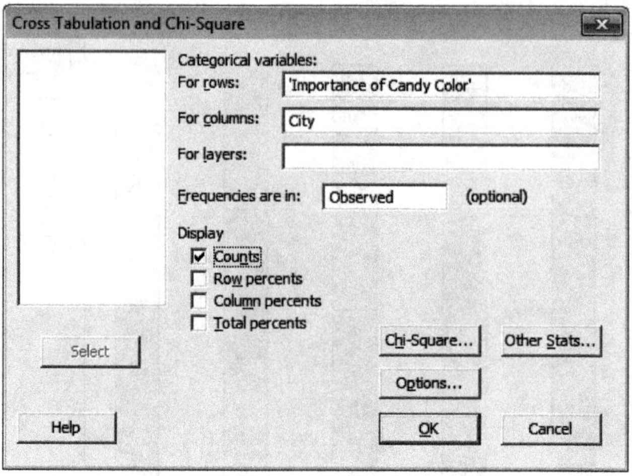

FIGURE 5.7
Creation of contingency table for chi-square analysis.

Tabulated statistics: Importance of Candy Color, City

Using frequencies in Observed

Rows: Importance of Candy Color Columns: City

	Boston	Chicago	Cleveland	New York City	San Francisco	All
1	92	29	38	53	62	274
2	63	58	46	36	70	273
3	98	100	82	41	125	446
4	480	478	326	263	486	2033
5	315	250	364	196	348	1473
6	257	203	405	304	252	1421
7	197	197	274	642	210	1520
All	1502	1315	1535	1535	1553	7440

Cell Contents: Count

FIGURE 5.8
Observed data.

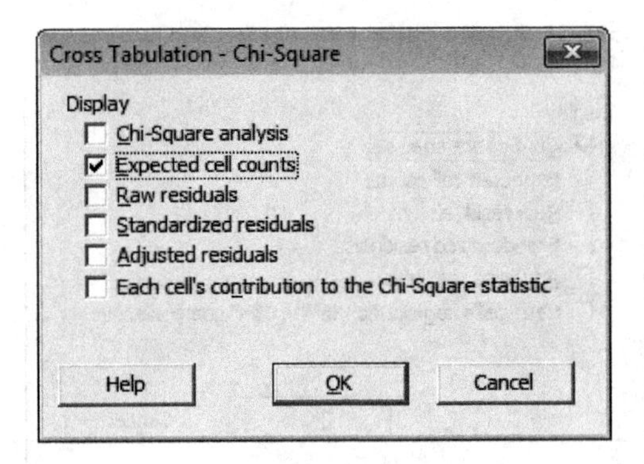

FIGURE 5.9
Selection of "Expected cell counts" option.

Tabulated statistics: Importance of Candy Color, City

Using frequencies in Observed

Rows: Importance of Candy Color   Columns: City

	Boston	Chicago	Cleveland	New York City	San Francisco	All
1	92	29	38	53	62	274
	55.3	48.4	56.5	56.5	57.2	274.0
2	63	58	46	36	70	273
	55.1	48.3	56.3	56.3	57.0	273.0
3	98	100	82	41	125	446
	90.0	78.8	92.0	92.0	93.1	446.0
4	480	478	326	263	486	2033
	410.4	359.3	419.4	419.4	424.4	2033.0
5	315	250	364	196	348	1473
	297.4	260.3	303.9	303.9	307.5	1473.0
6	257	203	405	304	252	1421
	286.9	251.2	293.2	293.2	296.6	1421.0
7	197	197	274	642	210	1520
	306.9	268.7	313.6	313.6	317.3	1520.0
All	1502	1315	1535	1535	1553	7440
	1502.0	1315.0	1535.0	1535.0	1553.0	7440.0

Cell Contents: Count
Expected count

FIGURE 5.10
Observed data and expected data.

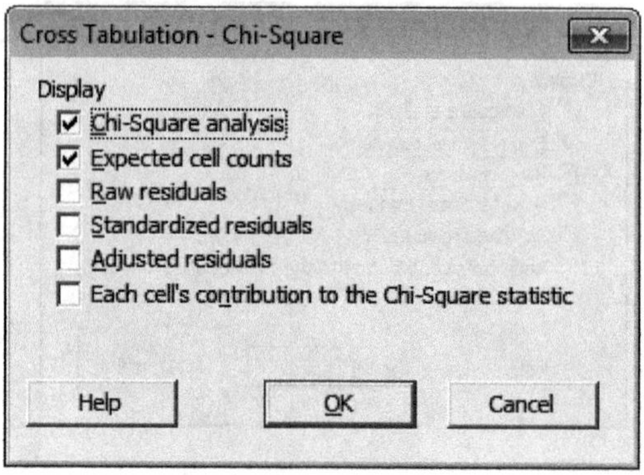

FIGURE 5.11
Selection of "Chi-Square analysis" option.

For the *P*-value of the chi-square analysis, click on "Chi-Square" in the dialog box shown in Figure 5.7, and the dialog box shown in Figure 5.11 opens. Check the box for "Chi-Square analysis" and click on "OK". It takes you back to the dialog box shown in Figure 5.7. Click on "OK" and the output shown in Figure 5.12 is the result. Because the Pearson chi-square *P*-value (0.000) is less than 0.05, there are significant differences among the ratings given by customers in the various cities. In order to view the chi-square probability distribution plot, select "Probability Distribution Plot" as shown in Figure 5.13. Doing so opens the dialog box in Figure 5.14. Select "View Probability" and click on "OK". This opens the dialog box shown in Figure 5.15. Select "Chi-Square" from the drop-down menu for "Distribution" and enter "24" for "Degrees of freedom" [The degrees of freedom are (number of ratings – 1) * (number of cities – 1) = (7 – 1) * (5 – 1) = 6 * 4 = 24]. Click on "OK" and the probability distribution plot shown in Figure 5.16 is the result. Notice that 36.42 is the critical value of the chi-square characteristic. In order to add a reference for the Pearson chi-square of 802.637 (refer to Figure 5.12), right-click on the plot shown in Figure 5.16 and select "Reference Lines" as shown in Figure 5.17. Doing so opens the dialog box shown in Figure 5.18. Enter "802.637" for "Show reference lines at X values" and click on "OK". The plot shown in Figure 5.19 is the result.

Now that it is clear that there are significant differences among the ratings given by customers in the various cities and that New York City seems to give a lot more importance to candy color than the other cities (see Figure 5.5), the company wants to check whether its claim of the following percentages in a packet are correct: 14% yellow candy, 13% red candy, 20% orange candy, 24% blue candy, 16% green candy, and 13% purple candy. To this end, a packet of candy is randomly selected, and the number of candies of each color is counted. The collected data are in the CHAPTER_5_2.MTW worksheet (the

Tabulated statistics: Importance of Candy Color, City

Using frequencies in Observed

Rows: Importance of Candy Color Columns: City

	Boston	Chicago	Cleveland	New York City	San Francisco	All
1	92	29	38	53	62	274
	55.3	48.4	56.5	56.5	57.2	274.0
2	63	58	46	36	70	273
	55.1	48.3	56.3	56.3	57.0	273.0
3	98	100	82	41	125	446
	90.0	78.8	92.0	92.0	93.1	446.0
4	480	478	326	263	486	2033
	410.4	359.3	419.4	419.4	424.4	2033.0
5	315	250	364	196	348	1473
	297.4	260.3	303.9	303.9	307.5	1473.0
6	257	203	405	304	252	1421
	286.9	251.2	293.2	293.2	296.6	1421.0
7	197	197	274	642	210	1520
	306.9	268.7	313.6	313.6	317.3	1520.0
All	1502	1315	1535	1535	1553	7440
	1502.0	1315.0	1535.0	1535.0	1553.0	7440.0

Cell Contents: Count
Expected count

Pearson Chi-Square = 802.637, DF = 24, P-Value = 0.000
Likelihood Ratio Chi-Square = 746.830, DF = 24, P-Value = 0.000

FIGURE 5.12
Output of chi-square analysis.

worksheet is available at the publisher's website; the data from the worksheet are also provided in the Appendix). A screenshot of the worksheet is shown in Figure 5.20. Label an empty column as "Expected", as shown in Figure 5.21. Figure 5.22 shows how to select "Calculator". Doing so opens the dialog box shown in Figure 5.23. Select the "Expected" column for "Store results in variable" and enter ""Percent Expected" * 106" in the "Expression" box (106 is the total number of candies in the packet). Then, check the box for "Assign as a formula" and click on "OK". This results in the "Expected" column values as shown in Figure 5.24. Figure 5.25 shows how to select "Chi-Square Goodness-of-Fit Test (One Variable)". This opens the dialog box shown in Figure 5.26. Select "Observed" for "Observed counts", and "Color" for "Category names

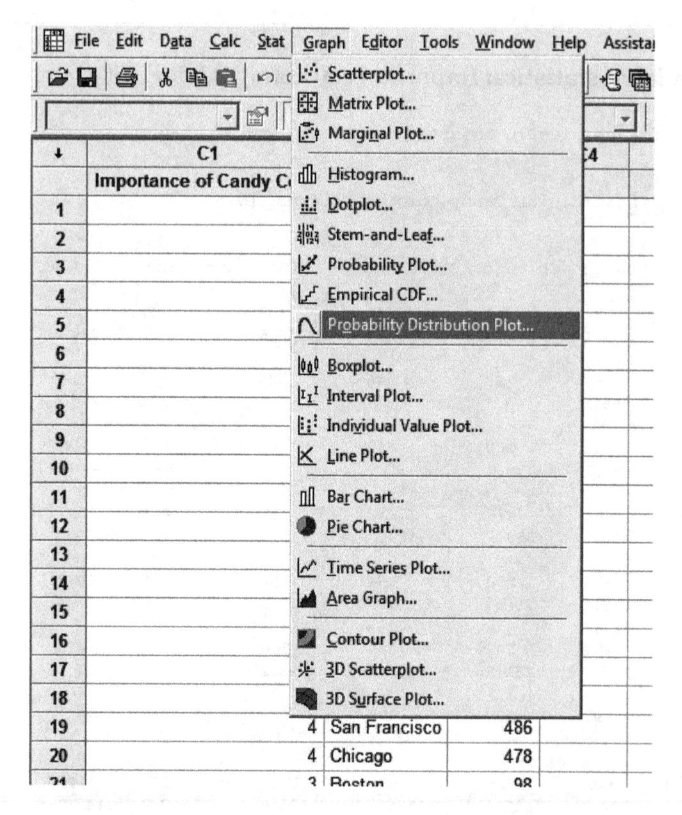

FIGURE 5.13
Selection of "Probability Distribution Plot".

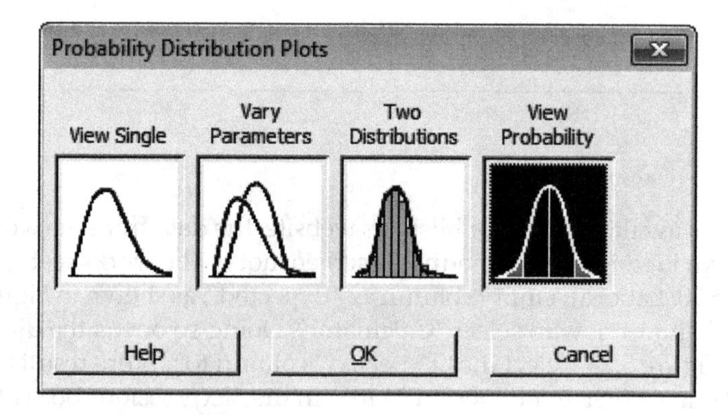

FIGURE 5.14
Selection of "View Probability".

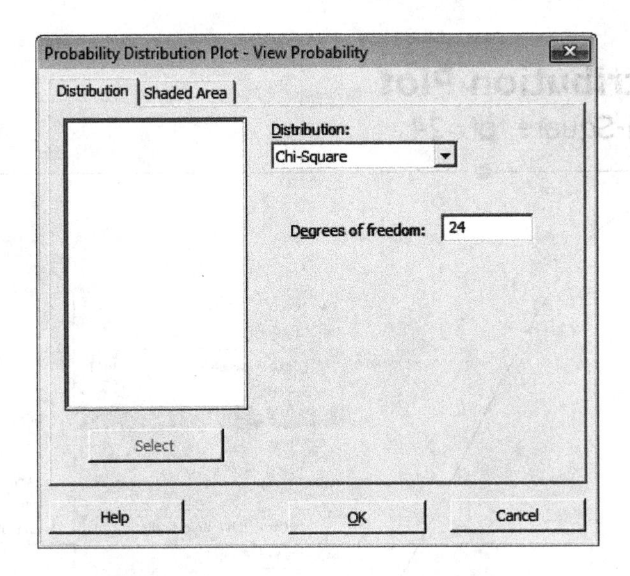

FIGURE 5.15
Entry of degrees of freedom.

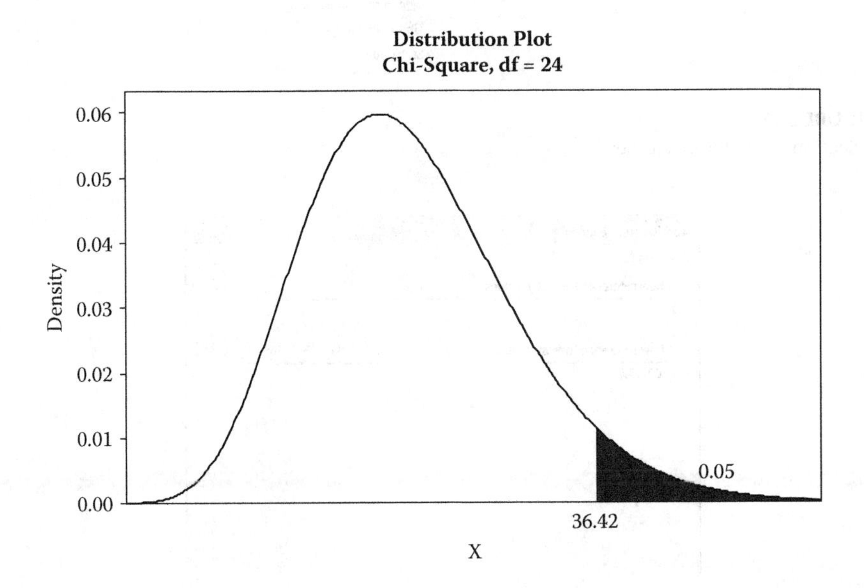

FIGURE 5.16
Chi-square probability distribution plot.

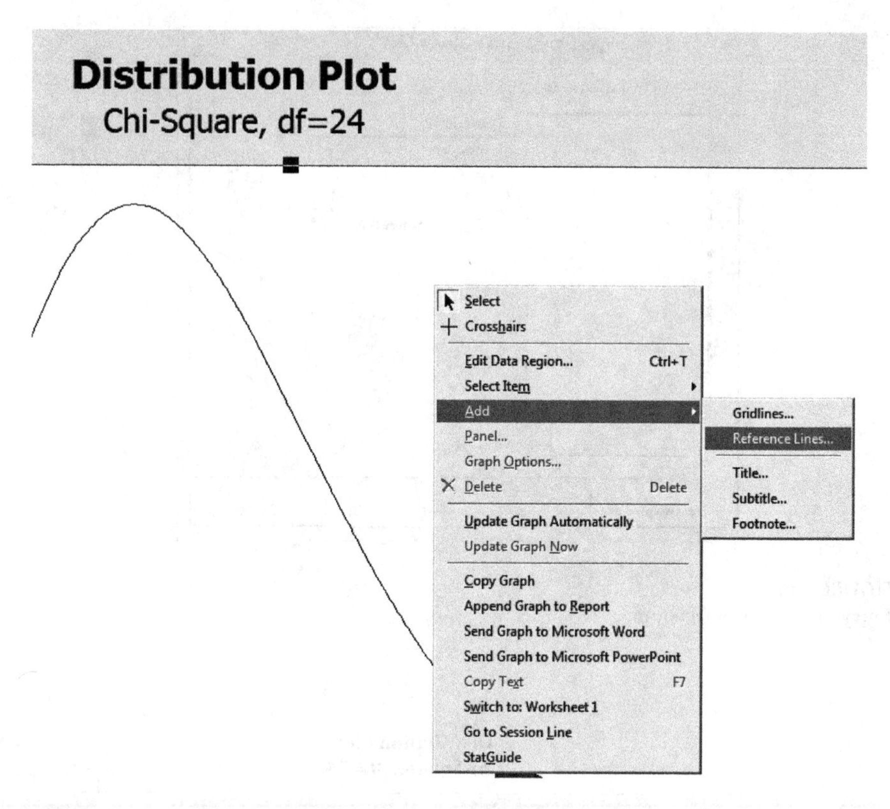

FIGURE 5.17
Selection of "Reference Lines".

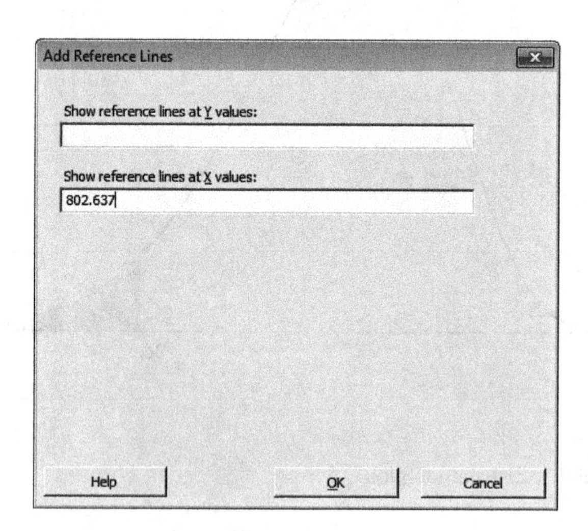

FIGURE 5.18
Entry of Pearson chi-square value as reference.

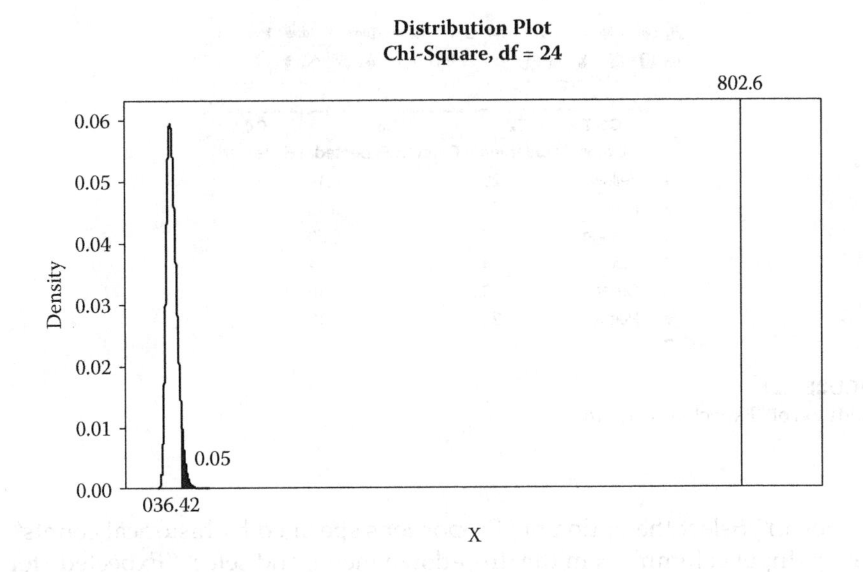

FIGURE 5.19
Critical chi-square value and Pearson chi-square value.

File Edit Data Calc Stat Graph Editor Tools Window

↓	C1-T	C2	C3
	Color	Observed	Percent Expected
1	Yellow	29	0.14
2	Red	23	0.13
3	Orange	12	0.20
4	Blue	14	0.24
5	Green	8	0.16
6	Purple	20	0.13

FIGURE 5.20
Observed counts and percent expected in a random packet.

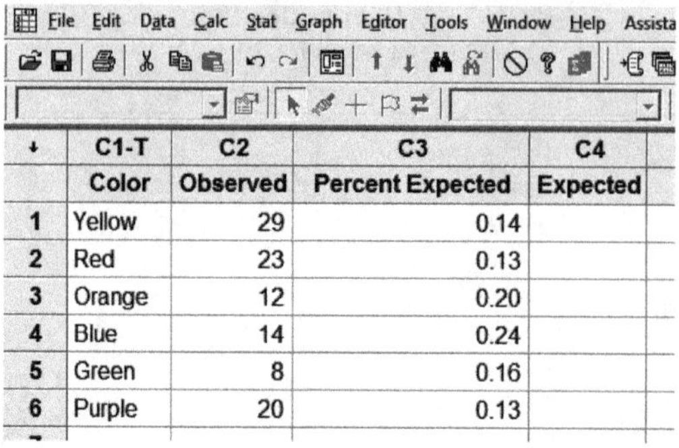

↓	C1-T	C2	C3	C4
	Color	Observed	Percent Expected	Expected
1	Yellow	29	0.14	
2	Red	23	0.13	
3	Orange	12	0.20	
4	Blue	14	0.24	
5	Green	8	0.16	
6	Purple	20	0.13	

FIGURE 5.21
Addition of "Expected" column.

(optional)". Select the option of "Proportions specified by historical counts", select "Input column" from the drop-down menu, and select "Expected" for the empty box. Click on "Graphs" and the dialog box shown in Figure 5.27 opens. Ensure that all three boxes are checked as shown in Figure 5.27. Click on "OK" and it takes you back to the dialog box shown in Figure 5.26. Click on "Results" and the dialog box shown in Figure 5.28 opens. Ensure that the box is checked as shown in Figure 5.28 and click on "OK". This takes you back to the dialog box shown in Figure 5.26. Click on "OK" and the outputs shown in Figures 5.29–5.31 are the results. Figure 5.29 shows the contribution of each color of candy to the chi-square value. Figure 5.30 shows the differences between the expected and observed counts for each color. Because the

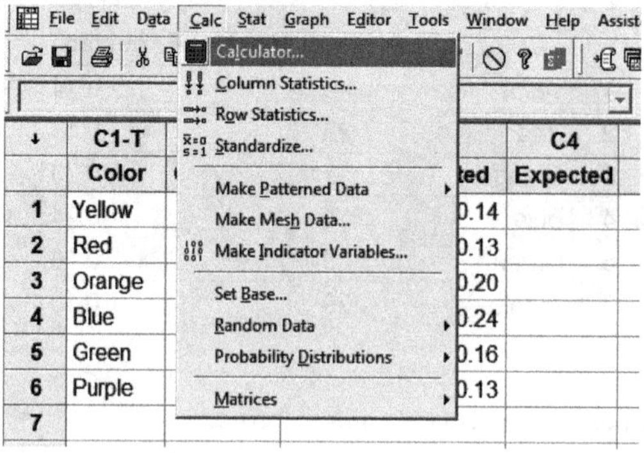

FIGURE 5.22
Use of "Calculator" for "Expected" column.

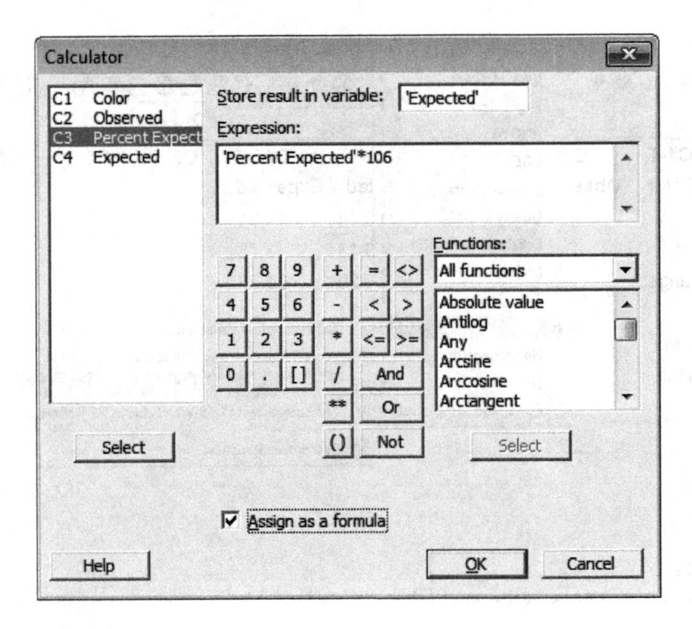

FIGURE 5.23
Calculation for "Expected" column.

File Edit Data Calc Stat Graph Editor Tools Window Help Assist

↓	C1-T	C2	C3	C4
	Color	Observed	Percent Expected	Expected
1	Yellow	29	0.14	14.84
2	Red	23	0.13	13.78
3	Orange	12	0.20	21.20
4	Blue	14	0.24	25.44
5	Green	8	0.16	16.96
6	Purple	20	0.13	13.78
7				

FIGURE 5.24
Data for "Expected" column.

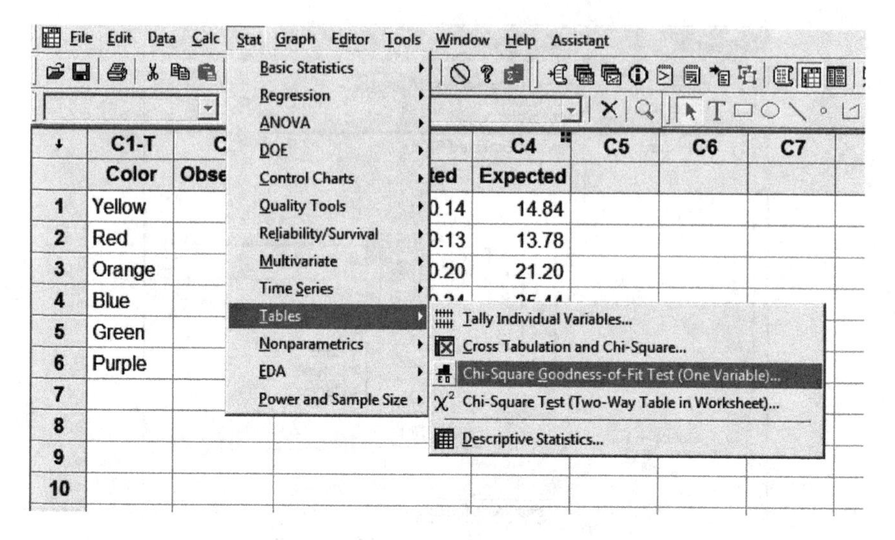

FIGURE 5.25
Selection of "Chi-Square Goodness-of-Fit Test (One Variable)".

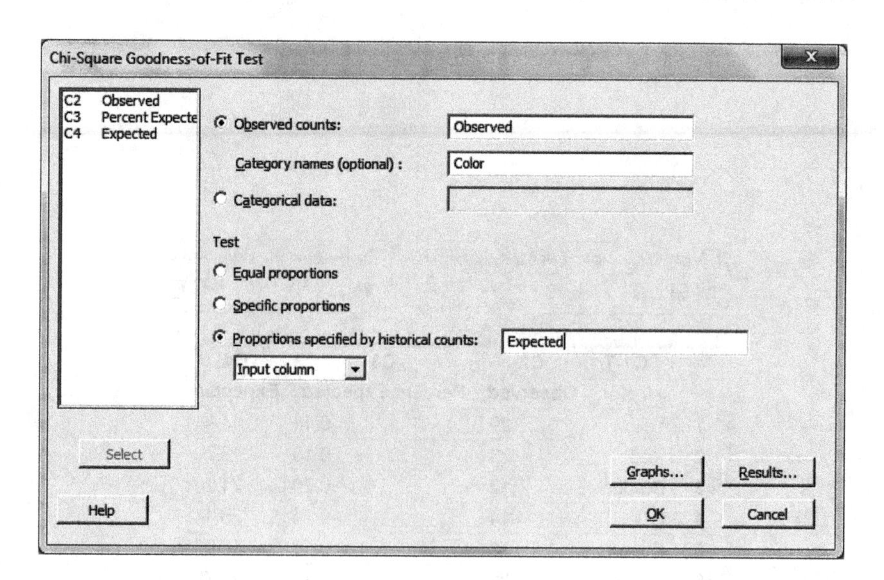

FIGURE 5.26
Entry of "Observed counts" and "Category names".

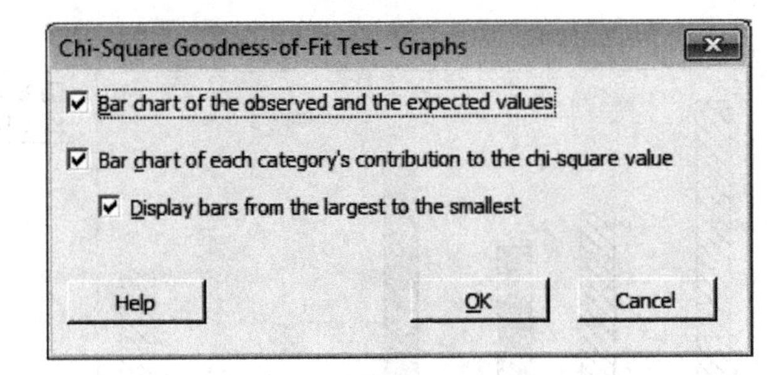

FIGURE 5.27
Selection of chi-square graphs.

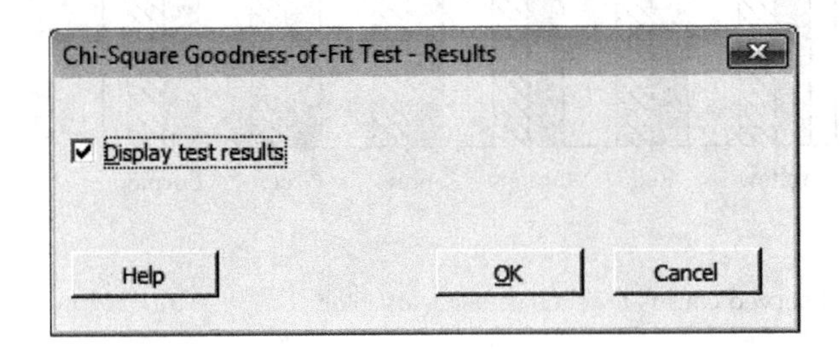

FIGURE 5.28
Selection of "Display test results".

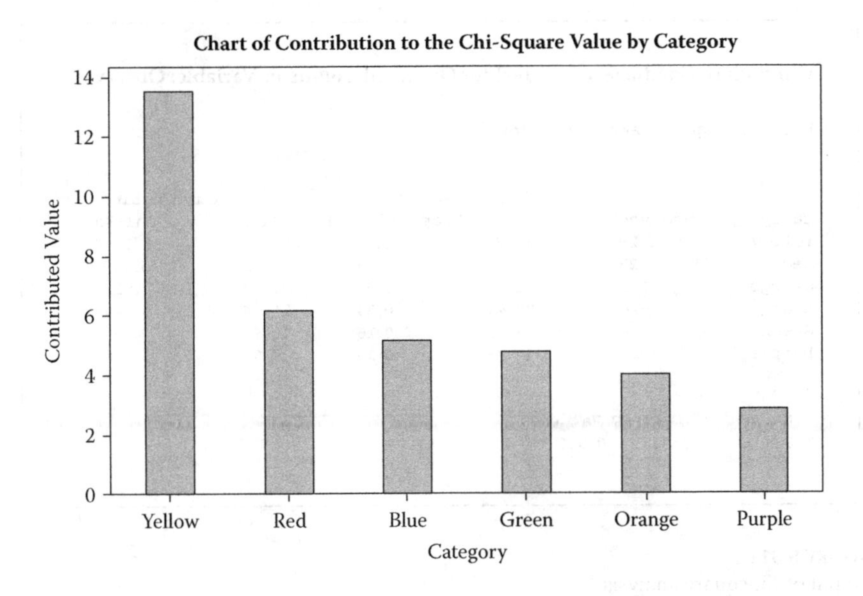

FIGURE 5.29
Contribution to chi-square value for each category of color.

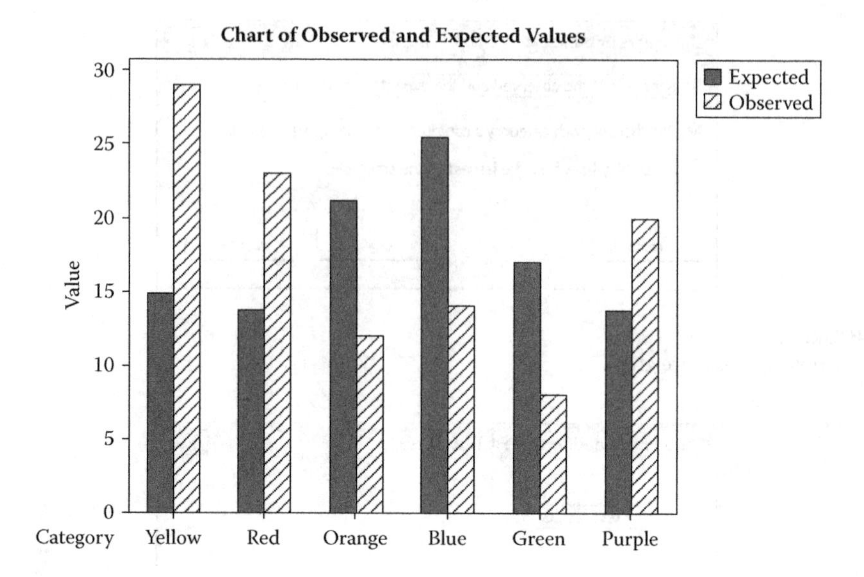

FIGURE 5.30
Expected and observed counts for each category of color.

Chi-Square Goodness-of-Fit Test for Observed Counts in Variable: Observed

Using category names in Color

Category	Observed	Historical Counts	Test Proportion	Expected	Contribution to Chi-Sq
Yellow	29	14.84	0.14	14.84	13.5112
Red	23	13.78	0.13	13.78	6.1690
Orange	12	21.20	0.20	21.20	3.9925
Blue	14	25.44	0.24	25.44	5.1444
Green	8	16.96	0.16	16.96	4.7336
Purple	20	13.78	0.13	13.78	2.8076

N	DF	Chi-Sq	P-Value
106	5	36.3581	0.000

FIGURE 5.31
Output of chi-square analysis.

P-value (0.000) is less than 0.05 in Figure 5.31, there are significant differences between what is observed in the packet and what is claimed by the company.

5.4 Improve and Control Phases

The company improves the production process to ensure that each packet contains 14% yellow candy, 13% red candy, 20% orange candy, 24% blue candy, 16% green candy, and 13% purple candy. The company continues to take random samples of candy packets in order to verify whether the improved process is in control.

6

Process Capability Analysis at a Manufacturing Company

This case study is about a Six Sigma project implemented by the production manager at a manufacturing firm that produces a critical automobile part used in cars produced by three major automobile companies. The production manager aims to improve the capability of the manufacturing process.

Recall the following process capability ratios from Chapter 2.

$$C_p = \frac{USL - LSL}{6\sigma}$$

$$C_{pl} = \frac{\mu - LSL}{3\sigma}$$

$$C_{pu} = \frac{USL - \mu}{3\sigma}$$

$$C_{pk} = \text{MIN}\{C_{pl}, C_{pu}\}$$

where

USL = Upper specification limit
LSL = Lower specification limit
μ = Process mean
σ = Process standard deviation

The higher the C_p and C_{pk} values are, the better the process is.

Section 6.1 gives a brief description of the define phase. Section 6.2 illustrates the measure phase with detailed instructions for using Minitab®. The analyze phase is briefly discussed in Section 6.3. Section 6.4 illustrates the improve phase with detailed instructions for using Minitab®. Finally, the control phase is briefly discussed in Section 6.5.

6.1 Define Phase

The production manager desires to increase the capability of the manufacturing process with a *USL* value of 60 units and an *LSL* value of 50 units for the part diameter. The problem statement is "to increase the C_p and C_{pk} values."

6.2 Measure Phase

Twenty samples, each containing 5 parts, are collected, and their diameters are measured. The data are shown in Table 6.1.

Before C_p and C_{pk} values are calculated, it is important to check whether the process data are normally distributed and in statistical control. The following is the approach to do so.

Open the CHAPTER_6_1.MTW worksheet containing the data from Table 6.1 in a single column (the worksheet is available at the publisher's website; the data from the worksheet are also provided in the Appendix). Figure 6.1 is a screenshot of the partial worksheet (it shows only 19 of the 100 numbers). Figures 6.2 and 6.3 illustrate how to check for normality and Figure 6.4 shows the normality test results. Because the *P*-value in Figure 6.4 is greater than 0.05, it is evident that the process data are normally distributed.

Figure 6.5 partially shows the data copied from Table 6.1 to the CHAPTER_6_1.MTW worksheet. In order to check whether the data are in statistical control, the data need to be transposed to have each sample in a single row. Figures 6.6 and 6.7 show how to transpose the data, and Figure 6.8 shows the transposed data in a new worksheet. (Do not delete the previous worksheet because you need it for process capability analysis later.) For clarity, the headings of the columns are revised, and the revised worksheet is shown in Figure 6.9.

Because the data are variable data and the sample size is 5, the appropriate control charts to construct are the $\overline{X}$ chart and R chart. Figures 6.10 and 6.11 show how to construct the R chart, and Figure 6.12 shows the R chart. The sample ranges are in statistical control, therefore check whether the sample means are in statistical control. Figures 6.13 and 6.14 show how to construct the $\overline{X}$ chart. It is evident from the $\overline{X}$ chart in Figure 6.15 that the sample means are also in statistical control.

Because the process data are normally distributed and are in statistical control, we can calculate the process capability ratios now. Figures 6.16 and 6.17 illustrate how to do so. Figure 6.18 shows that the *USL* and *LSL* are entered in the respective boxes. Click on "Options" in the dialog box shown in Figure 6.18, and the dialog box shown in Figure 6.19 opens. Uncheck the "Overall Analysis" box and enter the "Title" as shown in Figure 6.19. Click

TABLE 6.1
Production Data before Process Improvement

1	2	3	4	5	6	7	8	9	10	11	12	13	14	15	16	17	18	19	20
52.9	54.3	49.3	55.9	54.5	60.7	57.7	54.6	52.7	55.7	53.8	54.4	55.8	56	54.1	57.2	54.3	52.1	55	53.6
55	55.7	53.4	51.9	58.8	53.2	52.6	56	54.5	55.9	55.7	55	54.8	53.3	53.4	55.6	54.4	53.2	54.4	55.4
55.5	55.9	52.7	56.2	54.4	56.2	54.6	53	51.3	52.9	51.7	56.2	53.2	53.8	54.4	56	54.1	52.4	54.5	56.9
54.1	58.1	51.1	55.1	56.1	54.2	55.7	56.4	55.7	53.9	52.1	54	57	56.7	53.7	52	52.6	54.4	57.1	53.1
55.9	55.1	56.5	53	57.3	54.9	54.8	51.4	52.5	59.1	56.8	53.7	56.7	55.7	57.4	57.8	51.8	52.3	52.7	53.4

Minitab - Untitled -

File Edit Data

↓	C1
	Before
1	52.9
2	55.0
3	55.5
4	54.1
5	55.9
6	54.3
7	55.7
8	55.9
9	58.1
10	55.1
11	49.3
12	53.4
13	52.7
14	51.1
15	56.5
16	55.9
17	51.9
18	56.2
19	55.1

FIGURE 6.1
Data in one column in Minitab® worksheet before process improvement.

on "OK" and it takes you back to the dialog box shown in Figure 6.18. Click on "OK", and the graph shown in Figure 6.20 is the result. The current C_p and C_{pk} values are 0.85 and 0.79, respectively.

6.3 Analyze Phase

The production manager, along with her team, analyzes the process, and identifies that a couple of machine tools on the assembly line are not properly aligned.

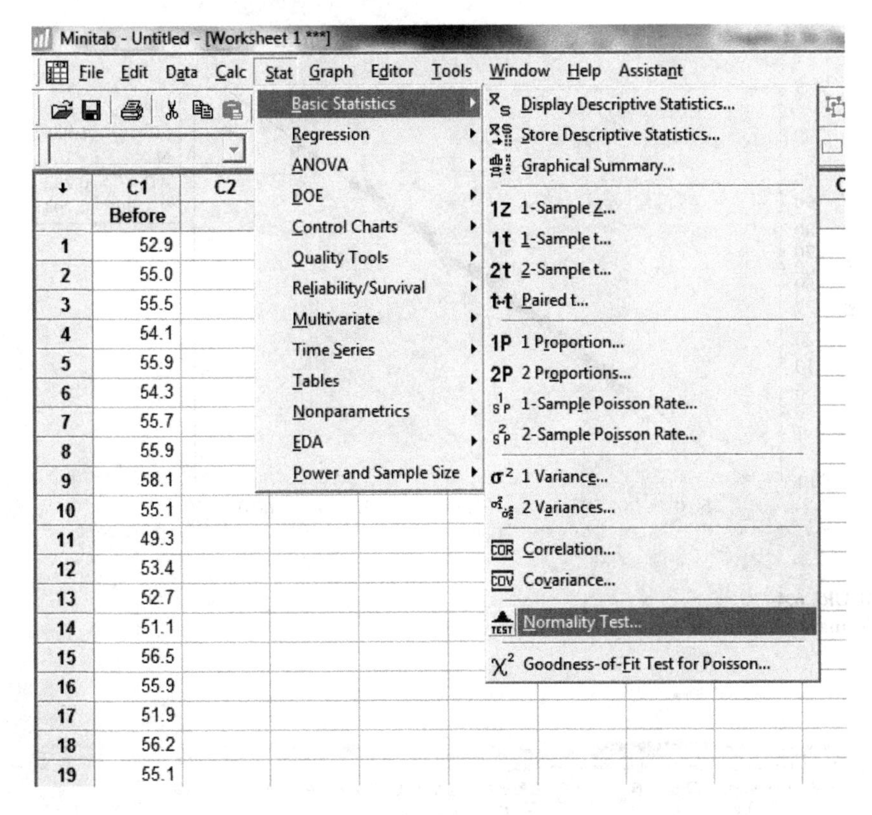

FIGURE 6.2
Approach to normality test before process improvement.

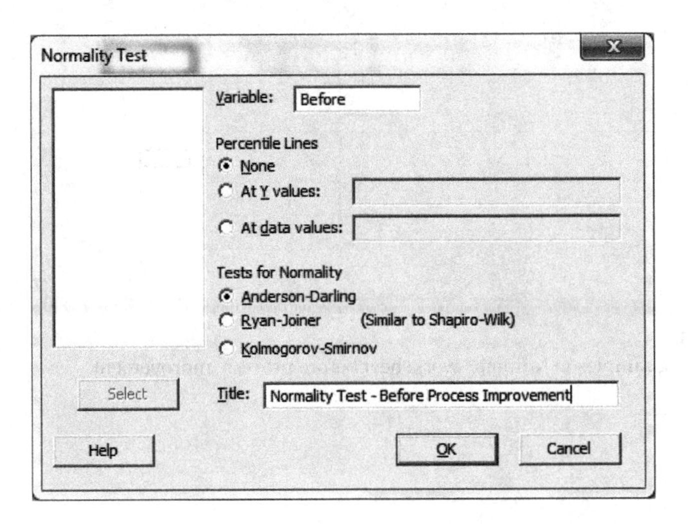

FIGURE 6.3
Normality test before process improvement.

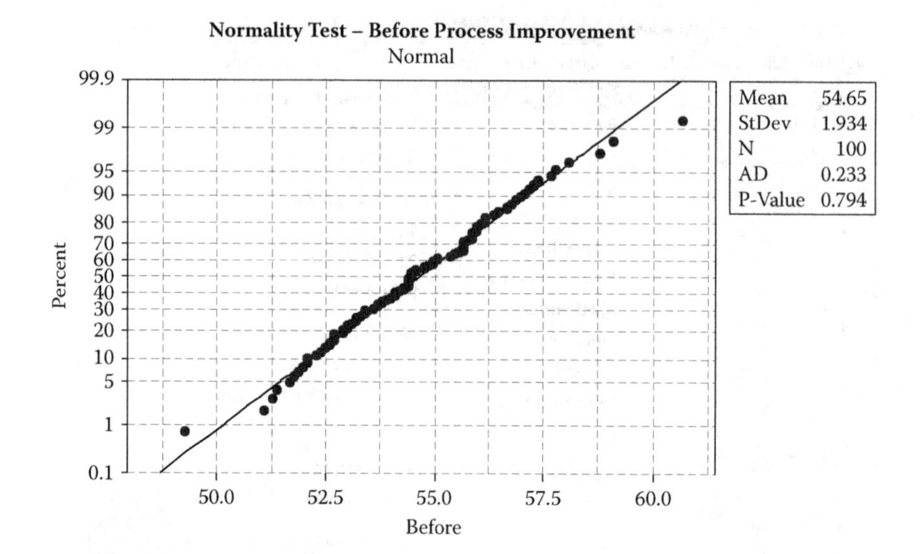

FIGURE 6.4
Normality test result before process improvement.

Minitab - Untitled - [Worksheet 1 ***]

File Edit Data Calc Stat Graph Editor Tools Window Help Assistant

↓	C1	C2	C3	C4	C5	C6	C7	C8	C9	C10	C11	C12	C13	C14	C15	C16
	Before		1	2	3	4	5	6	7	8	9	10	11	12	13	14
1	52.9		52.9	54.3	49.3	55.9	54.5	60.7	57.7	54.6	52.7	55.7	53.8	54.4	55.8	56.0
2	55.0		55.0	55.7	53.4	51.9	58.8	53.2	52.6	56.0	54.5	55.9	55.7	55.0	54.8	53.3
3	55.5		55.5	55.9	52.7	56.2	54.4	56.2	54.6	53.0	51.3	52.9	51.7	56.2	53.2	53.8
4	54.1		54.1	58.1	51.1	55.1	56.1	54.2	55.7	56.4	55.7	53.9	52.1	54.0	57.0	56.7
5	55.9		55.9	55.1	56.5	53.0	57.3	54.9	54.8	51.4	52.5	59.1	56.8	53.7	56.7	55.7
6	54.3															
7	55.7															
8	55.9															
9	58.1															
10	55.1															
11	49.3															
12	53.4															
13	52.7															
14	51.1															
15	56.5															
16	55.9															
17	51.9															
18	56.2															
19	55.1															

FIGURE 6.5
Data showing samples in Minitab® worksheet before process improvement.

Minitab - Untitled - [Worksheet 1 ***]

File Edit Data Calc Stat Graph Editor Tools Window Help Assistant

Subset Worksheet...
Split Worksheet...
Merge Worksheets...
Copy
Unstack Columns...
Stack
Transpose Columns...
Sort...
Rank...
Delete Rows...
Erase Variables...
Code
Change Data Type
Extract from Date/Time
Concatenate...
Display Data...

↓	C1	C4	C5	C6	C7	C8	C9
	Befor	2	3	4	5	6	7
1	52	54.3	49.3	55.9	54.5	60.7	57.7
2	55	55.7	53.4	51.9	58.8	53.2	52.6
3	55	55.9	52.7	56.2	54.4	56.2	54.6
4	54	58.1	51.1	55.1	56.1	54.2	55.7
5	55	55.1	56.5	53.0	57.3	54.9	54.8
6	54						
7	55						
8	55						
9	58						
10	55						
11	49						
12	53						
13	52						
14	51						
15	56.5						
16	55.9						
17	51.9						
18	56.2						
19	55.1						

FIGURE 6.6
Transpose of columns before process improvement.

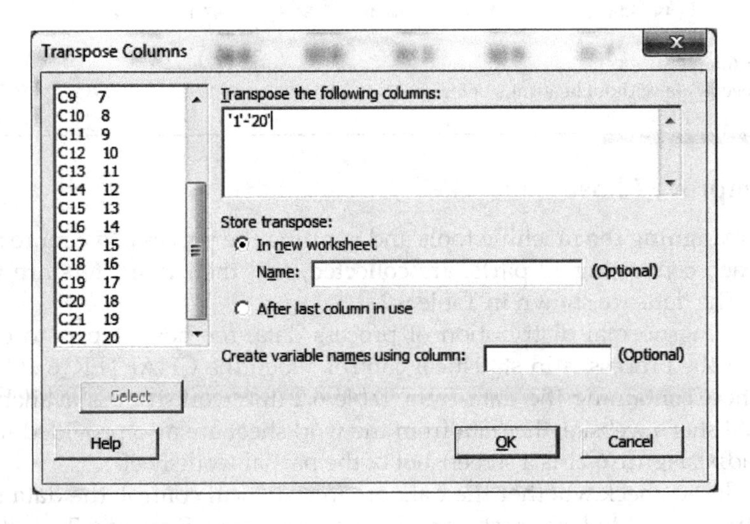

FIGURE 6.7
Storing transposed data in a new worksheet before process improvement.

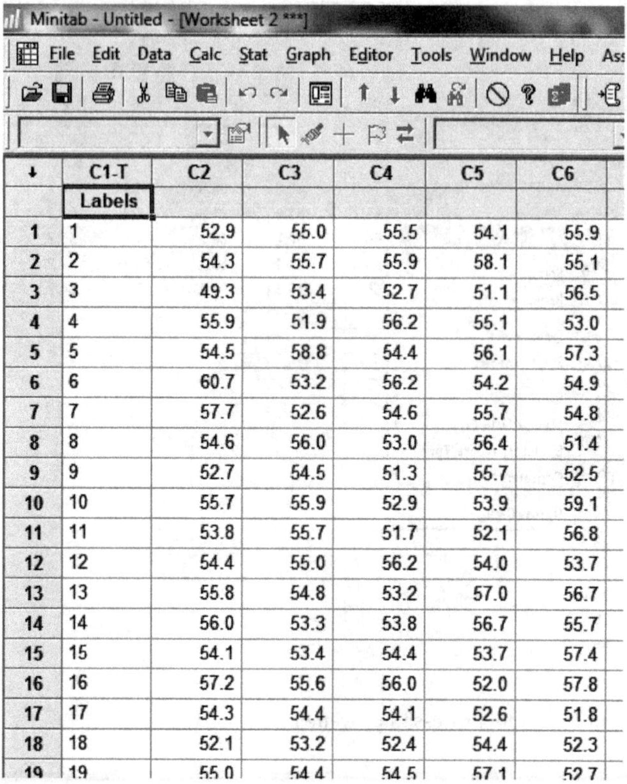

Minitab - Untitled - [Worksheet 2 ***]

File Edit Data Calc Stat Graph Editor Tools Window Help Ass

↓	C1-T	C2	C3	C4	C5	C6
	Labels					
1	1	52.9	55.0	55.5	54.1	55.9
2	2	54.3	55.7	55.9	58.1	55.1
3	3	49.3	53.4	52.7	51.1	56.5
4	4	55.9	51.9	56.2	55.1	53.0
5	5	54.5	58.8	54.4	56.1	57.3
6	6	60.7	53.2	56.2	54.2	54.9
7	7	57.7	52.6	54.6	55.7	54.8
8	8	54.6	56.0	53.0	56.4	51.4
9	9	52.7	54.5	51.3	55.7	52.5
10	10	55.7	55.9	52.9	53.9	59.1
11	11	53.8	55.7	51.7	52.1	56.8
12	12	54.4	55.0	56.2	54.0	53.7
13	13	55.8	54.8	53.2	57.0	56.7
14	14	56.0	53.3	53.8	56.7	55.7
15	15	54.1	53.4	54.4	53.7	57.4
16	16	57.2	55.6	56.0	52.0	57.8
17	17	54.3	54.4	54.1	52.6	51.8
18	18	52.1	53.2	52.4	54.4	52.3
19	19	55.0	54.4	54.5	57.1	52.7

FIGURE 6.8
Transposed data without headings before process improvement.

6.4 Improve Phase

Upon realigning the machine tools and running the process, five more samples, each containing 12 parts, are collected, and their diameters are measured. The data are shown in Table 6.2.

Assuming normal distribution of process data, the next step is to check whether the process is in statistical control. Open the CHAPTER_6_2.MTW worksheet containing the data from Table 6.2 (the worksheet is available at the publisher's website; the data from the worksheet are also provided in the Appendix). Figure 6.21 is a screenshot of the partial worksheet.

In order to check whether the data are in statistical control, the data need to be transposed to have each sample in a single row. Figures 6.22 and 6.23 show how to transpose the data, and Figure 6.24 shows the transposed data in a new worksheet. (Do not delete the previous worksheet because you need it for process capability analysis later.) For clarity, the headings of the columns are revised, and the revised worksheet is shown in Figure 6.25.

Minitab - Untitled - [Worksheet 2 ***]

File Edit Data Calc Stat Graph Editor Tools Window Help Assistant

↓	C1-T	C2	C3	C4	C5	C6
	Sample No.	Product 1	Product 2	Product 3	Product 4	Product 5
1	1	52.9	55.0	55.5	54.1	55.9
2	2	54.3	55.7	55.9	58.1	55.1
3	3	49.3	53.4	52.7	51.1	56.5
4	4	55.9	51.9	56.2	55.1	53.0
5	5	54.5	58.8	54.4	56.1	57.3
6	6	60.7	53.2	56.2	54.2	54.9
7	7	57.7	52.6	54.6	55.7	54.8
8	8	54.6	56.0	53.0	56.4	51.4
9	9	52.7	54.5	51.3	55.7	52.5
10	10	55.7	55.9	52.9	53.9	59.1
11	11	53.8	55.7	51.7	52.1	56.8
12	12	54.4	55.0	56.2	54.0	53.7
13	13	55.8	54.8	53.2	57.0	56.7
14	14	56.0	53.3	53.8	56.7	55.7
15	15	54.1	53.4	54.4	53.7	57.4
16	16	57.2	55.6	56.0	52.0	57.8
17	17	54.3	54.4	54.1	52.6	51.8
18	18	52.1	53.2	52.4	54.4	52.3
19	19	55.0	54.4	54.5	57.1	52.7

FIGURE 6.9
Transposed data with headings before process improvement.

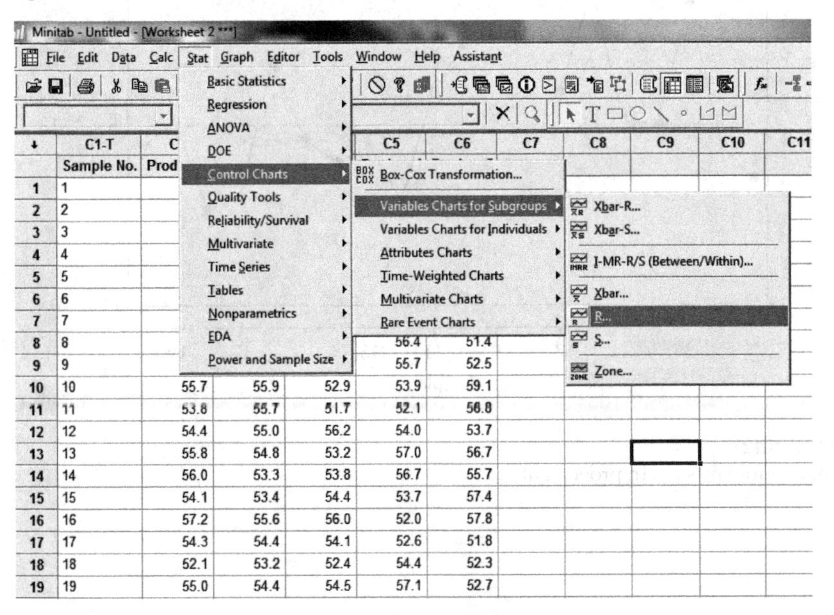

FIGURE 6.10
Approach to constructing R chart before process improvement.

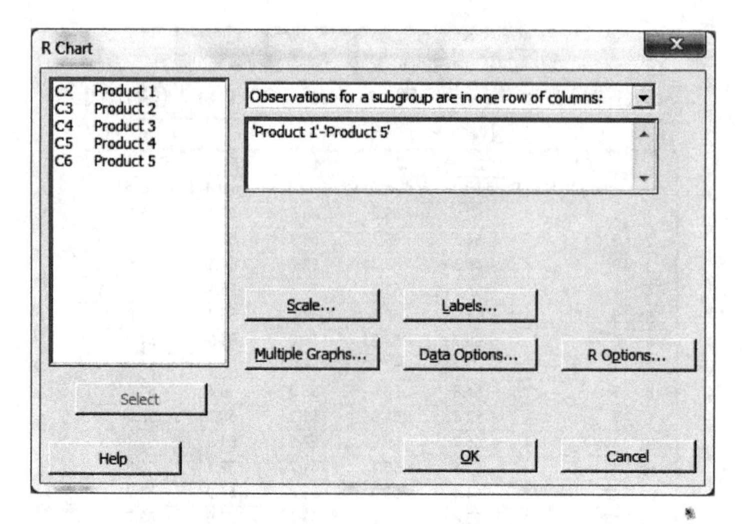

FIGURE 6.11
Selections for R chart before process improvement.

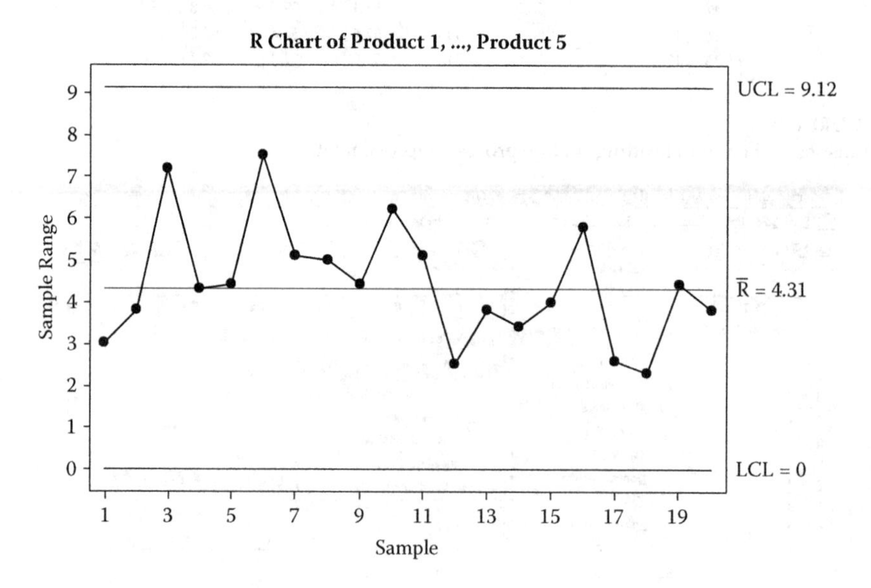

FIGURE 6.12
R chart before process improvement.

Minitab - Untitled - [Worksheet 2 ***]
File Edit Data Calc Stat Graph Editor Tools Window Help Assistant
Basic Statistics
Regression
ANOVA
DOE
Control Charts
Quality Tools
Reliability/Survival
Multivariate
Time Series
Tables
Nonparametrics
EDA
Power and Sample Size
Box-Cox Transformation...
Variables Charts for Subgroups
Variables Charts for Individuals
Attributes Charts
Time-Weighted Charts
Multivariate Charts
Rare Event Charts
Xbar-R...
Xbar-S...
I-MR-R/S (Between/Within)...
Xbar...
R...
S...
Zone...

↓	C1-T	C	C5	C6	C7	C8	C9	C10	C1
	Sample No.	Prod							
1	1								
2	2								
3	3								
4	4								
5	5								
6	6								
7	7								
8	8		56.4	51.4					
9	9		55.7	52.5					

↓	C1-T Sample No.	C2 Prod	C3	C4	C5	C6
10	10	55.7	55.9	52.9	53.9	59.1
11	11	53.8	55.7	51.7	52.1	56.8
12	12	54.4	55.0	56.2	54.0	53.7
13	13	55.8	54.8	53.2	57.0	56.7
14	14	56.0	53.3	53.8	56.7	55.7
15	15	54.1	53.4	54.4	53.7	57.4
16	16	57.2	55.6	56.0	52.0	57.8
17	17	54.3	54.4	54.1	52.6	51.8
18	18	52.1	53.2	52.4	54.4	52.3
19	19	55.0	54.4	54.5	57.1	52.7

FIGURE 6.13
Approach to constructing $\overline{X}$ chart before process improvement.

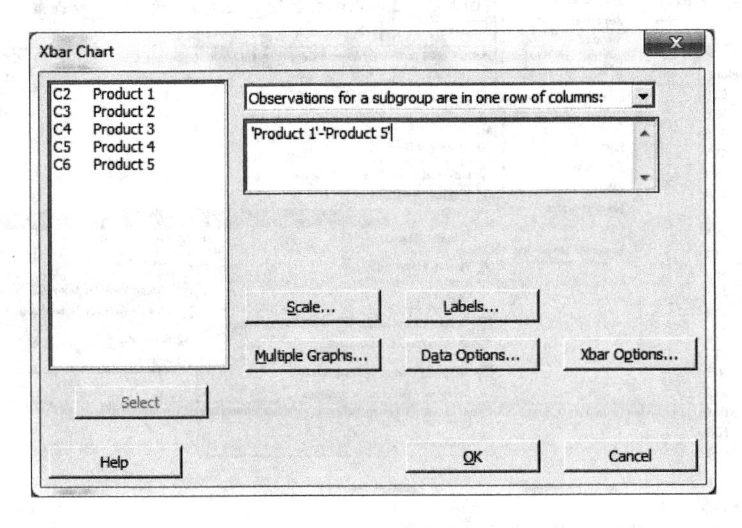

FIGURE 6.14
Selections for $\overline{X}$ chart before process improvement.

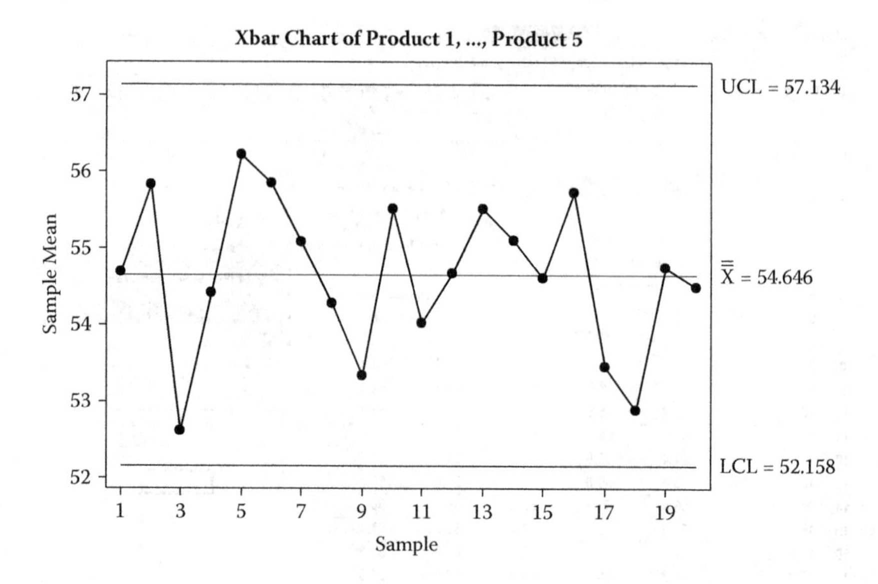

FIGURE 6.15
$\overline{X}$ chart before process improvement.

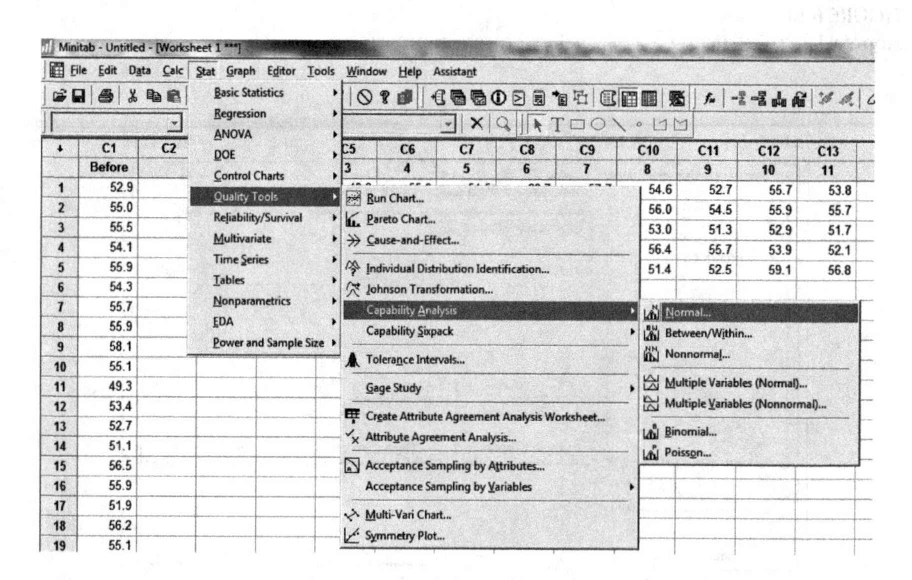

FIGURE 6.16
Approach to capability analysis before process improvement.

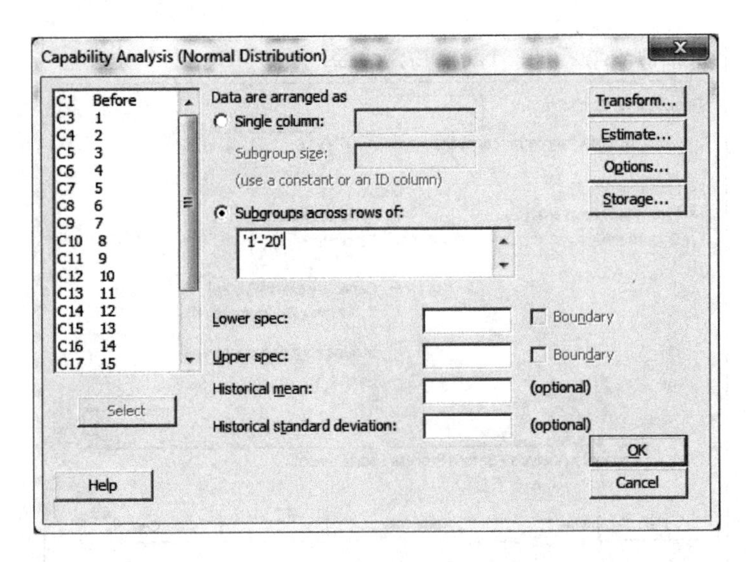

FIGURE 6.17
Subgroup selection for capability analysis before process improvement.

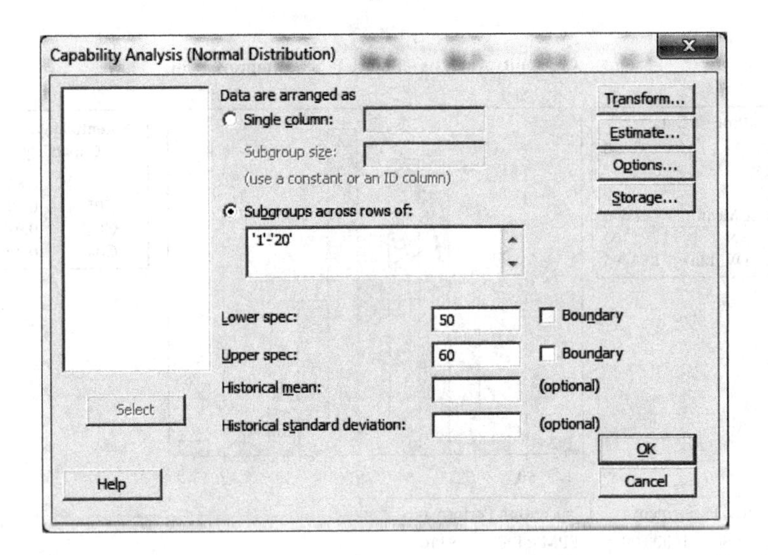

FIGURE 6.18
Specification limits for capability analysis before process improvement.

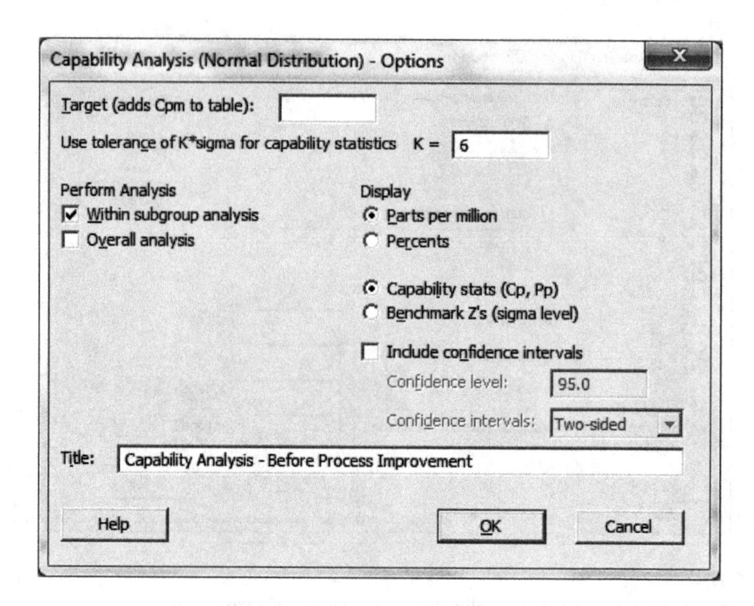

FIGURE 6.19
Options for capability analysis before process improvement.

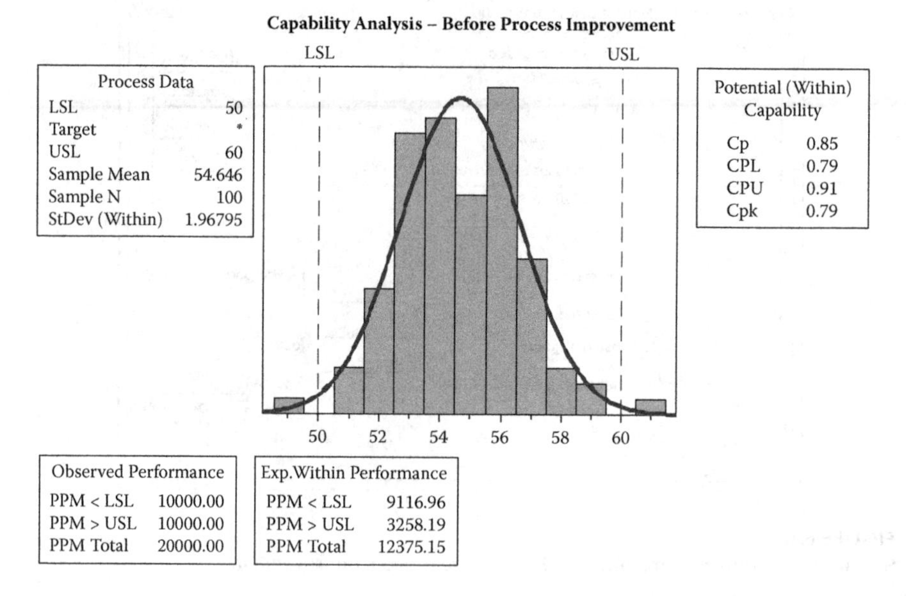

FIGURE 6.20
Capability analysis before process improvement.

TABLE 6.2
Production Data after Process Improvement

1	2	3	4	5
56	55	54	55.9	54.5
56	51	53.4	51.9	58.8
55.5	55.9	52.7	56.2	54.4
54.1	58.1	51.1	55.1	56.1
55.9	55.1	56.5	53	57.3
53.2	55.7	56	57.7	57.7
56.2	55.9	53.3	52.6	52.6
54.2	52.9	53.8	54.6	54.6
54.9	53.9	56.7	55.7	55.7
56	59.1	55.7	54.8	54.8
56	52.4	53.1	55	58
53	54.4	53.4	54.4	57

The data are variable data and the sample size is 12, therefore the appropriate control charts to construct are the $\overline{X}$ chart and S chart. Figures 6.26 and 6.27 show how to construct the S chart, and Figure 6.28 shows the S chart. Inasmuch as the sample standard deviations are in statistical control, check whether the sample means are in statistical control.

Minitab - Untitled - [Worksheet 1 ***]

File Edit Data Calc Stat Graph Editor Tools Window

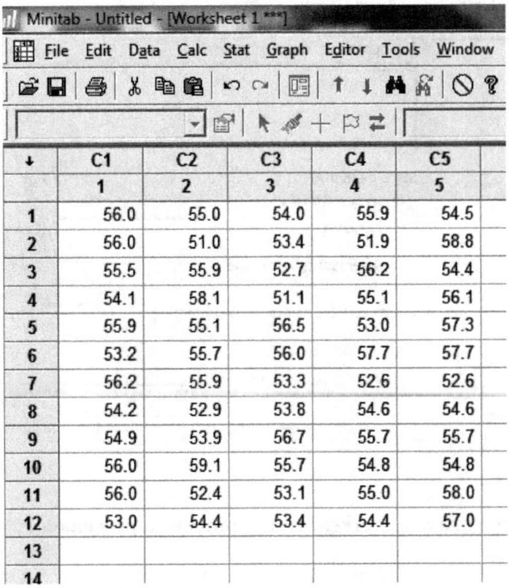

↓	C1	C2	C3	C4	C5
	1	2	3	4	5
1	56.0	55.0	54.0	55.9	54.5
2	56.0	51.0	53.4	51.9	58.8
3	55.5	55.9	52.7	56.2	54.4
4	54.1	58.1	51.1	55.1	56.1
5	55.9	55.1	56.5	53.0	57.3
6	53.2	55.7	56.0	57.7	57.7
7	56.2	55.9	53.3	52.6	52.6
8	54.2	52.9	53.8	54.6	54.6
9	54.9	53.9	56.7	55.7	55.7
10	56.0	59.1	55.7	54.8	54.8
11	56.0	52.4	53.1	55.0	58.0
12	53.0	54.4	53.4	54.4	57.0
13					
14					

FIGURE 6.21
Data showing samples in Minitab® worksheet after process improvement.

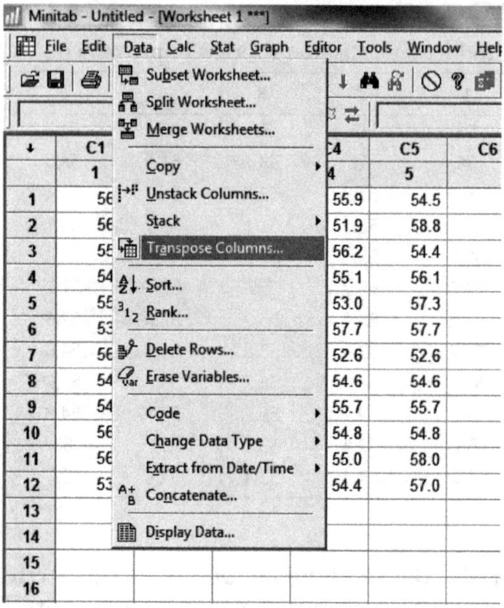

FIGURE 6.22
Transpose of columns after process improvement.

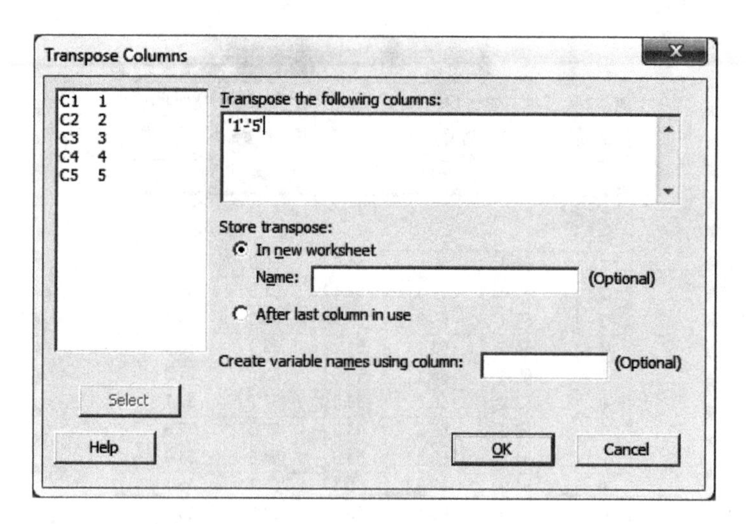

FIGURE 6.23
Storing transposed data in a new worksheet after process improvement.

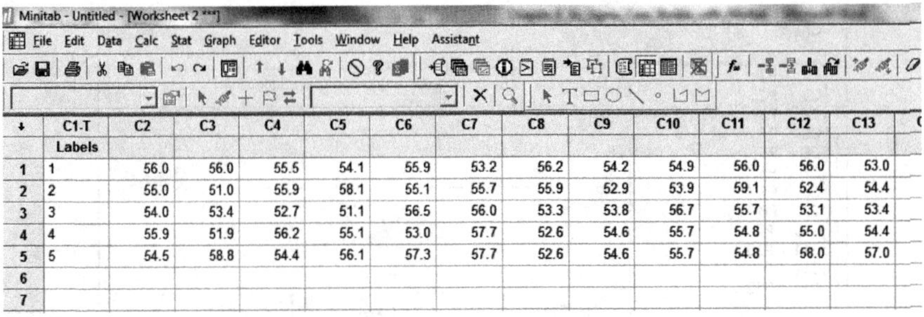

	C1-T Labels	C2	C3	C4	C5	C6	C7	C8	C9	C10	C11	C12	C13
1	1	56.0	56.0	55.5	54.1	55.9	53.2	56.2	54.2	54.9	56.0	56.0	53.0
2	2	55.0	51.0	55.9	58.1	55.1	55.7	55.9	52.9	53.9	59.1	52.4	54.4
3	3	54.0	53.4	52.7	51.1	56.5	56.0	53.3	53.8	56.7	55.7	53.1	53.4
4	4	55.9	51.9	56.2	55.1	53.0	57.7	52.6	54.6	55.7	54.8	55.0	54.4
5	5	54.5	58.8	54.4	56.1	57.3	57.7	52.6	54.6	55.7	54.8	58.0	57.0

FIGURE 6.24
Transposed data without headings after process improvement.

	C1-T Sample No.	C2 P1	C3 P2	C4 P3	C5 P4	C6 P5	C7 P6	C8 P7	C9 P8	C10 P9	C11 P10	C12 P11	C13 P12
1	1	56.0	56.0	55.5	54.1	55.9	53.2	56.2	54.2	54.9	56.0	56.0	53.0
2	2	55.0	51.0	55.9	58.1	55.1	55.7	55.9	52.9	53.9	59.1	52.4	54.4
3	3	54.0	53.4	52.7	51.1	56.5	56.0	53.3	53.8	56.7	55.7	53.1	53.4
4	4	55.9	51.9	56.2	55.1	53.0	57.7	52.6	54.6	55.7	54.8	55.0	54.4
5	5	54.5	58.8	54.4	56.1	57.3	57.7	52.6	54.6	55.7	54.8	58.0	57.0

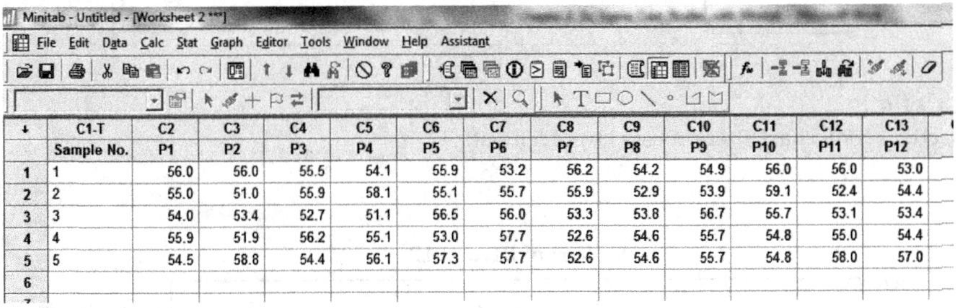

FIGURE 6.25
Transposed data with headings after process improvement.

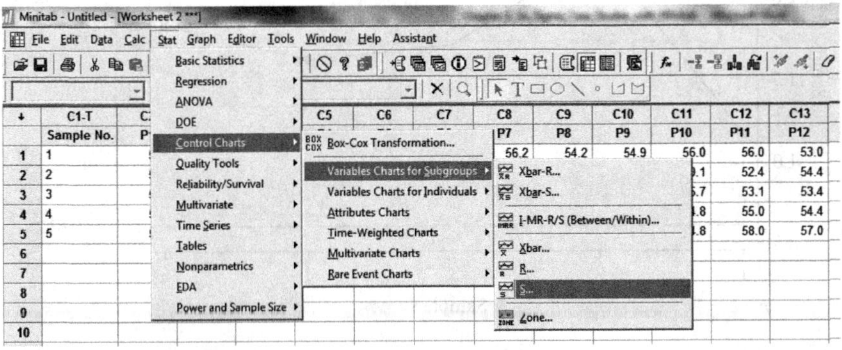

FIGURE 6.26
Approach to constructing *S* chart after process improvement.

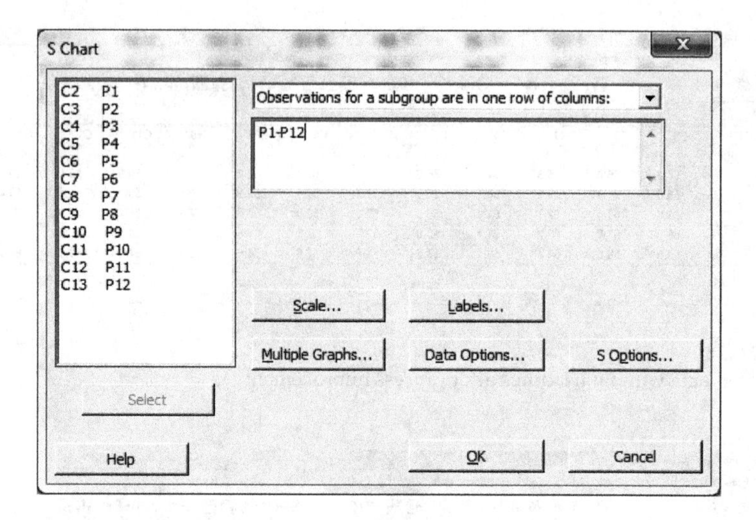

FIGURE 6.27
Selections for *S* chart after process improvement.

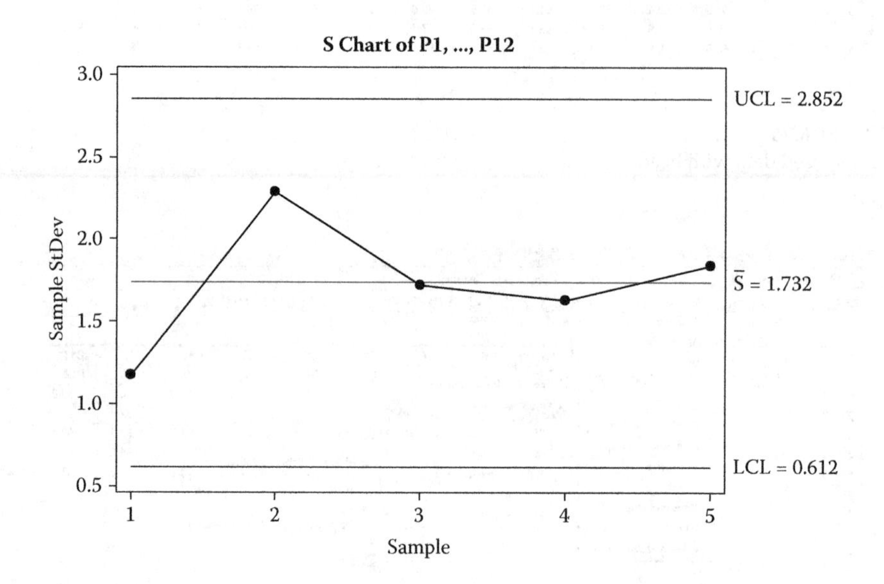

FIGURE 6.28
S chart after process improvement.

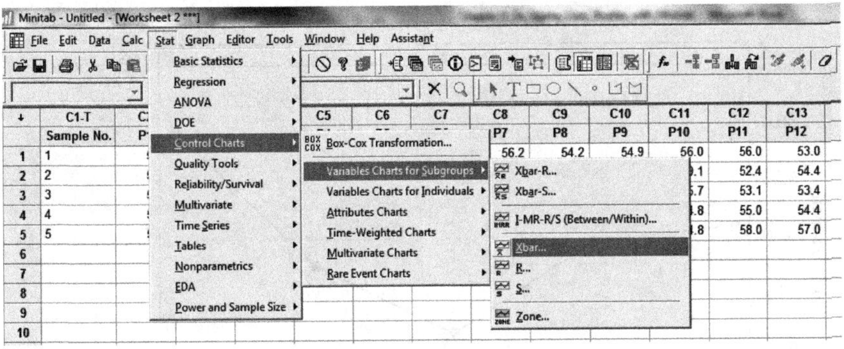

FIGURE 6.29
Approach to constructing $\overline{X}$ chart after process improvement.

Figures 6.29 and 6.30 show how to construct the $\overline{X}$ chart. It is evident from the $\overline{X}$ chart in Figure 6.31 that the sample means are also in statistical control.

Because the process data are normally distributed and are in statistical control, we can calculate the process capability ratios now. Figures 6.32 and 6.33 illustrate how to do so. Figure 6.34 shows that the *USL* and *LSL* are entered in the respective boxes. Click on "Options" in the dialog box shown in Figure 6.34, and the dialog box shown in Figure 6.35 opens. Uncheck the "Overall Analysis" box and enter the "Title" as shown in Figure 6.35. Click on "OK" and it takes you back to the dialog box shown in Figure 6.34. Click on "OK" and the graph shown in Figure 6.36 is the result. The new C_p and C_{pk} values are 0.89 and 0.89, respectively. As is obvious, the process is improved.

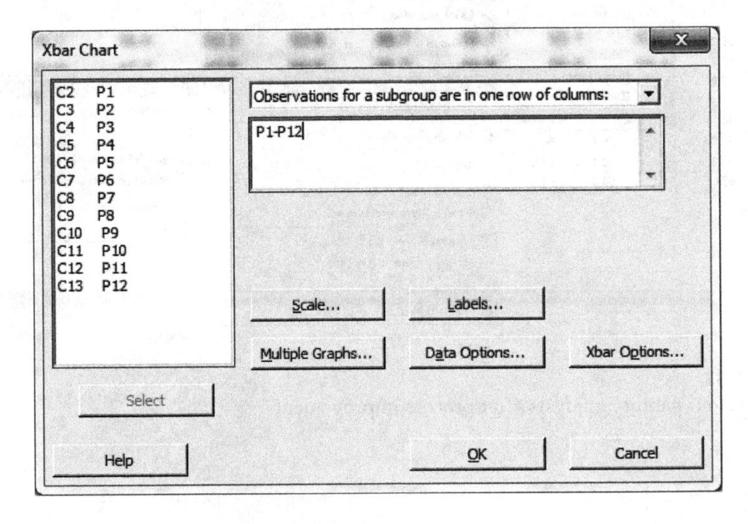

FIGURE 6.30
Selections for $\overline{X}$ chart after process improvement.

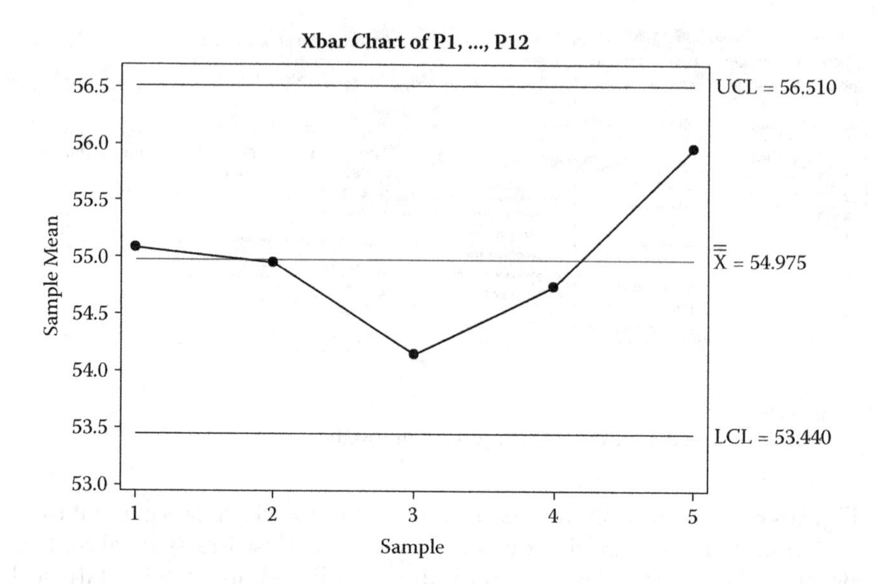

FIGURE 6.31
$\overline{X}$ chart after process improvement.

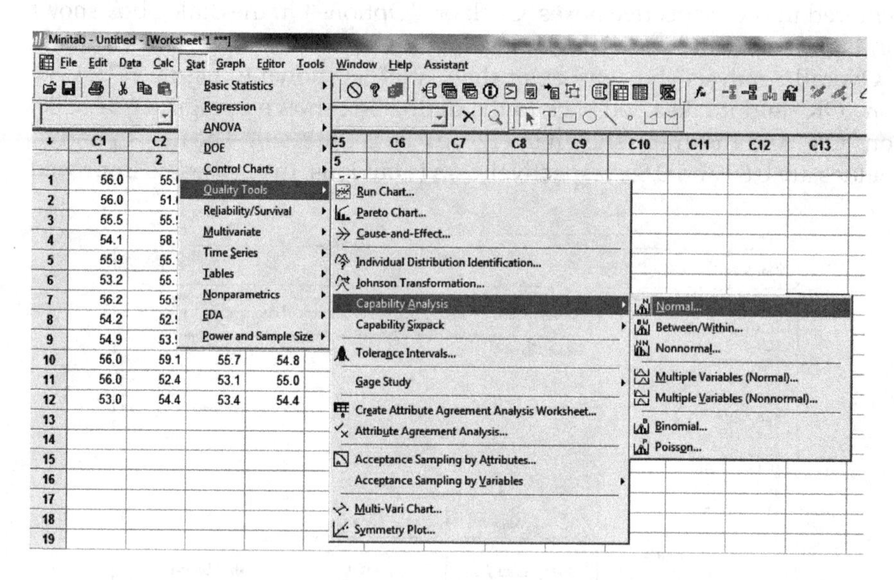

FIGURE 6.32
Approach to capability analysis after process improvement.

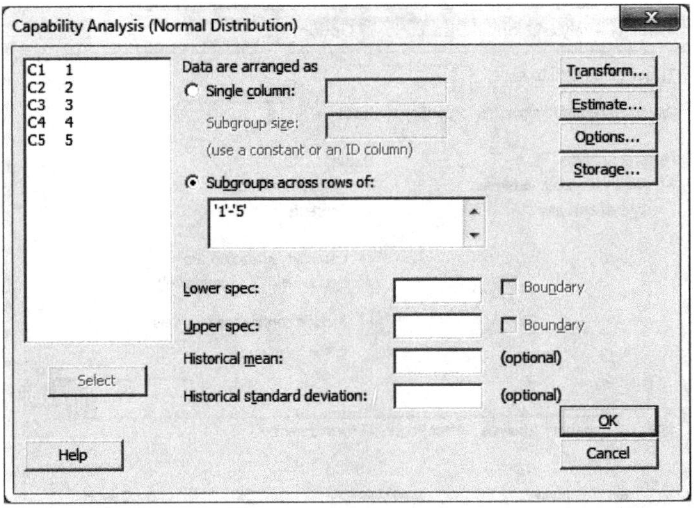

Capability Analysis (Normal Distribution)

C1 1
C2 2
C3 3
C4 4
C5 5

Data are arranged as
Single column:
Subgroup size:
(use a constant or an ID column)
Subgroups across rows of:
'1'-'5'

Lower spec: Boundary
Upper spec: Boundary
Historical mean: (optional)
Historical standard deviation: (optional)

Transform... Estimate... Options... Storage...
Select Help OK Cancel

FIGURE 6.33
Subgroup selection for capability analysis after process improvement.

Capability Analysis (Normal Distribution)

Data are arranged as
Single column:
Subgroup size:
(use a constant or an ID column)
Subgroups across rows of:
'1'-'5'

Lower spec: 50 Boundary
Upper spec: 60 Boundary
Historical mean: (optional)
Historical standard deviation: (optional)

Transform... Estimate... Options... Storage...
Select Help OK Cancel

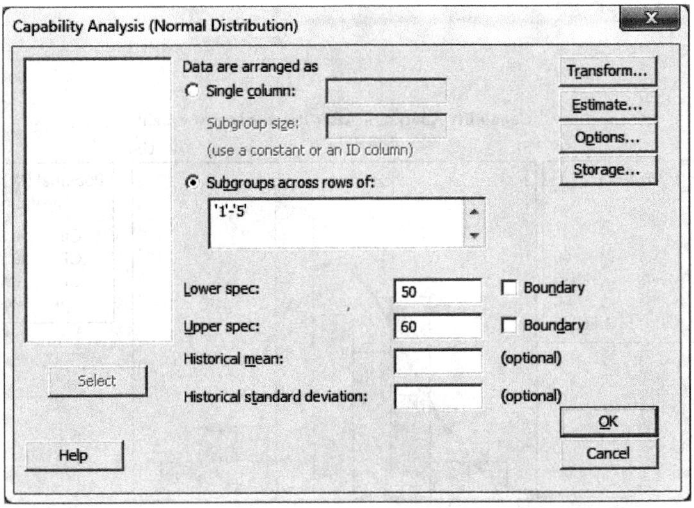

FIGURE 6.34
Specification limits for capability analysis after process improvement.

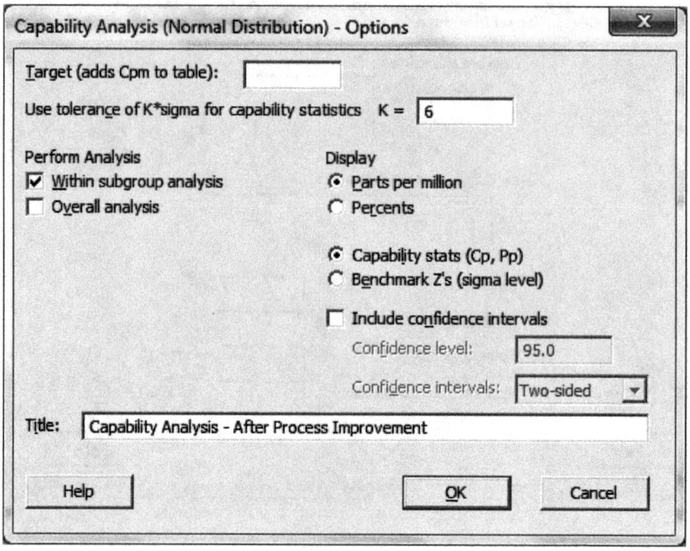

Capability Analysis (Normal Distribution) - Options

Target (adds Cpm to table):

Use tolerance of K*sigma for capability statistics K = 6

Perform Analysis
- [x] Within subgroup analysis
- [ ] Overall analysis

Display
- (•) Parts per million
- ( ) Percents

- (•) Capability stats (Cp, Pp)
- ( ) Benchmark Z's (sigma level)

- [ ] Include confidence intervals

Confidence level: 95.0

Confidence intervals: Two-sided

Title: Capability Analysis - After Process Improvement

Help | OK | Cancel

FIGURE 6.35
Options for capability analysis after process improvement.

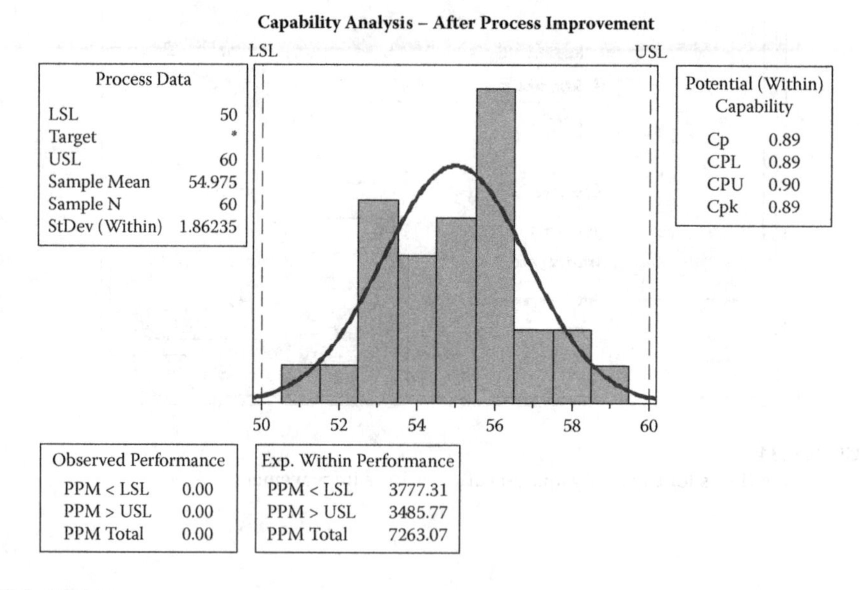

FIGURE 6.36
Capability analysis after process improvement.

6.5 Control Phase

With the help of the supplier of the machines and their tools, the production manager installs a fail-safe mechanism that prevents misalignment of the two tools.

7

Binary Logistic Regression to Predict Customer Satisfaction at a Restaurant

The owners of a busy restaurant in downtown Boston wish to improve customer satisfaction. To this end, the restaurant manager is asked to collect customer satisfaction data and suggest feasible recommendations to the owners.

Sections 7.1 and 7.2 give brief descriptions of the define phase and the measure phase, respectively. Section 7.3 illustrates the analyze phase with detailed instructions for using Minitab®. The improve phase and the control phase are briefly explained in Section 7.4.

7.1 Define Phase

The objective is to improve the proportion of satisfied customers at the restaurant. Inasmuch as customer satisfaction is a binary ("yes" or "no") event, the restaurant manager uses binary logistic regression that might help to predict the probability of satisfying a given customer.

If P is the probability of satisfying a customer, then the binary regression model is

$$\text{log(odds of satisfying a customer)} = \log\left(\frac{P}{1-P}\right) = \beta_0 + \beta_1 X_1 + \beta_2 X_2 + \cdots$$

where β_0 is a constant and $\beta_1, \beta_2, \ldots$ are coefficients of independent factors $X_1, X_2, \ldots$ In other words,

$$P = \frac{e^{(\beta_0 + \beta_1 X_1 + \beta_2 X_2 + \cdots)}}{1 + e^{(\beta_0 + \beta_1 X_1 + \beta_2 X_2 + \cdots)}}$$

The independent factors that the restaurant manager wishes to consider are the variety of drinks on the menu, the number of waiters, and whether a free refill is offered.

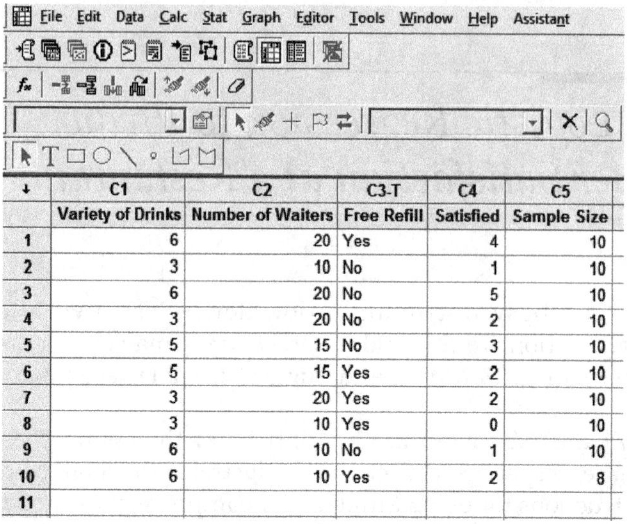

↓	C1	C2	C3-T	C4	C5
	Variety of Drinks	Number of Waiters	Free Refill	Satisfied	Sample Size
1	6	20	Yes	4	10
2	3	10	No	1	10
3	6	20	No	5	10
4	3	20	No	2	10
5	5	15	No	3	10
6	5	15	Yes	2	10
7	3	20	Yes	2	10
8	3	10	Yes	0	10
9	6	10	No	1	10
10	6	10	Yes	2	8
11					

FIGURE 7.1
Data collected to measure customer satisfaction.

7.2 Measure Phase

Random samples of customers are taken on different days with different levels of these factors. These customers are asked whether they are satisfied with the service.

Open the CHAPTER_7.MTW worksheet for the collected data (the worksheet is available at the publisher's website; the data from the worksheet are also provided in the Appendix). Figure 7.1 shows a screenshot of the worksheet. Notice that the first row in the worksheet is on a day when 6 different drinks are on the menu, 20 waiters are working, a free refill is offered, and 4 out of the 10 sampled customers are satisfied with the service.

7.3 Analyze Phase

Figure 7.2 shows how to select "Binary Logistic Regression". Doing so opens the dialog box shown in Figure 7.3. Select "Response in event/trial format", select the "Satisfied" column for "Number of events", and "Sample Size" column for "Number of trials". For the "Model", select "Variety of Drinks", "Number of Waiters", and "Free Refill". Because "Free Refill" is a categorical variable (as opposed to a numerical variable), select "Free Refill" for "Factors (optional)". Click on "Options" and the dialog box shown in Figure 7.4 opens.

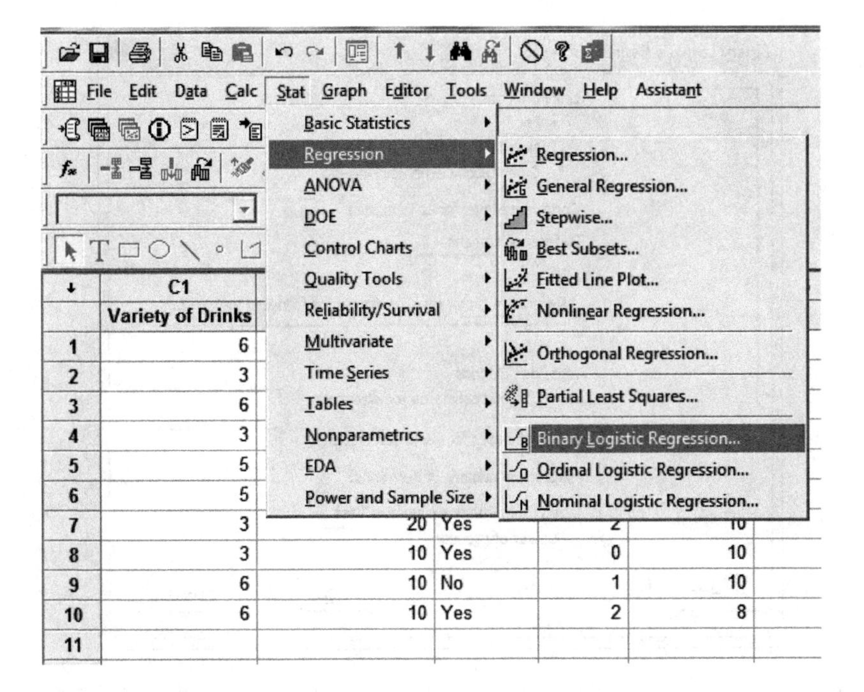

FIGURE 7.2
Selection of "Binary Logistic Regression".

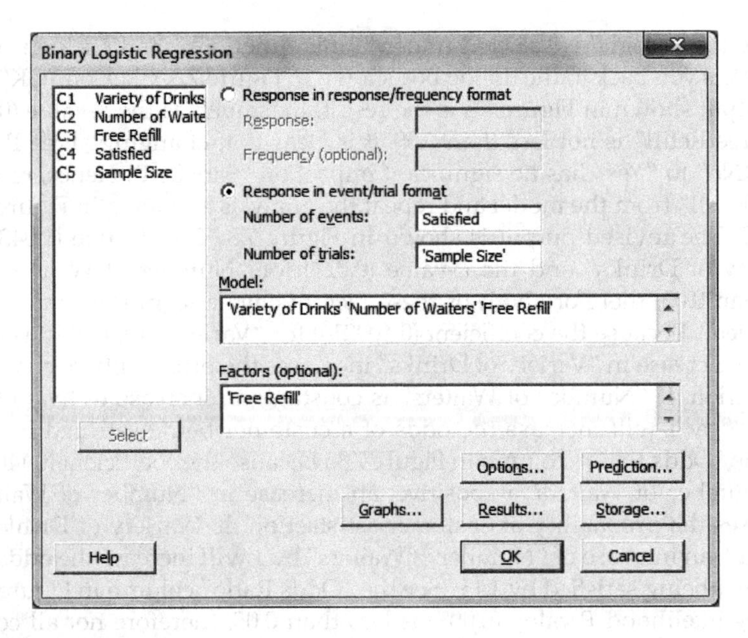

FIGURE 7.3
Selection of variables for binary logistic regression model.

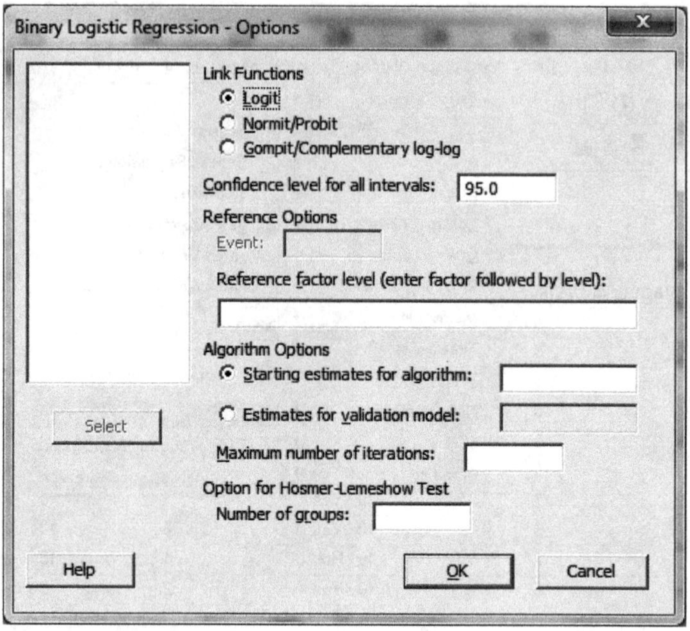

FIGURE 7.4
Selection of "Logit".

Ensure that "Logit" is selected under "Link Functions" and click on "OK". This takes you back to the dialog box shown in Figure 7.3. Click on "OK" and the output shown in Figure 7.5 is the result. Inasmuch as the *P*-value (0.673) for "Free Refill" is not less than 0.05, it is clear that changing "Free Refill" from "No" to "Yes" has no significant impact on "Satisfied". Hence, remove "Free Refill" from the model and repeat the analysis as shown in Figures 7.6 and 7.7. The revised output is shown in Figure 7.8. The *P*-value (0.043) for "Variety of Drinks" and the *P*-value (0.023) for "Number of Waiters" are less than 0.05, therefore both of these factors have a significant impact on "Satisfied". Because the coefficient (0.409216) for "Variety of Drinks" is positive, an increase in "Variety of Drinks" increases the probability of customer satisfaction. If "Number of Waiters" is constant, an increase of "Variety of Drinks" by 1 will increase the odds of a customer being satisfied by 1.51. (See the "Odds Ratio" column in Figure 7.8.) Because the coefficient (0.140952) for "Number of Waiters" is positive, an increase in "Number of Waiters" increases the probability of customer satisfaction. If "Variety of Drinks" is constant, an increase of "Number of Waiters" by 1 will increase the odds of a customer being satisfied by 1.15. (See the "Odds Ratio" column in Figure 7.8.) The log-likelihood *P*-value (0.007) is less than 0.05, therefore not all coefficients in the binary logistics regression model are zero. Also, because the

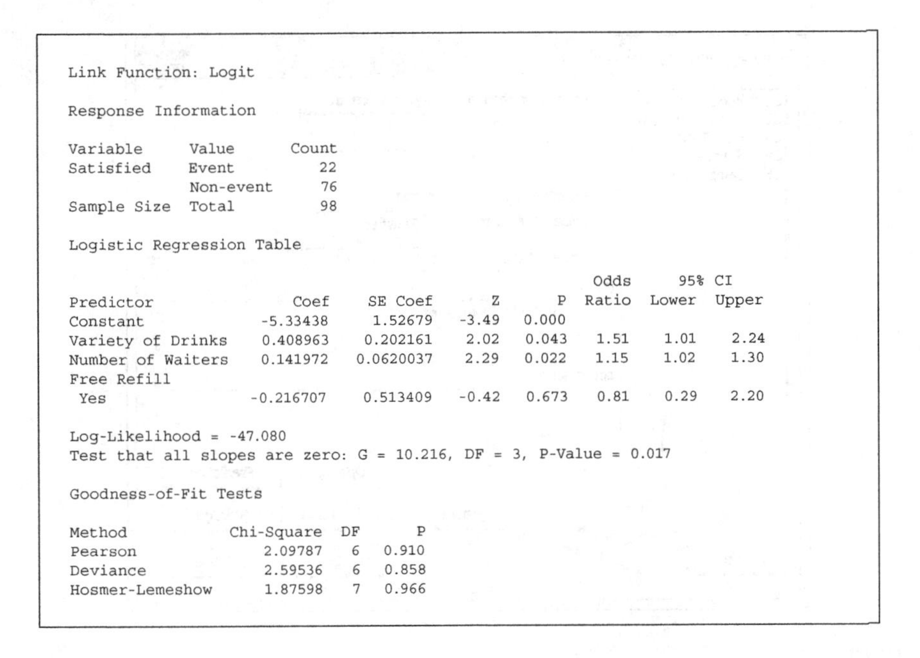

Link Function: Logit

Response Information

Variable	Value	Count
Satisfied	Event	22
	Non-event	76
Sample Size	Total	98

Logistic Regression Table

					Odds	95% CI	
Predictor	Coef	SE Coef	Z	P	Ratio	Lower	Upper
Constant	-5.33438	1.52679	-3.49	0.000			
Variety of Drinks	0.408963	0.202161	2.02	0.043	1.51	1.01	2.24
Number of Waiters	0.141972	0.0620037	2.29	0.022	1.15	1.02	1.30
Free Refill							
Yes	-0.216707	0.513409	-0.42	0.673	0.81	0.29	2.20

Log-Likelihood = -47.080

Test that all slopes are zero: G = 10.216, DF = 3, P-Value = 0.017

Goodness-of-Fit Tests

Method	Chi-Square	DF	P
Pearson	2.09787	6	0.910
Deviance	2.59536	6	0.858
Hosmer-Lemeshow	1.87598	7	0.966

FIGURE 7.5
Output of binary logistic regression.

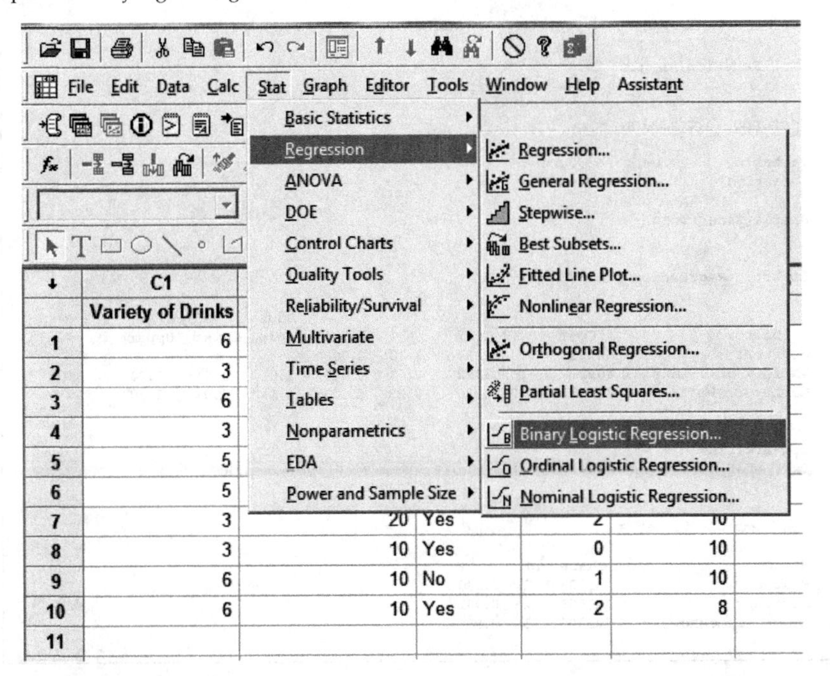

FIGURE 7.6
Repeat of analysis.

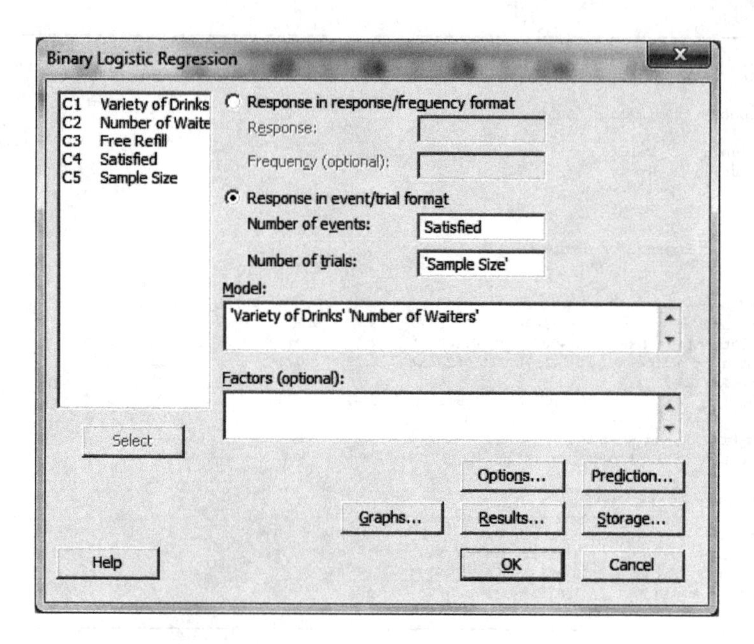

FIGURE 7.7
Removal of "Free Refill" from the model.

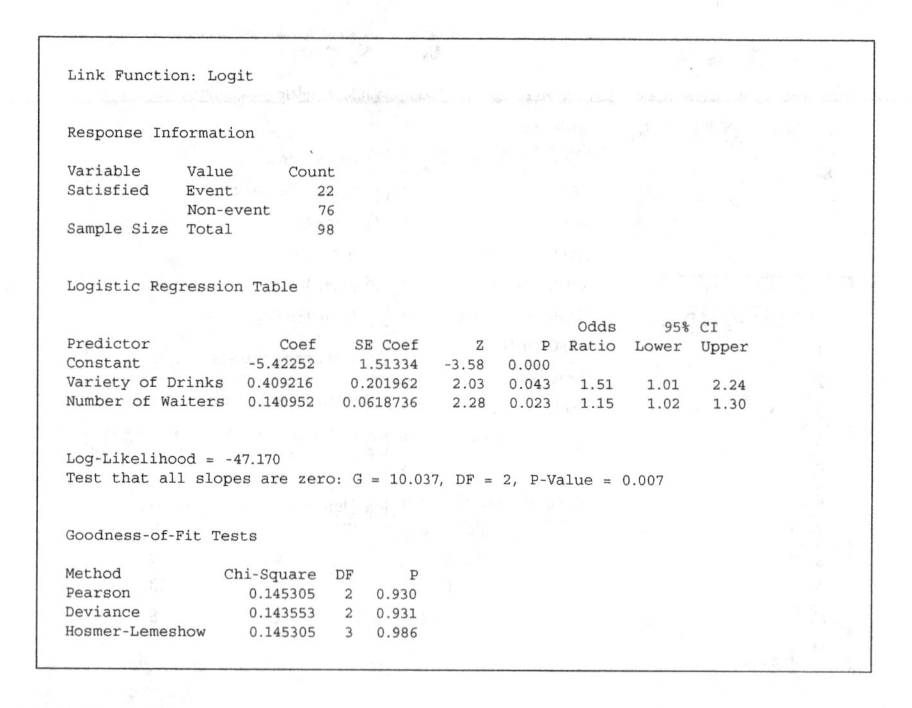

Link Function: Logit

Response Information

Variable	Value	Count
Satisfied	Event	22
	Non-event	76
Sample Size	Total	98

Logistic Regression Table

Predictor	Coef	SE Coef	Z	P	Odds Ratio	95% CI Lower	95% CI Upper
Constant	-5.42252	1.51334	-3.58	0.000			
Variety of Drinks	0.409216	0.201962	2.03	0.043	1.51	1.01	2.24
Number of Waiters	0.140952	0.0618736	2.28	0.023	1.15	1.02	1.30

Log-Likelihood = -47.170
Test that all slopes are zero: G = 10.037, DF = 2, P-Value = 0.007

Goodness-of-Fit Tests

Method	Chi-Square	DF	P
Pearson	0.145305	2	0.930
Deviance	0.143553	2	0.931
Hosmer-Lemeshow	0.145305	3	0.986

FIGURE 7.8
Revised output of binary logistic regression.

File Edit Data Calc Stat Graph Editor Tools Window Help Assistant

↓	C1	C2	C3-T	C4	C5	C6	C7
	Variety of Drinks	Number of Waiters	Free Refill	Satisfied	Sample Size	Variety 2	Waiters 2
1	6	20	Yes	4	10		
2	3	10	No	1	10		
3	6	20	No	5	10		
4	3	20	No	2	10		
5	5	15	No	3	10		
6	5	15	Yes	2	10		
7	3	20	Yes	2	10		
8	3	10	Yes	0	10		
9	6	10	No	1	10		
10	6	10	Yes	2	8		
11							

FIGURE 7.9
Creation of new columns, "Variety 2" and "Waiters 2".

P-value of each of the three goodness-of-fit tests is not less than 0.05, the binary logistics regression model is a good fit to the data collected.

The probability of customer satisfaction can be predicted as follows, with the equation:

$$P = \frac{e^{(-5.42252 + 0.409216 * \text{Variety of Drinks} + 0.140952 * \text{Number of Waiters})}}{1 + e^{(-5.42252 + 0.409216 * \text{Variety of Drinks} + 0.140952 * \text{Number of Waiters})}}$$

For prediction of customer satisfaction for different levels of "Variety of Drinks" and "Number of Waiters", label two empty columns as "Variety 2" and "Waiters 2", as shown in Figure 7.9. Then, as shown in Figure 7.10, select "Make Mesh Data". Doing so opens the dialog box shown in Figure 7.11. For "X", select "Variety 2" for "Store in", enter "3" for "From", enter "6" for "To", and enter "4" for "Number of positions". (This is done in order to consider the following 4 values for "Variety of Drinks": 3, 4, 5, 6.) For "Y", select "Waiters 2" for "Store in", enter "10" for "From", enter "20" for "To", and enter "3" for "Number of positions". (This is done in order to consider the following three values for "Number of Waiters": 10, 15, 20.) Click on "OK", and all possible combinations of "Variety of Drinks" and "Number of Waiters" appear in "Variety 2" and "Waiters 2" columns, as shown in Figure 7.12. In order to calculate the probability of customer satisfaction for each of these combinations, label an empty column as "Prob Satistifed" as shown in Figure 7.12. Then, select "Calculator" as shown in Figure 7.13.

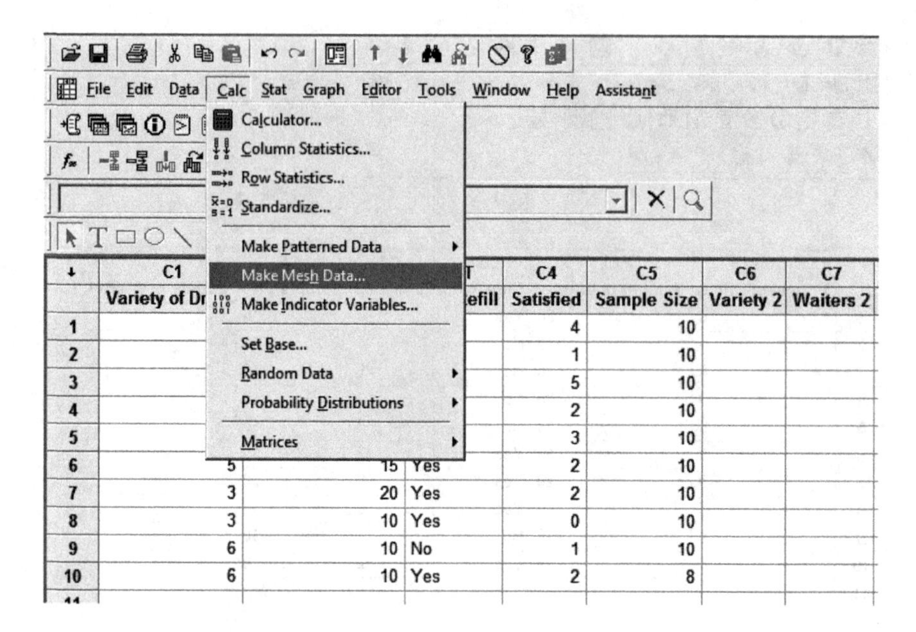

FIGURE 7.10
Selection of "Make Mesh Data".

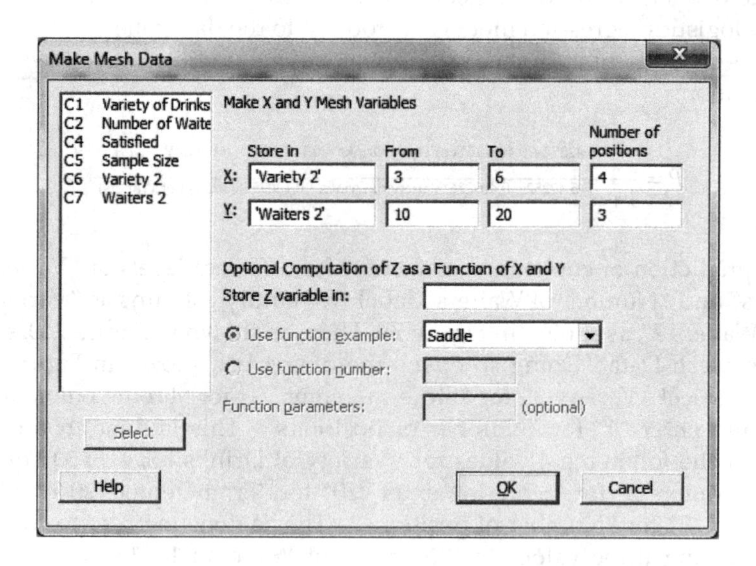

FIGURE 7.11
Entry of positions for mesh data.

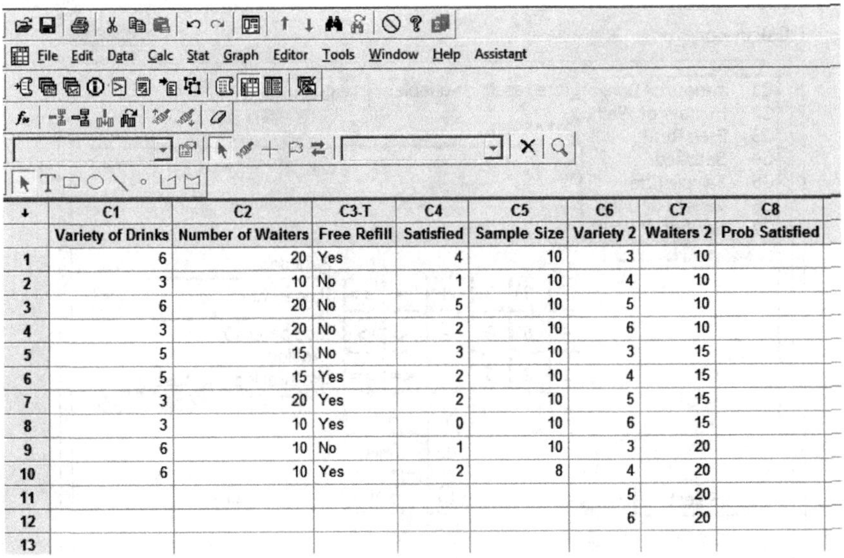
File Edit Data Calc Stat Graph Editor Tools Window Help Assistant

↓	C1	C2	C3-T	C4	C5	C6	C7	C8
	Variety of Drinks	Number of Waiters	Free Refill	Satisfied	Sample Size	Variety 2	Waiters 2	Prob Satisfied
1	6	20	Yes	4	10	3	10	
2	3	10	No	1	10	4	10	
3	6	20	No	5	10	5	10	
4	3	20	No	2	10	6	10	
5	5	15	No	3	10	3	15	
6	5	15	Yes	2	10	4	15	
7	3	20	Yes	2	10	5	15	
8	3	10	Yes	0	10	6	15	
9	6	10	No	1	10	3	20	
10	6	10	Yes	2	8	4	20	
11						5	20	
12						6	20	
13								

FIGURE 7.12
Meshed data.

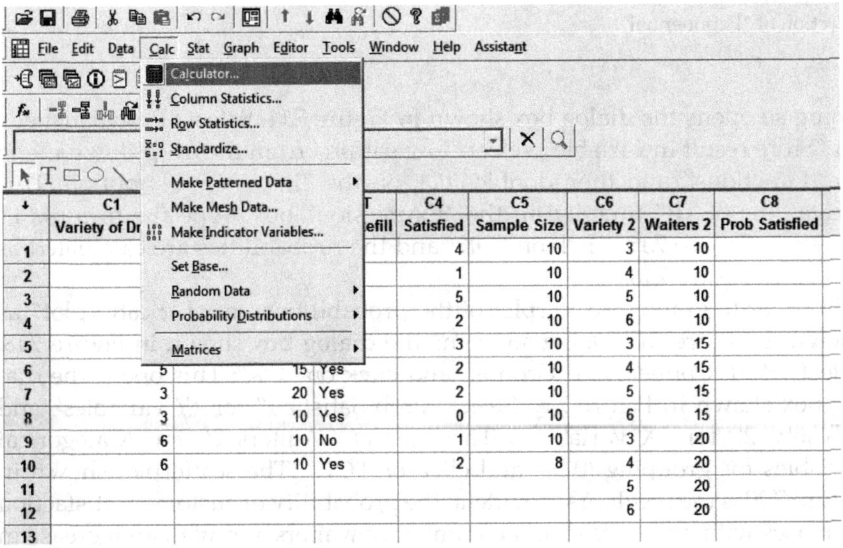
File Edit Data Calc Stat Graph Editor Tools Window Help Assistant
Calculator...
Column Statistics...
Row Statistics...
Standardize...
Make Patterned Data
Make Mesh Data...
Make Indicator Variables...
Set Base...
Random Data
Probability Distributions
Matrices

↓	C1	C2	C3-T	C4	C5	C6	C7	C8
	Variety of Dr		efill	Satisfied	Sample Size	Variety 2	Waiters 2	Prob Satisfied
1				4	10	3	10	
2				1	10	4	10	
3				5	10	5	10	
4				2	10	6	10	
5				3	10	3	15	
6	5	15	Yes	2	10	4	15	
7	3	20	Yes	2	10	5	15	
8	3	10	Yes	0	10	6	15	
9	6	10	No	1	10	3	20	
10	6	10	Yes	2	8	4	20	
11						5	20	
12						6	20	
13								

FIGURE 7.13
Selection of "Calculator".

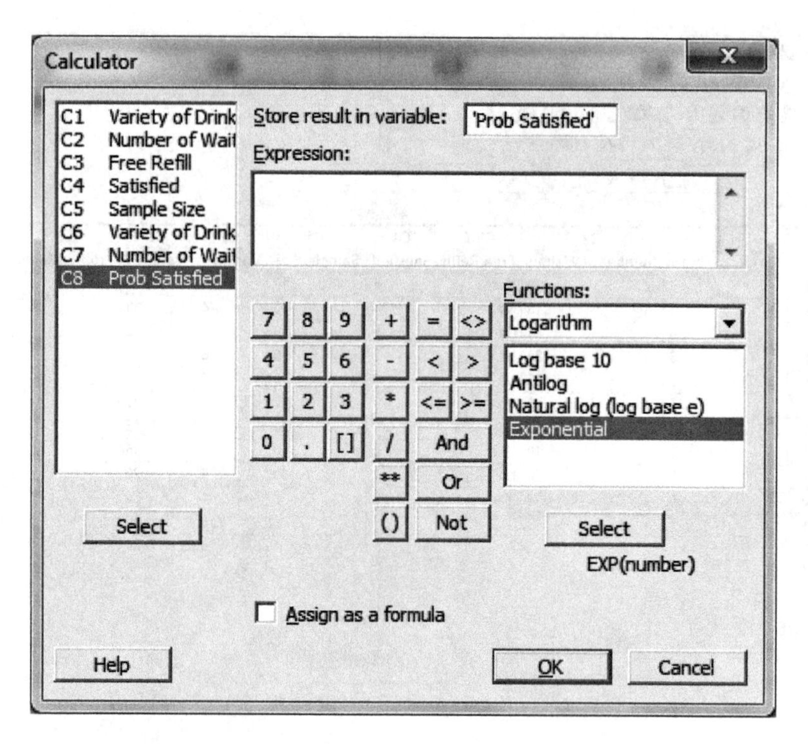

FIGURE 7.14
Selection of "Exponential".

Doing so opens the dialog box shown in Figure 7.14. Select "Prob Satisfied" for "Store result in variable", select "Logarithm" from the drop-down menu for "Functions", and then double-click on the "Exponential" option. This creates the "EXP" function in the "Expression" box. Type the formula as shown in Figure 7.15. Click on "OK" and the probabilities are calculated as shown in Figure 7.16.

In order to graph a scatterplot of the probabilities, select "Scatterplot" as shown in Figure 7.17. Doing so opens the dialog box shown in Figure 7.18. Select "With Connect and Groups" and click on "OK". This opens the dialog box shown in Figure 7.19. Select "Prob Satisfied" for "Y variables" and "Variety 2" for "X variables". Then, select "Waiters 2" for "Categorical variables for grouping (0-3)" and click on "OK". The scatterplot shown in Figure 7.20 is the result. As is evident, the probability of customer satisfaction increases with an increase in the number of waiters and with an increase in the number of different drinks on the menu.

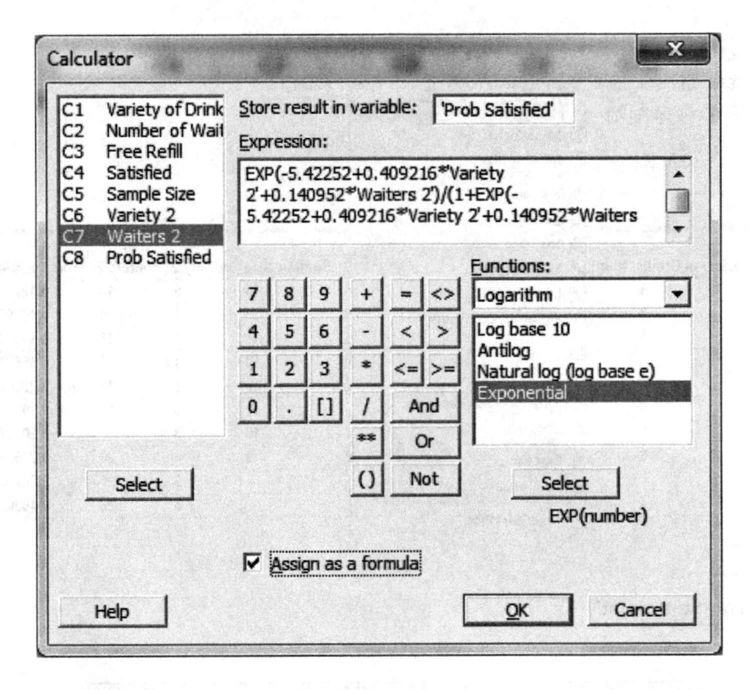

FIGURE 7.15
Entry of formula for probability of customer satisfaction.

↓	C1	C2	C3-T	C4	C5	C6	C7	C8
	Variety of Drinks	Number of Waiters	Free Refill	Satisfied	Sample Size	Variety 2	Waiters 2	Prob Satisfied
3	6	20	No	5	10	5	10	0.122720
4	3	20	No	2	10	6	10	0.173976
5	5	15	No	3	10	3	15	0.110998
6	5	15	Yes	2	10	4	15	0.158241
7	3	20	Yes	2	10	5	15	0.220602
8	3	10	Yes	0	10	6	15	0.290016
9	6	10	No	1	10	3	20	0.201679
10	6	10	Yes	2	8	4	20	0.275556
11						5	20	0.364149
12						6	20	0.463022

FIGURE 7.16
Probabilities of customer satisfaction.

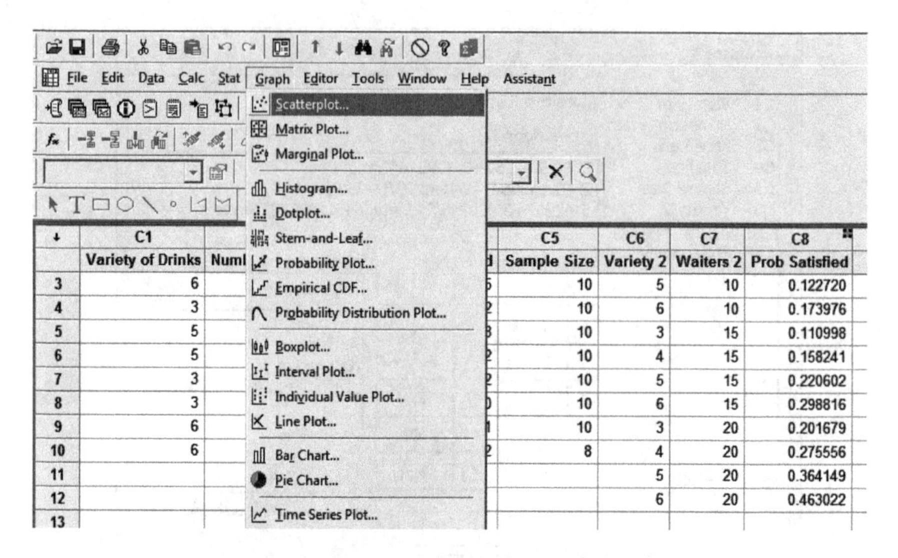

FIGURE 7.17
Selection of "Scatterplot".

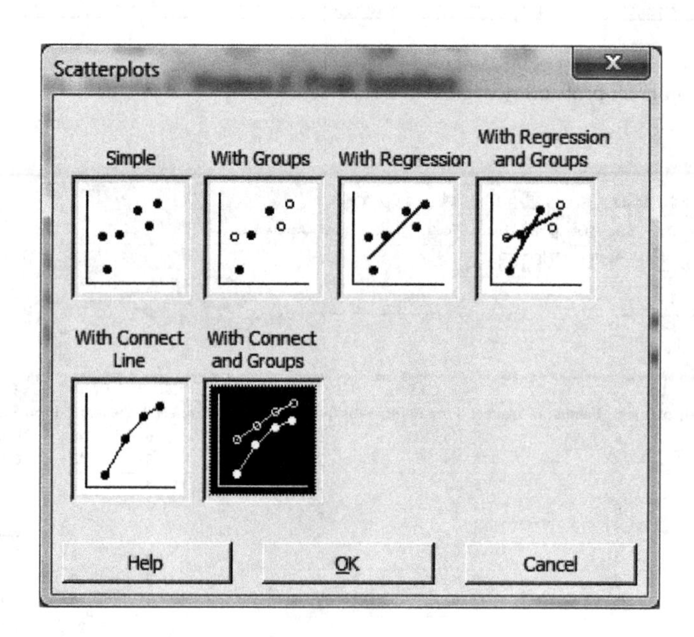

FIGURE 7.18
Selection of "With Connect and Groups".

Scatterplot - With Connect and Groups

C1 Variety of Drinks
C2 Number of Waiters
C3 Free Refill
C4 Satisfied
C5 Sample Size
C6 Variety 2
C7 Waiters 2
C8 Prob Satisfied

	Y variables	X variables
1	'Prob Satisfie	'Variety 2'
2		
3		
4		
5		
6		

Categorical variables for grouping (0-3):
'Waiters 2'

X-Y pairs form groups

Select | Scale... | Labels... | Data View...

Multiple Graphs... | Data Options...

Help | OK | Cancel

FIGURE 7.19
Selection of variables for scatterplot.

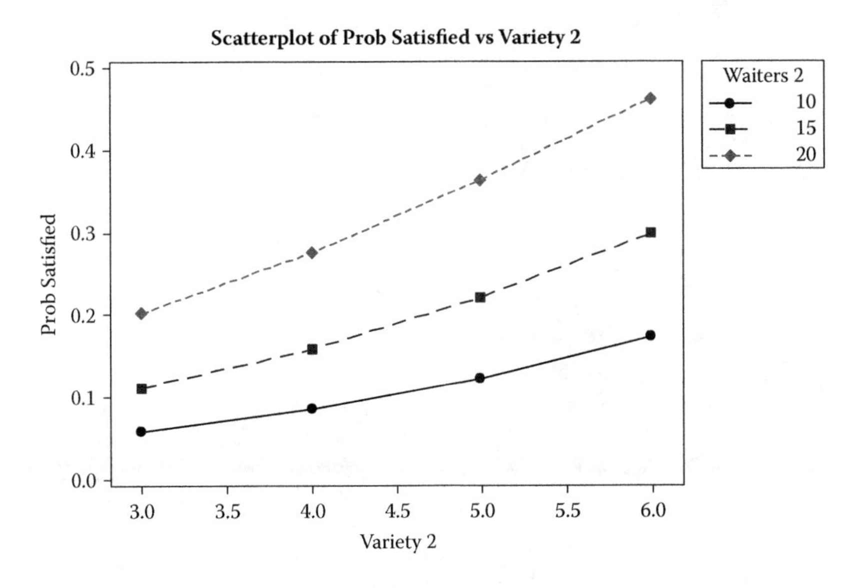

FIGURE 7.20
Scatterplot.

7.4 Improve and Control Phases

The manager recommends to the owners to stop offering refills on drinks, increase the variety of drinks on the menu, and increase the number of waiters working in the restaurant. The owners consider implementing these recommendations and encourage the manager to continue collecting feedback from the customers in the future so that quality is continuously improved and controlled.

8

Item Analysis and Cluster Analysis to Gather "Voice of the Customer" (VOC) Data from Employees at a Service Firm

A service firm decides to improve employee satisfaction via a Six Sigma project. To this end, the management brainstorms to identify three factors to which employees might give importance: empowerment, ease of commute to work, and challenging work.

Section 8.1 gives a brief description of the define phase. Sections 8.2 and 8.3 illustrate the measure and analyze phases, respectively, with detailed instructions for using Minitab®. The improve and control phases are briefly explained in Section 8.4.

8.1 Define Phase

The objective is to improve the proportion of satisfied employees in the company. Seven "questions" (Q1–Q7) are designed for employee empowerment and all 63 employees in the firm are asked to rate their answers to each question on a 1–5 scale, where 1 is "Strongly Disagree", 2 is "Disagree", 3 is "Neutral", 4 is "Agree", and 5 is "Strongly Agree". The questions are as follows:

Q1. I like to have the firm's goals shared with me.

Q2. I do not like to be trusted.

Q3. I do not like to have the responsibility to solve problems.

Q4. I like to have authority to make decisions.

Q5. I like a leadership role.

Q6. I do not like to be recognized for empowered behavior.

Q7. I like to have opportunities to affect results.

8.2 Measure Phase

Open the CHAPTER_8_1.MTW worksheet for the data collected (the worksheet is available at the publisher's website; the data from the worksheet are also provided in the Appendix). Figure 8.1 is a screenshot of the partial worksheet. Q2, Q3, and Q6 are "negative" questions (high ratings for which mean empowerment is undesired), therefore it is necessary to convert them into "positive" questions (high ratings for which mean empowerment is desired) similar to Q1, Q4, Q5, and Q7, before averaging the ratings for each question for empowerment. To this end, as shown in Figure 8.2, label three empty columns as Q2P, Q3P, and Q6P. Then, select "Calculator" as shown in Figure 8.3. Doing so opens the dialog box shown in Figure 8.4. Select "Q2P" for "Store result in variable", and enter "6- 'Q2' " for "Expression". (This is so that a rating of 1 for Q2 becomes a rating of 5 for Q2P, and so on.) Check the box for "Assign as a formula" and click on "OK". Repeat this process for Q3P and Q6P, as shown in Figures 8.5 and 8.6, respectively. The calculated ratings for Q2P, Q3P, and Q6P can be seen in Figure 8.7.

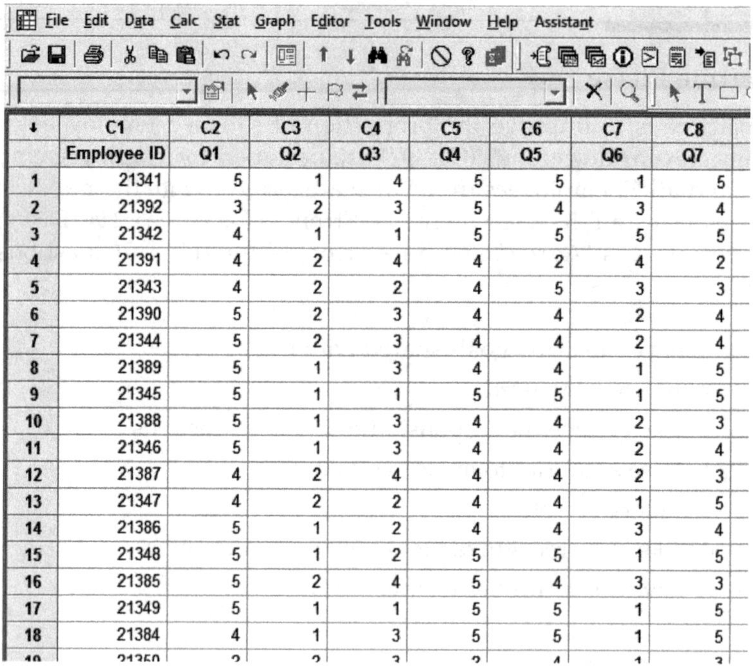

File Edit Data Calc Stat Graph Editor Tools Window Help Assistant

↓	C1	C2	C3	C4	C5	C6	C7	C8
	Employee ID	Q1	Q2	Q3	Q4	Q5	Q6	Q7
1	21341	5	1	4	5	5	1	5
2	21392	3	2	3	5	4	3	4
3	21342	4	1	1	5	5	5	5
4	21391	4	2	4	4	2	4	2
5	21343	4	2	2	4	5	3	3
6	21390	5	2	3	4	4	2	4
7	21344	5	2	3	4	4	2	4
8	21389	5	1	3	4	4	1	5
9	21345	5	1	1	5	5	1	5
10	21388	5	1	3	4	4	2	3
11	21346	5	1	3	4	4	2	4
12	21387	4	2	4	4	4	2	3
13	21347	4	2	2	4	4	1	5
14	21386	5	1	2	4	4	3	4
15	21348	5	1	2	5	5	1	5
16	21385	5	2	4	5	4	3	3
17	21349	5	1	1	5	5	1	5
18	21384	4	1	3	5	5	1	5
19	21350	2	2	3	2	4	1	3

FIGURE 8.1
Data collected for employee empowerment.

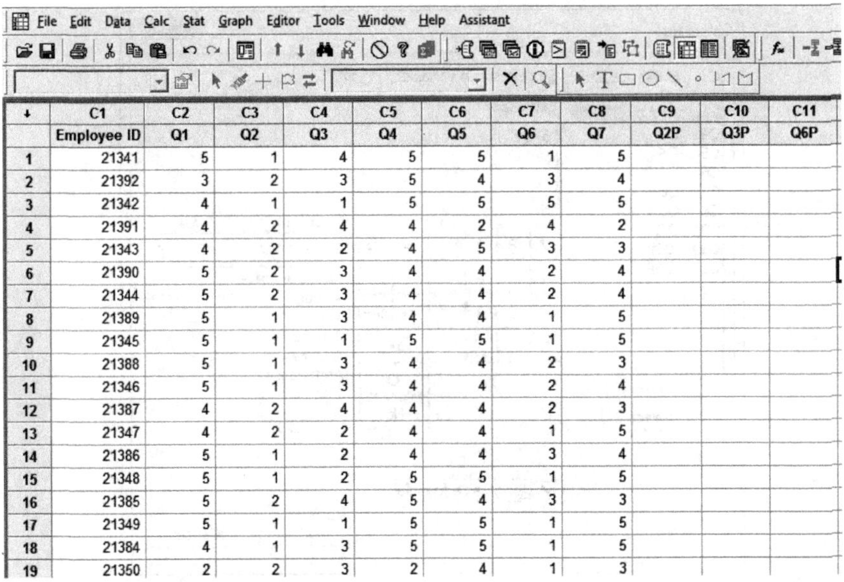

File Edit Data Calc Stat Graph Editor Tools Window Help Assistant

↓	C1	C2	C3	C4	C5	C6	C7	C8	C9	C10	C11
	Employee ID	Q1	Q2	Q3	Q4	Q5	Q6	Q7	Q2P	Q3P	Q6P
1	21341	5	1	4	5	5	1	5			
2	21392	3	2	3	5	4	3	4			
3	21342	4	1	1	5	5	5	5			
4	21391	4	2	4	4	2	4	2			
5	21343	4	2	2	4	5	3	3			
6	21390	5	2	3	4	4	2	4			
7	21344	5	2	3	4	4	2	4			
8	21389	5	1	3	4	4	1	5			
9	21345	5	1	1	5	5	1	5			
10	21388	5	1	3	4	4	2	3			
11	21346	5	1	3	4	4	2	4			
12	21387	4	2	4	4	4	2	3			
13	21347	4	2	2	4	4	1	5			
14	21386	5	1	2	4	4	3	4			
15	21348	5	1	2	5	5	1	5			
16	21385	5	2	4	5	4	3	3			
17	21349	5	1	1	5	5	1	5			
18	21384	4	1	3	5	5	1	5			
19	21350	2	2	3	2	4	1	3			

FIGURE 8.2
Conversion of negative questions to positive.

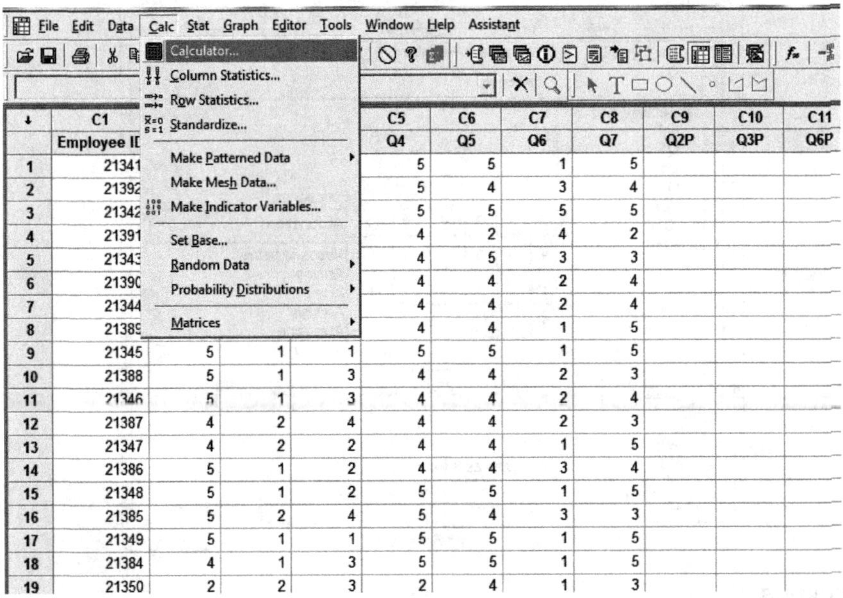

File Edit Data Calc Stat Graph Editor Tools Window Help Assistant

Calculator...
Column Statistics...
Row Statistics...
Standardize...
Make Patterned Data
Make Mesh Data...
Make Indicator Variables...
Set Base...
Random Data
Probability Distributions
Matrices

↓	C1	C2	C3	C4	C5	C6	C7	C8	C9	C10	C11
	Employee ID				Q4	Q5	Q6	Q7	Q2P	Q3P	Q6P
1	21341				5	5	1	5			
2	21392				5	4	3	4			
3	21342				5	5	5	5			
4	21391				4	2	4	2			
5	21343				4	5	3	3			
6	21390				4	4	2	4			
7	21344				4	4	2	4			
8	21389				4	4	1	5			
9	21345	5	1	1	5	5	1	5			
10	21388	5	1	3	4	4	2	3			
11	21346	5	1	3	4	4	2	4			
12	21387	4	2	4	4	4	2	3			
13	21347	4	2	2	4	4	1	5			
14	21386	5	1	2	4	4	3	4			
15	21348	5	1	2	5	5	1	5			
16	21385	5	2	4	5	4	3	3			
17	21349	5	1	1	5	5	1	5			
18	21384	4	1	3	5	5	1	5			
19	21350	2	2	3	2	4	1	3			

FIGURE 8.3
Selection of "Calculator".

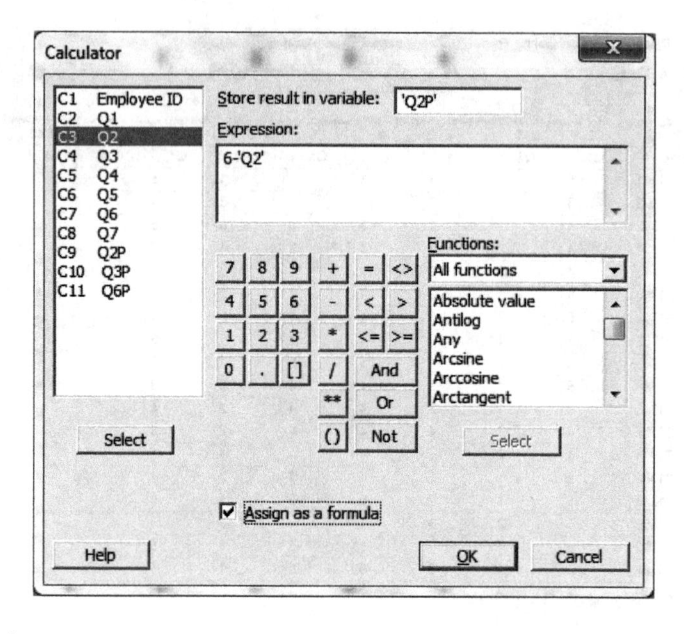

FIGURE 8.4
Conversion of Q2 to Q2P.

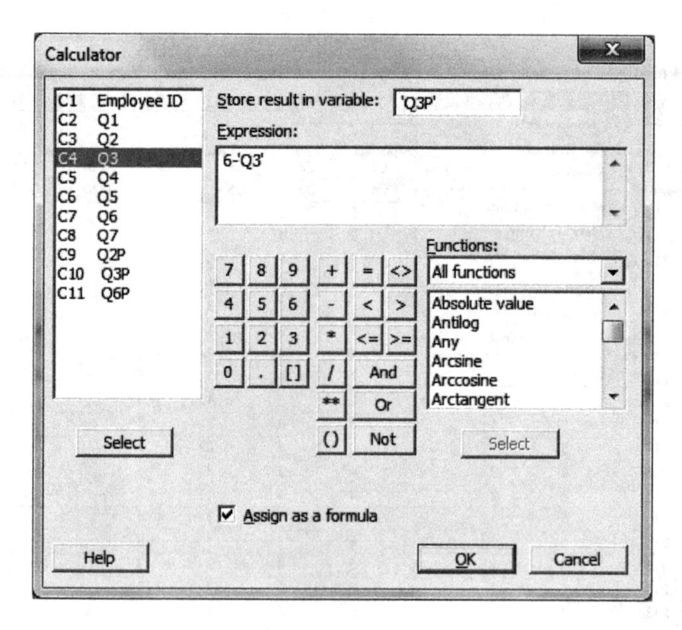

FIGURE 8.5
Conversion of Q3 to Q3P.

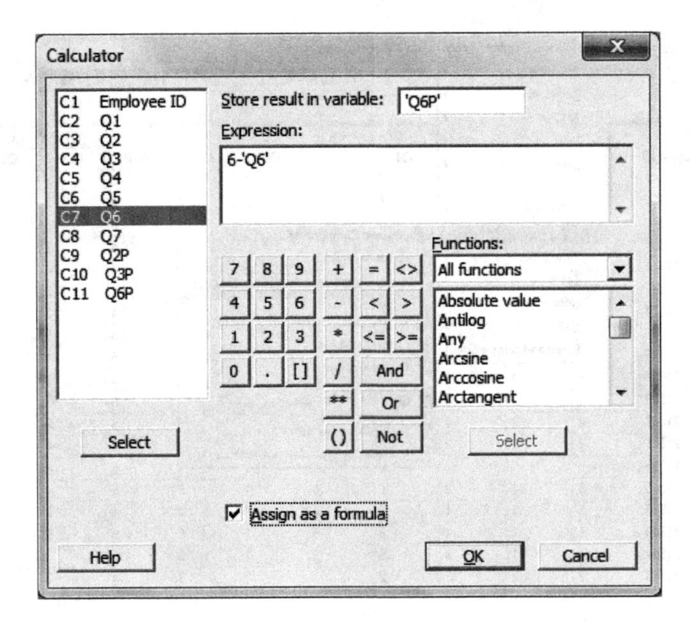

FIGURE 8.6
Conversion of Q6 to Q6P.

File Edit Data Calc Stat Graph Editor Tools Window Help Assistant

↓	C1	C2	C3	C4	C5	C6	C7	C8	C9	C10	C11
	Employee ID	Q1	Q2	Q3	Q4	Q5	Q6	Q7	Q2P	Q3P	Q6P
1	21341	5	1	4	5	5	1	5	5	2	5
2	21392	3	2	3	5	4	3	4	4	3	3
3	21342	4	1	1	5	5	5	5	5	5	1
4	21391	4	2	4	4	2	4	2	4	2	2
5	21343	4	2	2	4	5	3	3	4	4	3
6	21390	5	2	3	4	4	2	4	4	3	4
7	21344	5	2	3	4	4	2	4	4	3	4
8	21389	5	1	3	4	4	1	5	5	3	5
9	21345	5	1	1	5	5	1	5	5	5	5
10	21388	5	1	3	4	4	2	3	5	3	4
11	21346	5	1	3	4	4	2	4	5	3	4
12	21387	4	2	4	4	4	2	3	4	2	4
13	21347	4	2	2	4	4	1	5	4	4	5
14	21386	5	1	2	4	4	3	4	5	4	3
15	21348	5	1	2	5	5	1	5	5	4	5
16	21385	5	2	4	5	4	3	3	4	2	3
17	21349	5	1	1	5	5	1	5	5	5	5
18	21384	4	1	3	5	5	1	5	5	3	5
19	21350	2	2	3	2	4	1	3	4	3	5

FIGURE 8.7
Positive questions Q2P, Q3P, and Q6P.

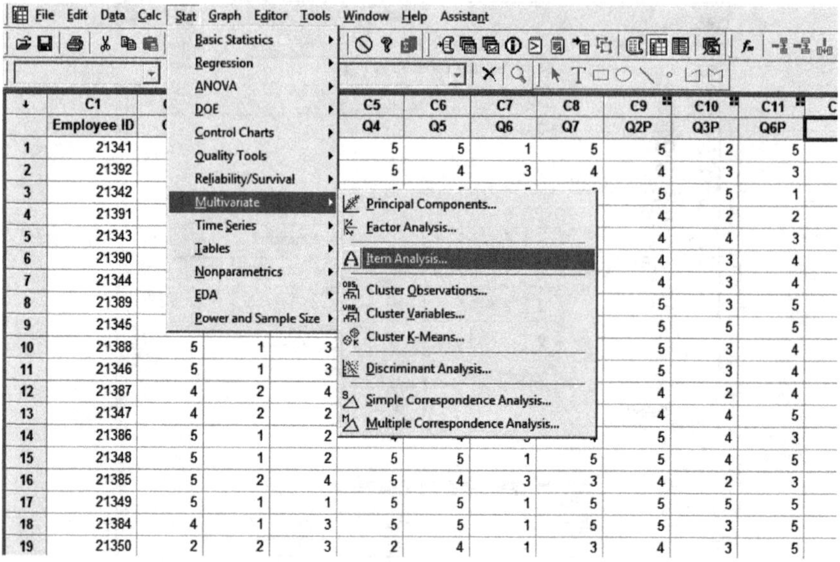

FIGURE 8.8
Selection of "Item Analysis".

8.3 Analyze Phase

Select "Item Analysis" as shown in Figure 8.8. Doing so opens the dialog box shown in Figure 8.9. Then, select "Q1 Q2P Q3P Q4 Q5 Q6P Q7" columns for "Variables". Click on "Graphs" and the dialog box shown in Figure 8.10 opens. Check the box for "Matrix plot for data with smoother" if it is not already checked, and click on "OK". This takes you back to the dialog box shown in Figure 8.9. Click on "OK" and the outputs shown in Figures 8.11 and 8.12 are the results. Figure 8.11 is a matrix plot for all possible pairs of questions. It shows for each pair, how, with an increase in the rating for one question, the rating for the other question changes. Essentially, it shows whether there is any correlation between the ratings of the two questions in each pair. A better understanding of this can be obtained by looking at the output shown in Figure 8.12. Notice that Q6P has a negative correlation with Q4. In addition, it has a very small positive correlation with the other questions. Also, Cronback's Alpha (0.7142) is satisfactory (>0.70) but not ideal (at least 0.80). Hence, the management decides to remove Q6P and repeat the item analysis (see Figure 8.13). The revised outputs are shown in Figures 8.14 and 8.15. Cronback's Alpha (0.7352), although not ideal, is higher than before. Because there are no negative correlations in the revised output, the management decides to average the ratings for Q1, Q2P, Q3P, Q4, Q5, and Q7, and call it the rating for "Empowerment". Label a blank column as "Empowerment", as shown in Figure 8.16. Figure 8.17

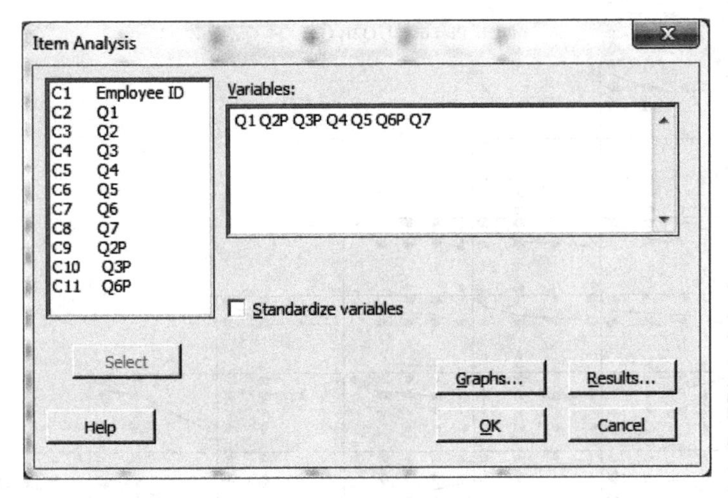

FIGURE 8.9
Selection of variables for item analysis.

shows how to select "Calculator". Doing so opens the dialog box shown in Figure 8.18. Select the "Empowerment" column for "Store result in variable". In the list of "All Functions", double-click on "Means (rows)". This will add the "RMEAN" function to the "Expression" box. As shown in Figure 8.18, select the Q1, Q2P, Q3P, Q4, Q5, and Q7 columns for the "RMEAN" function. Check the box for "Assign as a formula" and click on "OK". This calculates the average "Empowerment" ratings (see Figure 8.19).

A similar approach is used for the other two factors: ease of commute to work, and challenging work. In other words, questionnaires are developed for these two factors and analysis is carried out in a manner similar to "Empowerment" factor. Open the CHAPTER_8_2.MTW worksheet for average ratings of all three factors (the worksheet is available at the publisher's website; the data from the worksheet are also provided in the Appendix). Figure 8.20 is a screenshot of the partial worksheet. Label an empty column as "Cluster", as shown in Figure 8.21. Figure 8.22 shows to how to select "Cluster Observations". Doing so opens the dialog box

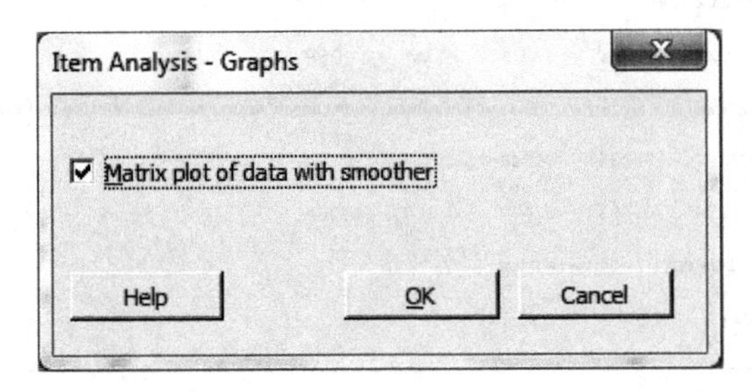

FIGURE 8.10
Selection of "Matrix plot".

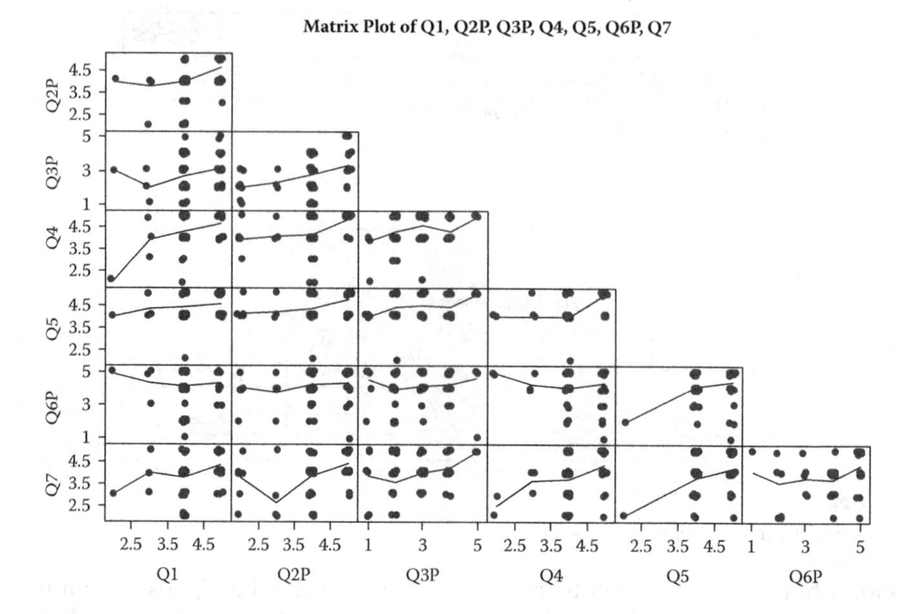

FIGURE 8.11
Matrix plot.

shown in Figure 8.23. As shown, select the "Empowerment", "Ease of Commute", and "Challenge" columns for "Variables or distance width", enter "5" for number clusters, and check the box for "Dendogram". Click on "Customize"; the dialog box shown in Figure 8.24 opens. Enter "Clusters of Employees" for the "Title" and click on "OK". This takes you back to the dialog box shown in Figure 8.23. Click on "Storage" and the dialog box shown in Figure 8.25 opens. Select the "Cluster" column for "Cluster membership column" and click on "OK". This takes you back to the dialog box shown in Figure 8.23. Click on "OK" and the outputs shown in Figures 8.26–8.28 are the results. The dendogram in Figure 8.26 explains

Correlation Matrix

	Q1	Q2P	Q3P	Q4	Q5	Q6P
Q2P	0.395					
Q3P	0.220	0.396				
Q4	0.429	0.313	0.298			
Q5	0.157	0.292	0.285	0.544		
Q6P	0.038	0.073	0.020	-0.049	0.169	
Q7	0.263	0.294	0.353	0.407	0.360	0.268

Cell Contents: Pearson correlation

Cronbach's Alpha = 0.7142

FIGURE 8.12
Output of item analysis.

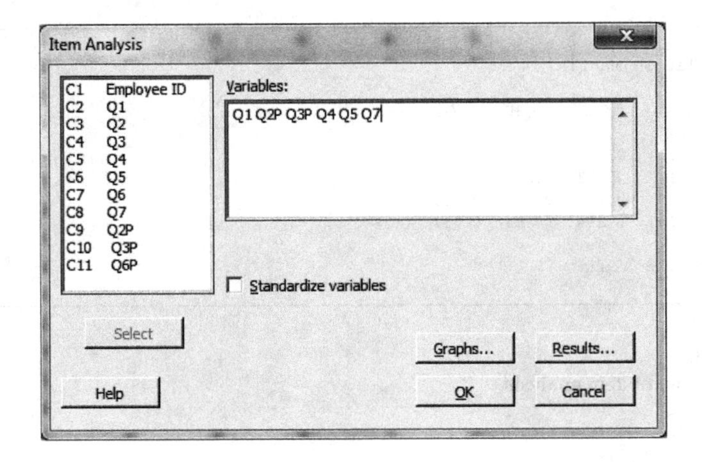

FIGURE 8.13
Revised selection of variables for item analysis.

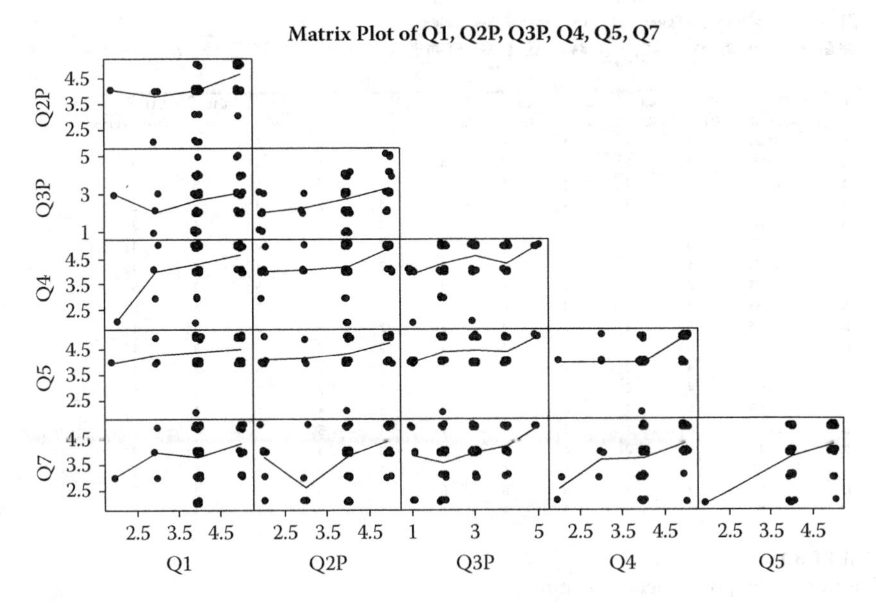

FIGURE 8.14
Revised matrix plot.

```
Correlation Matrix

       Q1     Q2P    Q3P     Q4     Q5
Q2P   0.395
Q3P   0.220  0.396
Q4    0.429  0.313  0.298
Q5    0.157  0.292  0.285  0.544
Q7    0.263  0.294  0.353  0.407  0.360

Cronbach's Alpha = 0.7352
```

FIGURE 8.15
Revised output of item analysis.

the process in which Minitab® has considered Euclidean distances among employees and then combined employees into clusters. The "Cluster" column in Figure 8.27 shows which employee belongs to which of the five clusters created. Figure 8.28 shows how many employees are in each cluster (see the "Number of observations" column).

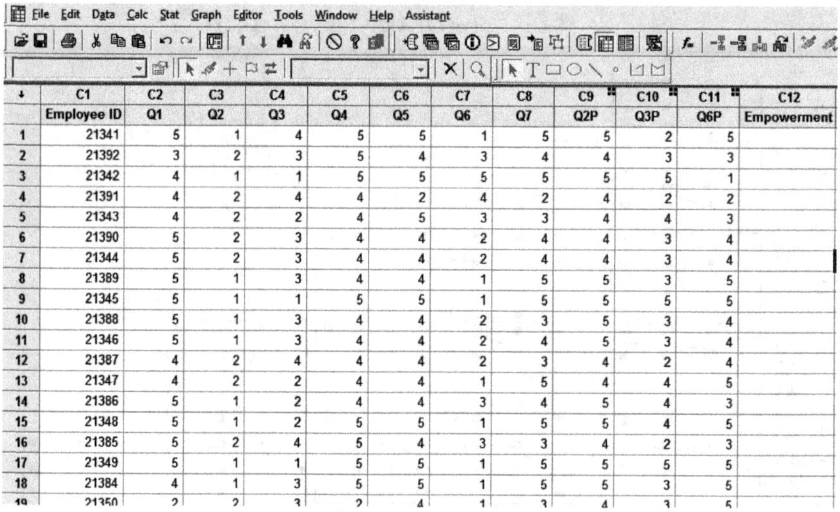

File Edit Data Calc Stat Graph Editor Tools Window Help Assistant

↓	C1	C2	C3	C4	C5	C6	C7	C8	C9	C10	C11	C12
	Employee ID	Q1	Q2	Q3	Q4	Q5	Q6	Q7	Q2P	Q3P	Q6P	Empowerment
1	21341	5	1	4	5	5	1	5	5	2	5	
2	21392	3	2	3	5	4	3	4	4	3	3	
3	21342	4	1	1	5	5	5	5	5	5	1	
4	21391	4	2	4	4	2	4	2	4	2	2	
5	21343	4	2	2	4	5	3	3	4	4	3	
6	21390	5	2	3	4	4	2	4	4	3	4	
7	21344	5	2	3	4	4	2	4	4	3	4	
8	21389	5	1	3	4	4	1	5	5	3	5	
9	21345	5	1	1	5	5	1	5	5	5	5	
10	21388	5	1	3	4	4	2	3	5	3	4	
11	21346	5	1	3	4	4	2	4	5	3	4	
12	21387	4	2	4	4	4	2	3	4	2	4	
13	21347	4	2	2	4	4	1	5	4	4	5	
14	21386	5	1	2	4	4	3	4	5	4	3	
15	21348	5	1	2	5	5	1	5	5	4	5	
16	21385	5	2	4	5	4	3	3	4	2	3	
17	21349	5	1	1	5	5	1	5	5	5	5	
18	21384	4	1	3	5	5	1	5	5	3	5	
19	21350	2	2	3	2	4	1	3	4	3	5	

FIGURE 8.16
Creation of "Empowerment" column.

File Edit Data Calc Stat Graph Editor Tools Window Help Assistant

Calculator...
Column Statistics...
Row Statistics...
Standardize...
Make Patterned Data
Make Mesh Data...
Make Indicator Variables...
Set Base...
Random Data
Probability Distributions
Matrices

↓	C1				C5	C6	C7	C8	C9	C10	C11	C12
	Employee ID				Q4	Q5	Q6	Q7	Q2P	Q3P	Q6P	Empowerment
1	21341				5	5	1	5	5	2	5	
2	21392				5	4	3	4	4	3	3	
3	21342				5	5	5	5	5	5	1	
4	21391				4	2	4	2	4	2	2	
5	21343				4	5	3	3	4	4	3	
6	21390				4	4	2	4	4	3	4	
7	21344				4	4	2	4	4	3	4	
8	21389				4	4	1	5	5	3	5	
9	21345	5	1	1	5	5	1	5	5	5	5	
10	21388	5	1	3	4	4	2	3	5	3	4	
11	21346	5	1	3	4	4	2	4	5	3	4	
12	21387	4	2	4	4	4	2	3	4	2	4	
13	21347	4	2	2	4	4	1	5	4	4	5	
14	21386	5	1	2	4	4	3	4	5	4	3	
15	21348	5	1	2	5	5	1	5	5	4	5	
16	21385	5	2	4	5	4	3	3	4	2	3	
17	21349	5	1	1	5	5	1	5	5	5	5	
18	21384	4	1	3	5	5	1	5	5	3	5	
19	21350	2	2	3	2	4	1	3	4	3	5	

FIGURE 8.17
Use of "Calculator" for "Empowerment".

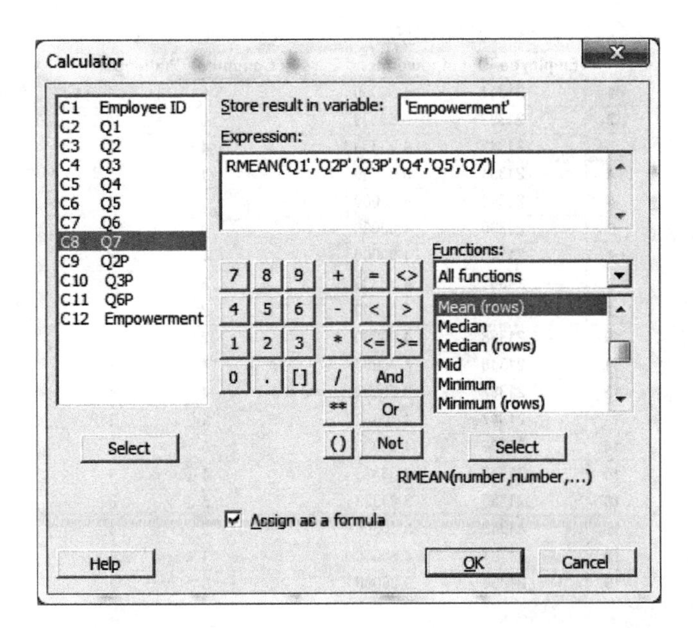

FIGURE 8.18
Calculation of "Empowerment".

File Edit Data Calc Stat Graph Editor Tools Window Help Assistant

↓	C1	C2	C3	C4	C5	C6	C7	C8	C9	C10	C11	C12
	Employee ID	Q1	Q2	Q3	Q4	Q5	Q6	Q7	Q2P	Q3P	Q6P	Empowerment
1	21341	5	1	4	5	5	1	5	5	2	5	4.50000
2	21392	3	2	3	5	4	3	4	4	3	3	3.83333
3	21342	4	1	1	5	5	5	5	5	5	1	4.83333
4	21391	4	2	4	4	2	4	2	4	2	2	3.00000
5	21343	4	2	2	4	5	3	3	4	4	3	4.00000
6	21390	5	2	3	4	4	2	4	4	3	4	4.00000
7	21344	5	2	3	4	4	2	4	4	3	4	4.00000
8	21389	5	1	3	4	4	1	5	5	3	5	4.33333
9	21345	5	1	1	5	5	1	5	5	5	5	5.00000
10	21388	5	1	3	4	4	2	3	5	3	4	4.00000
11	21346	5	1	3	4	4	2	4	5	3	4	4.16667
12	21387	4	2	4	4	4	2	3	4	2	4	3.50000
13	21347	4	2	2	4	4	1	5	4	4	5	4.16667
14	21386	5	1	2	4	4	3	4	5	4	3	4.33333
15	21348	5	1	2	5	5	1	5	5	4	5	4.83333
16	21385	5	2	4	5	4	3	3	4	2	3	3.83333
17	21349	5	1	1	5	5	1	5	5	5	5	5.00000
18	21384	4	1	3	5	5	1	5	5	3	5	4.50000
19	21350	2	2	3	2	4	1	3	4	3	5	3.00000

FIGURE 8.19
Empowerment ratings.

↓	C1	C2	C3	C4
	Employee ID	Empowerment	Ease of Commute	Challenge
1	21341	4.50000	5	1
2	21392	3.83333	5	1
3	21342	4.83333	4	3
4	21391	3.00000	1	2
5	21343	4.00000	1	4
6	21390	4.00000	3	3
7	21344	4.00000	4	3
8	21389	4.33333	3	5
9	21345	5.00000	3	3
10	21388	4.00000	2	1
11	21346	4.16667	4	5
12	21387	3.50000	4	4
13	21347	4.16667	1	3
14	21386	4.33333	5	4
15	21348	4.83333	4	1
16	21385	3.83333	3	3
17	21349	5.00000	2	4
18	21384	4.50000	1	4
19	21350	3.00000	5	1

FIGURE 8.20
Ratings for empowerment, ease of commute, and challenge.

↓	C1 Employee ID	C2 Empowerment	C3 Ease of Commute	C4 Challenge	C5 Cluster
1	21341	4.50000	5	1	
2	21392	3.83333	5	1	
3	21342	4.83333	4	3	
4	21391	3.00000	1	2	
5	21343	4.00000	1	4	
6	21390	4.00000	3	3	
7	21344	4.00000	4	3	
8	21389	4.33333	3	5	
9	21345	5.00000	3	3	
10	21388	4.00000	2	1	
11	21346	4.16667	4	5	
12	21387	3.50000	4	4	
13	21347	4.16667	1	3	
14	21386	4.33333	5	4	
15	21348	4.83333	4	1	
16	21385	3.83333	3	3	
17	21349	5.00000	2	4	
18	21384	4.50000	1	4	
19	21350	3.00000	5	1	

FIGURE 8.21
Creation of "Cluster" column.

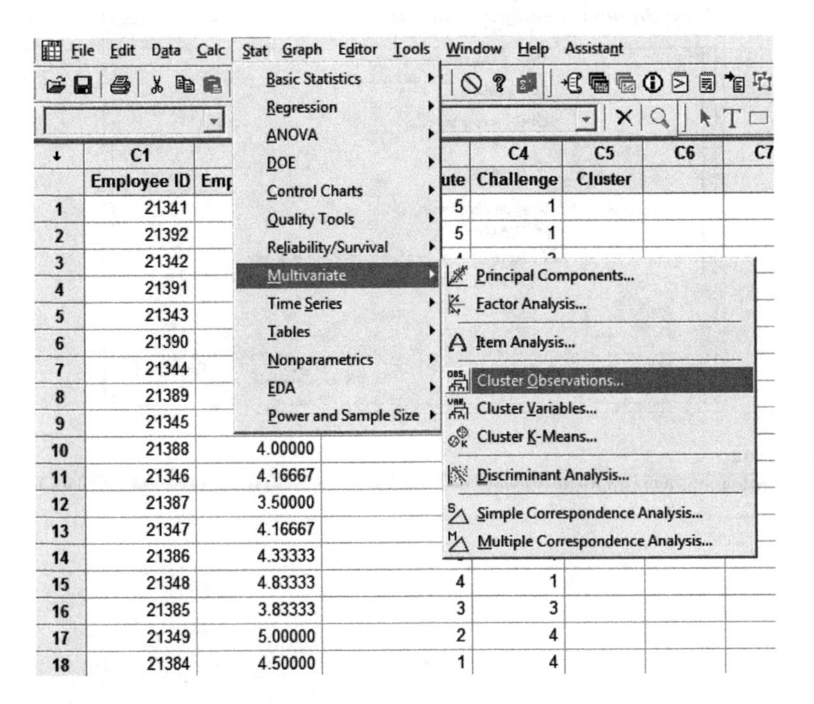

FIGURE 8.22
Selection of "Cluster Observations".

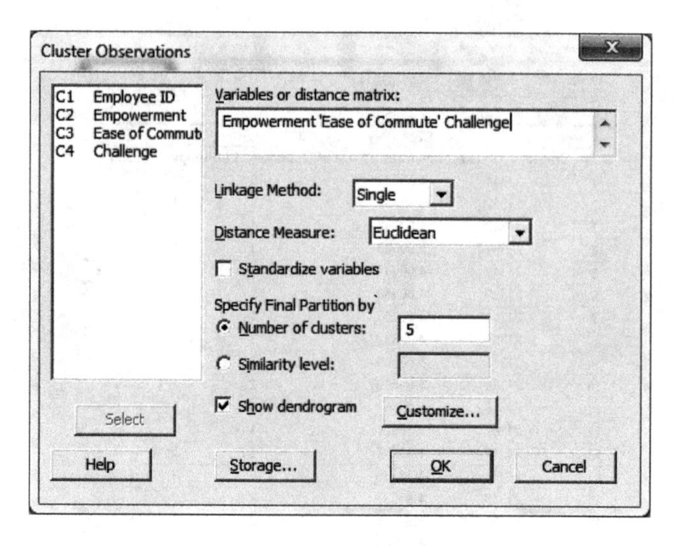

FIGURE 8.23
Selection of variables for cluster analysis.

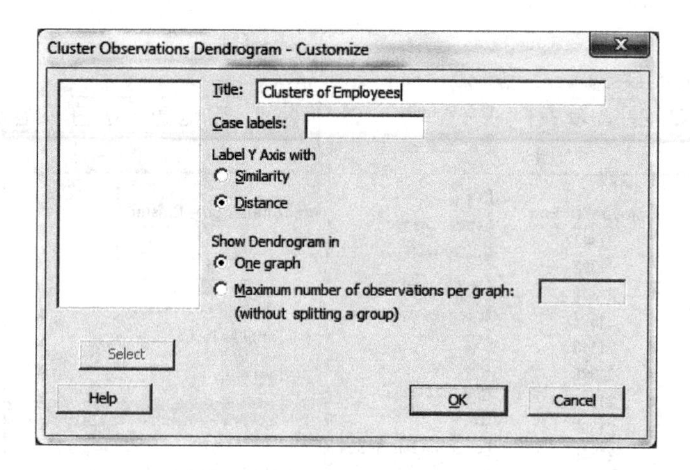

FIGURE 8.24
Entry of title for cluster analysis output.

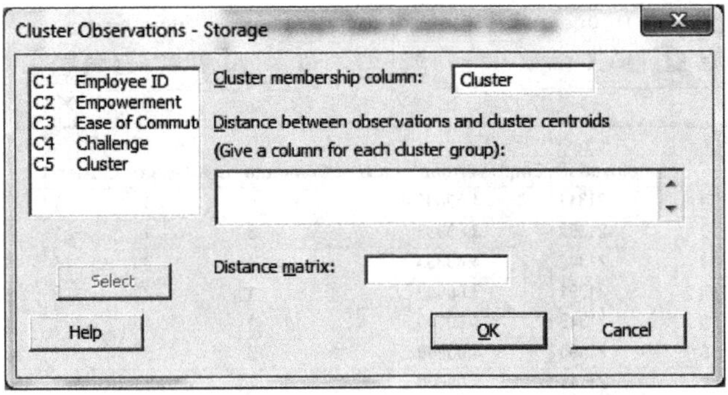

Cluster Observations - Storage

C1 Employee ID
C2 Empowerment
C3 Ease of Commut
C4 Challenge
C5 Cluster

Cluster membership column: Cluster

Distance between observations and cluster centroids
(Give a column for each cluster group):

Distance matrix:

Select

Help OK Cancel

FIGURE 8.25
Storage of clusters in "Cluster" column.

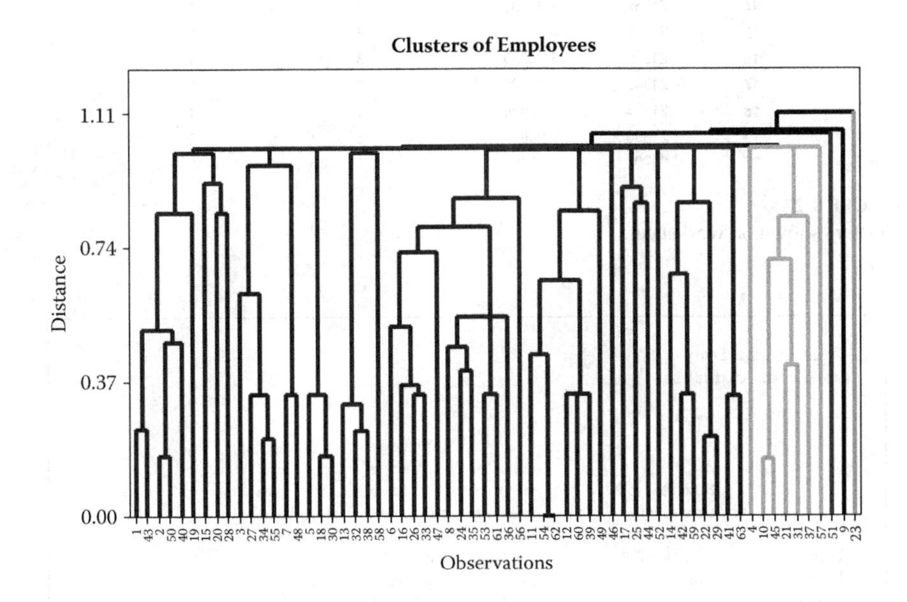

FIGURE 8.26
Dendrogram.

↓	C1 Employee ID	C2 Empowerment	C3 Ease of Commute	C4 Challenge	C5 Cluster
1	21341	4.50000	5	1	1
2	21392	3.83333	5	1	1
3	21342	4.83333	4	3	1
4	21391	3.00000	1	2	2
5	21343	4.00000	1	4	1
6	21390	4.00000	3	3	1
7	21344	4.00000	4	3	1
8	21389	4.33333	3	5	1
9	21345	5.00000	3	3	3
10	21388	4.00000	2	1	2
11	21346	4.16667	4	5	1
12	21387	3.50000	4	4	1
13	21347	4.16667	1	3	1
14	21386	4.33333	5	4	1
15	21348	4.83333	4	1	1
16	21385	3.83333	3	3	1
17	21349	5.00000	2	4	1
18	21384	4.50000	1	4	1
19	21350	3.00000	5	1	1

FIGURE 8.27
Clusters shown on worksheet.

```
Final Partition
Number of clusters: 5


                                         Average    Maximum
                              Within    distance   distance
               Number of  cluster sum       from       from
            observations   of squares   centroid   centroid
Cluster1              53      213.794    1.88158    3.20743
Cluster2               7        4.202    0.73654    1.16163
Cluster3               1        0.000    0.00000    0.00000
Cluster4               1        0.000    0.00000    0.00000
Cluster5               1        0.000    0.00000    0.00000
```

FIGURE 8.28
Count of employees in each cluster.

8.4 Improve and Control Phases

Because most (53) out of the 63 employees fall in the same cluster, the management decides to brainstorm with all the employees, about potential improvements with respect to the three factors (viz., empowerment, ease of commute to work, and challenging work) and implement the economically viable options. The management also decides to continue collecting feedback from the employees in the future so that employee satisfaction is continuously improved and controlled.

9

Mixture Designs to Optimize Pollution Level and Temperature of Fuels

This case study is about a laboratory using Six Sigma's define-measure-analyze-design-verify (DMADV) approach to design the compositions of two fuel mixtures: Neo and Zeo.

Sections 9.1 and 9.2 give brief descriptions of the define phase and the measure phase, respectively. Section 9.3 illustrates the analyze phase with detailed instructions for using Minitab®. The design and verify phases are briefly explained in Section 9.4.

9.1 Define Phase

The laboratory wishes to decide the components and their proportions for two fuel mixtures: Neo and Zeo.

9.2 Measure Phase

The objective for Neo is to minimize the pollution level, and the objective for Zeo is to optimize the temperature generated by the fuel.

9.3 Analyze Phase

The laboratory is considering three components for fuel Neo: P, Q, and R. The total percentage of the proportions of these three components must be 100, and each component may range between 0% and 100%. To design the fuel mixture, select "Create Mixture Design" as shown in Figure 9.1. Doing so opens the dialog box shown in Figure 9.2. Select "Simplex centroid" for "Type of Design" and "3" for "Number of components". Click on "Designs"

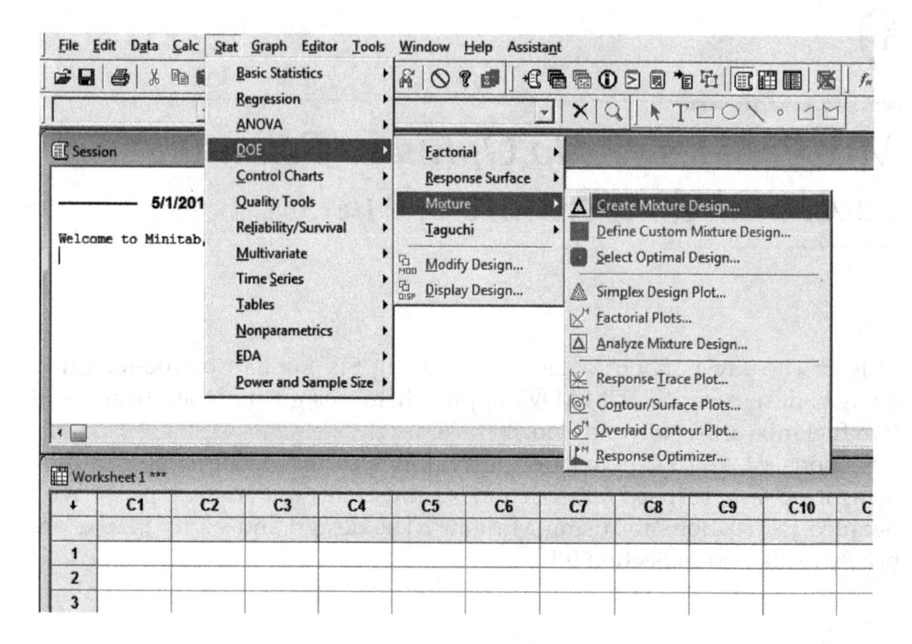

FIGURE 9.1
Selection of "Create Mixture Design" for fuel Neo.

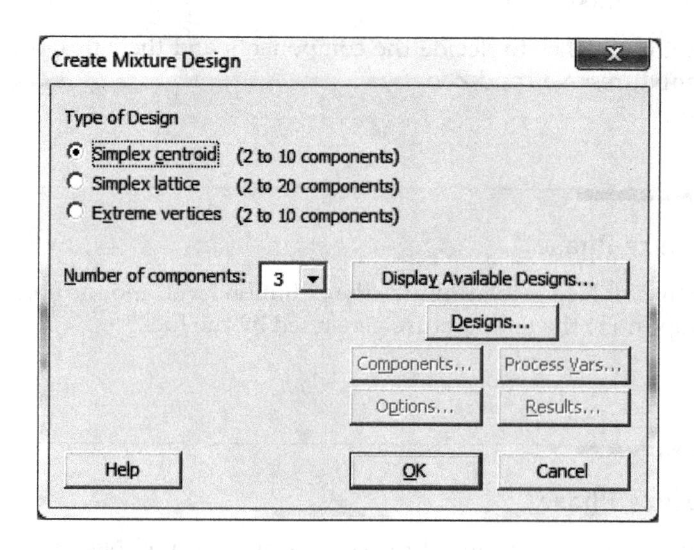

FIGURE 9.2
Selection of "Simplex centroid".

and the dialog box shown in Figure 9.3 opens. Select "3" for "Number of replicates for the whole design" and click on "OK". This takes you back to the dialog box shown in Figure 9.2. Click on "Components" and the dialog box shown in Figure 9.4 opens. Enter "100" for "Single total". Also, name the three components in the "Name" column, enter "0" for "Lower", and enter "100" for "Upper", for each of the three components. Click on "OK" and it takes you back to the dialog box shown in Figure 9.2. Click on "Options" and the dialog box shown in Figure 9.5 opens. Uncheck the "Randomize runs" box so that it is easier for you to replicate the results in this case study. Click on "OK" and it takes you back to the dialog box shown in Figure 9.2. Click on "Results" and the dialog box shown in Figure 9.6 opens. Ensure that "Detailed description" is selected, and click on "OK". This takes you back to the dialog box shown in Figure 9.2. Click on "OK" and the partial mixture design shown in Figure 9.7 is the result. In order to see the simplex design plot, select "Simplex Design Plot" as shown in Figure 9.8. Doing so opens the dialog box shown in Figure 9.9. Click on "OK" and the simplex design plot shown in Figure 9.10 is the result. As shown in Figure 9.11, label an empty column as "Sulfation" (a measure of pollution level). The data from the experiment are shown in Figure 9.11. (For example, in the first run, the sulfation level is found to be "3" when P = 100%, Q = 0%, and R = 0%.) Open the CHAPTER_9_NEO.MTW worksheet to analyze the data (the worksheet

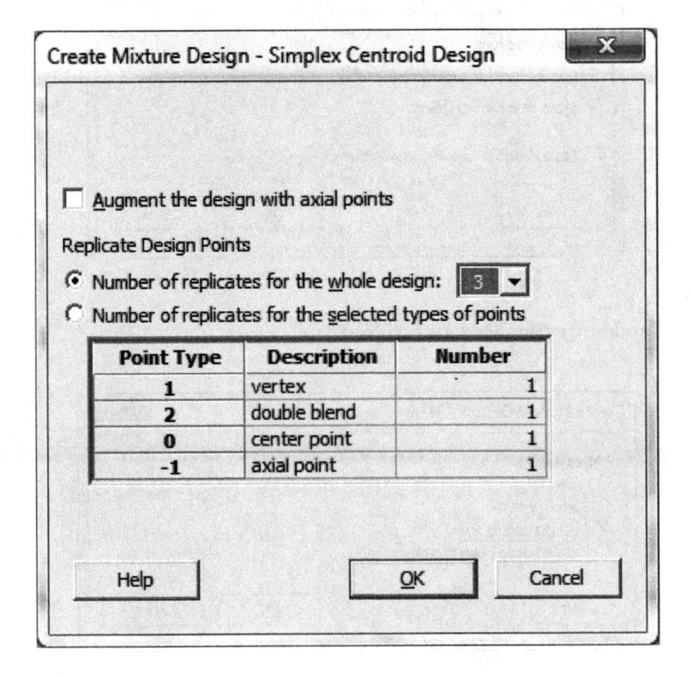

FIGURE 9.3
Entry of replicates.

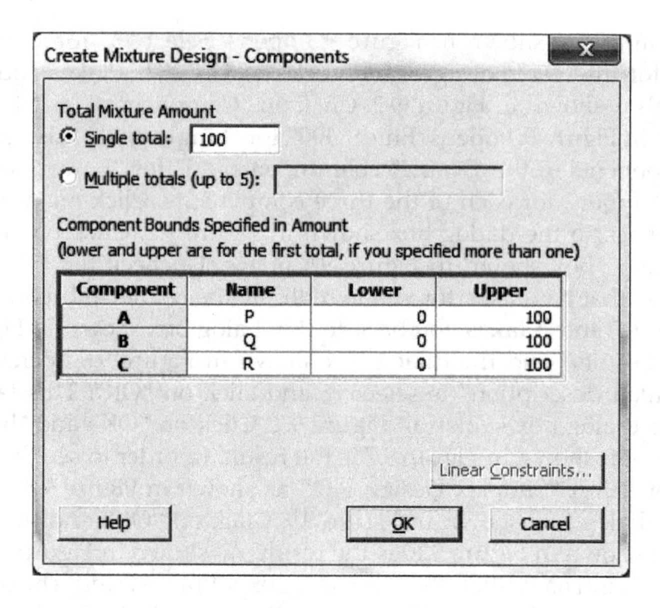

FIGURE 9.4
Entry of lower and upper bounds for components of fuel Neo.

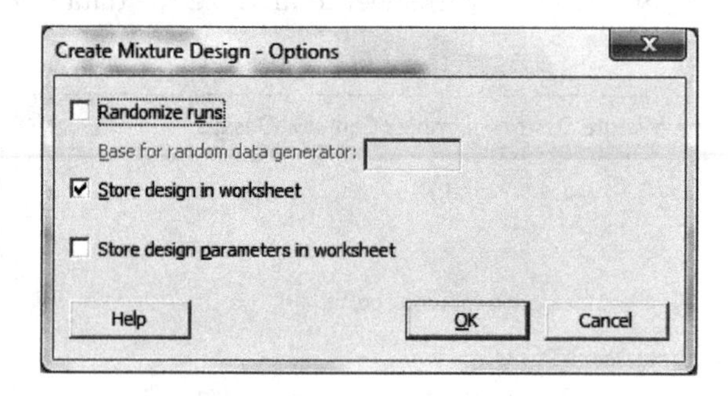

FIGURE 9.5
Unchecking "Randomize runs" box for fuel Neo.

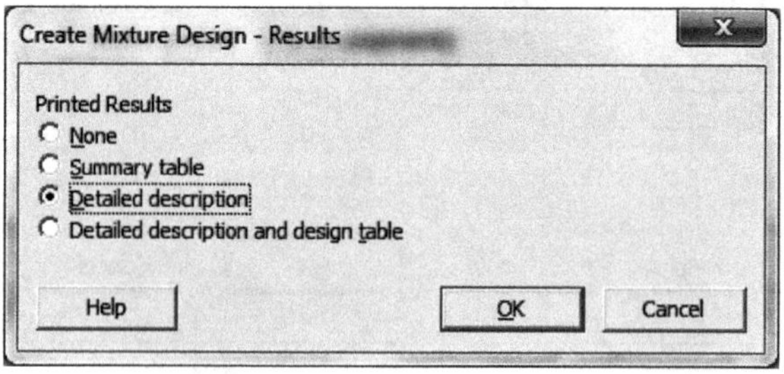

Create Mixture Design - Results
Printed Results
None
Summary table
Detailed description
Detailed description and design table
Help
OK
Cancel

FIGURE 9.6
Selection of "Detailed description" for fuel Neo.

	C1	C2	C3	C4	C5	C6	C7
	StdOrder	RunOrder	PtType	Blocks	P	Q	R
1	1	1	1	1	100.000	0.000	0.000
2	2	2	1	1	0.000	100.000	0.000
3	3	3	1	1	0.000	0.000	100.000
4	4	4	2	1	50.000	50.000	0.000
5	5	5	2	1	50.000	0.000	50.000
6	6	6	2	1	0.000	50.000	50.000
7	7	7	0	1	33.333	33.333	33.333
8	8	8	1	1	100.000	0.000	0.000
9	9	9	1	1	0.000	100.000	0.000
10	10	10	1	1	0.000	0.000	100.000
11	11	11	2	1	50.000	50.000	0.000
12	12	12	2	1	50.000	0.000	50.000
13	13	13	2	1	0.000	50.000	50.000
14	14	14	0	1	33.333	33.333	33.333
15	15	15	1	1	100.000	0.000	0.000
16	16	16	1	1	0.000	100.000	0.000
17	17	17	1	1	0.000	0.000	100.000
18	18	18	2	1	50.000	50.000	0.000
19	19	19	2	1	50.000	0.000	50.000
20	20	20	2	1	0.000	50.000	50.000

FIGURE 9.7
Partial mixture design for fuel Neo.

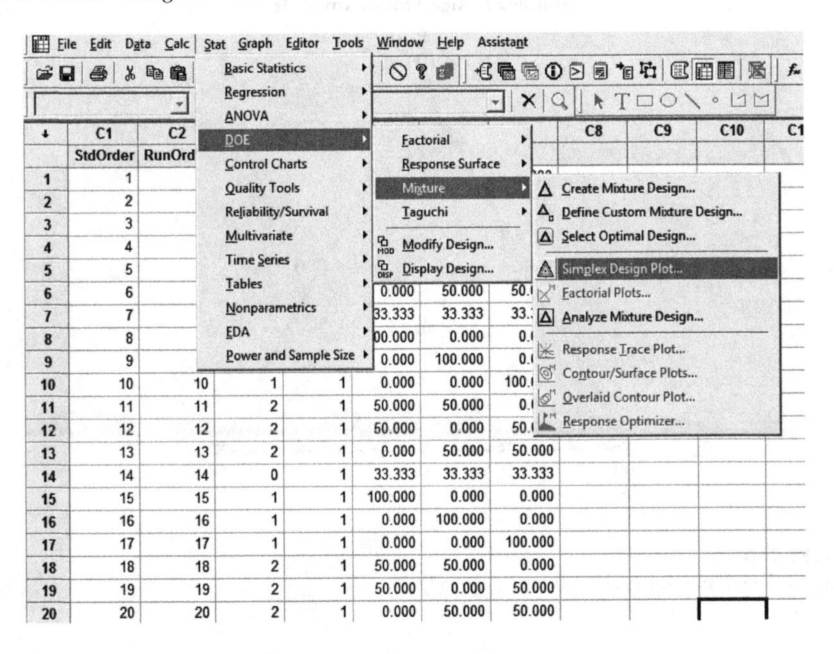

FIGURE 9.8
Selection of "Simplex Design Plot" for fuel Neo.

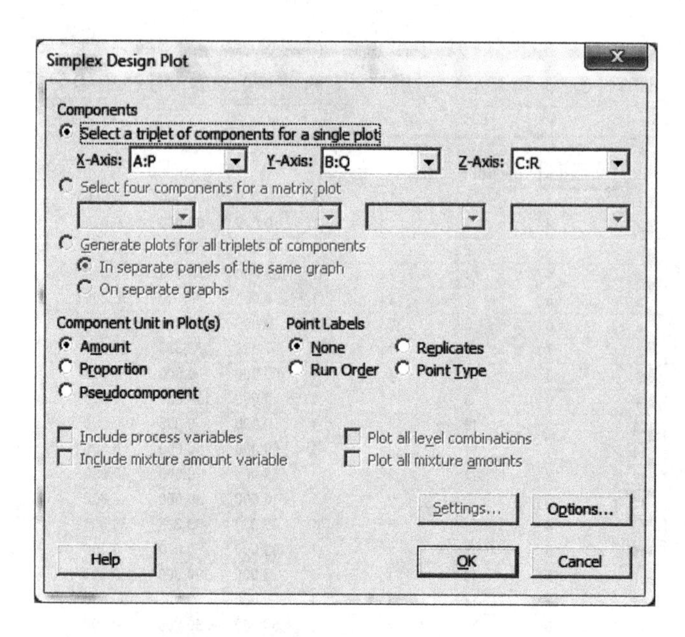

FIGURE 9.9
Options for "Simplex Design Plot" for fuel Neo.

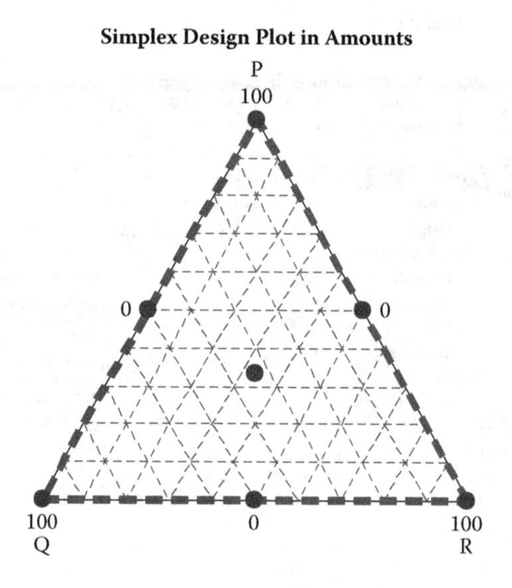

FIGURE 9.10
Simplex design plot for fuel Neo.

is available at the publisher's website; the data from the worksheet are also provided in the Appendix).

To analyze the design created for fuel Neo, select "Analyze Mixture Design" as shown in Figure 9.12. Doing so opens the dialog box shown in Figure 9.13. Select "Sulfation" for "Responses" and click on "Terms". This opens the dialog box shown in Figure 9.14. Choose "Quadratic" for "Include component terms for model" and choose the "Selected Terms" as shown in Figure 9.14. Click on "OK" and it takes you back to the dialog box shown in Figure 9.13. Click on "OK" and the output shown in Figure 9.15 is the result. Starting from the term with the highest *P*-value that is greater than 0.05, eliminate one term at a time from the analysis by repeating the above steps (beginning with what is shown in Figure 9.12). Figure 9.16 shows the elimination of the QR term from the analysis and Figure 9.17 shows the corresponding output. Figure 9.18 shows the elimination of the PR term (as well) from the analysis and Figure 9.19 shows the corresponding output. Figure 9.20 shows the elimination of the PQ term (as well) from the analysis and Figure 9.21 shows the corresponding output. Notice that the *P*-value (0.048) in Figure 9.21 is less than 0.05, and hence this model in Figure 9.21 can be considered now for optimization.

File Edit Data Calc Stat Graph Editor Tools Window Help Assistant

↓	C1	C2	C3	C4	C5	C6	C7	C8
	StdOrder	RunOrder	PtType	Blocks	P	Q	R	Sulfation
1	1	1	1	1	100.000	0.000	0.000	3
2	2	2	1	1	0.000	100.000	0.000	2
3	3	3	1	1	0.000	0.000	100.000	3
4	4	4	2	1	50.000	50.000	0.000	9
5	5	5	2	1	50.000	0.000	50.000	6
6	6	6	2	1	0.000	50.000	50.000	1
7	7	7	0	1	33.333	33.333	33.333	4
8	8	8	1	1	100.000	0.000	0.000	35
9	9	9	1	1	0.000	100.000	0.000	11
10	10	10	1	1	0.000	0.000	100.000	10
11	11	11	2	1	50.000	50.000	0.000	10
12	12	12	2	1	50.000	0.000	50.000	12
13	13	13	2	1	0.000	50.000	50.000	12
14	14	14	0	1	33.333	33.333	33.333	16
15	15	15	1	1	100.000	0.000	0.000	17
16	16	16	1	1	0.000	100.000	0.000	12
17	17	17	1	1	0.000	0.000	100.000	1
18	18	18	2	1	50.000	50.000	0.000	13
19	19	19	2	1	50.000	0.000	50.000	11
20	20	20	2	1	0.000	50.000	50.000	2
21	21	21	0	1	33.333	33.333	33.333	3

FIGURE 9.11
Mixture design for fuel Neo.

File Edit Data Calc Stat Graph Editor Tools Window Help Assistant

Basic Statistics
Regression
ANOVA
DOE
Control Charts
Quality Tools
Reliability/Survival
Multivariate
Time Series
Tables
Nonparametrics
EDA
Power and Sample Size

Factorial
Response Surface
Mixture
Taguchi
Modify Design...
Display Design...

Create Mixture Design...
Define Custom Mixture Design...
Select Optimal Design...
Simplex Design Plot...
Factorial Plots...
Analyze Mixture Design...
Response Trace Plot...
Contour/Surface Plots...
Overlaid Contour Plot...
Response Optimizer...

C1 StdOrder, C2 RunOrd, C8 Sulfation, C9, C10

FIGURE 9.12
Selection of "Analyze Mixture Design" for fuel Neo.

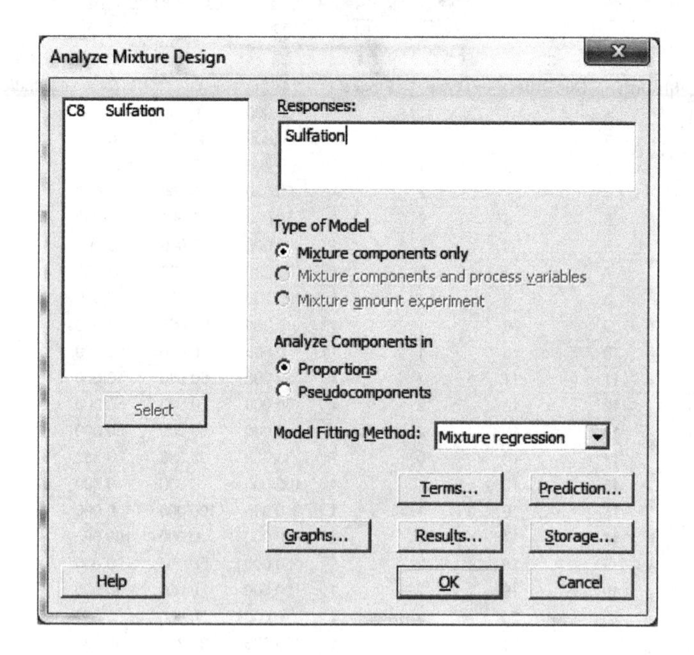

FIGURE 9.13
Selection of "Sulfation" column for "Responses".

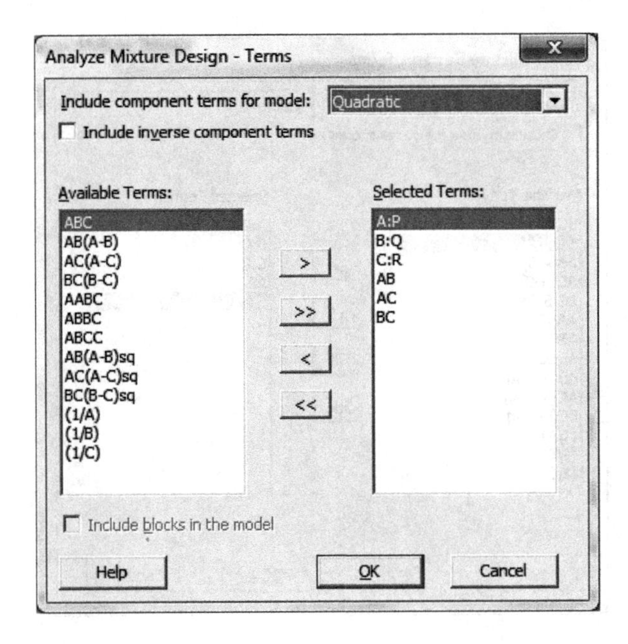

FIGURE 9.14
Quadratic terms for fuel Neo mixture design.

For optimization (minimization) of sulfation level, select "Response Optimizer" as shown in Figure 9.22. Doing so opens the dialog box shown in Figure 9.23. Move the "Sulfation" column from "Available" to "Selected", as shown in Figure 9.24. Click on "Setup" and the dialog box shown in Figure 9.25 opens. Select "Minimize" for "Goal", enter "20" for "Target" and "25" for "Upper". (The laboratory aims for a target sulfation level as low as 20, and is not willing to go above an upper bound of 25.) Click on "OK" and the optimal solution (P: 0%, Q: 0%, and R: 100%) shown in Figure 9.26 is the result.

Analysis of Variance for Sulfation (component proportions)

Source	DF	Seq SS	Adj SS	Adj MS	F	P
Regression	5	372.25	372.252	74.450	1.34	0.300
Linear	2	344.53	293.947	146.973	2.65	0.103
Quadratic	3	27.72	27.723	9.241	0.17	0.917
P*Q	1	15.87	17.601	17.601	0.32	0.582
P*R	1	6.66	7.195	7.195	0.13	0.724
Q*R	1	5.19	5.194	5.194	0.09	0.764
Residual Error	15	831.96	831.961	55.464		
Lack-of-Fit	1	0.00	0.001	0.001	0.00	0.997
Pure Error	14	831.96	831.960	59.426		
Total	20	1204.21				

FIGURE 9.15
ANOVA output for fuel Neo.

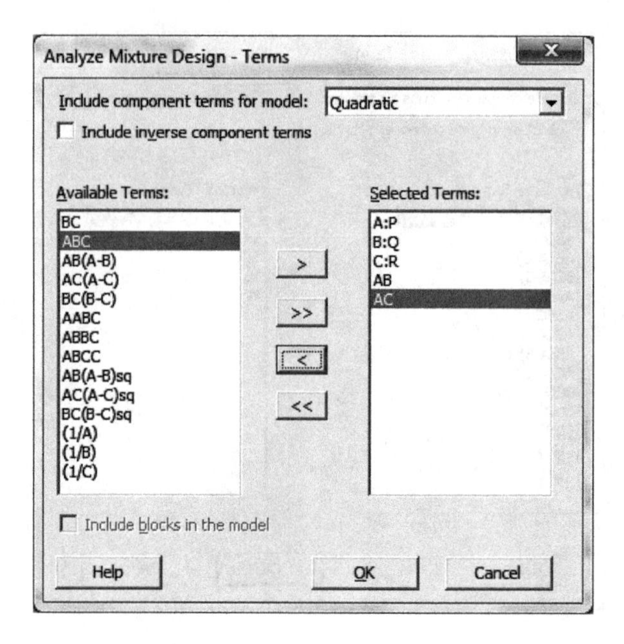

FIGURE 9.16
Removal of QR term.

The laboratory is considering three components for fuel Zeo: X, Y, and Z. The total percentage of the proportions of these three components must be 100, and each component must be at least 20%. To design the fuel mixture, select "Create Mixture Design" as shown in Figure 9.27. Doing so opens the dialog box shown in Figure 9.28. Select "Simplex lattice" for "Type of Design" and "3" for "Number of components". Click on "Designs" and the dialog box shown in Figure 9.29 opens. Select "2" for "Degree of lattice" and click on "OK". This takes you back to the dialog box shown in Figure 9.28. Click on

Analysis of Variance for Sulfation (component proportions)

Source	DF	Seq SS	Adj SS	Adj MS	F	P
Regression	4	367.06	367.058	91.765	1.75	0.188
Linear	2	344.53	365.404	182.702	3.49	0.055
Quadratic	2	22.53	22.530	11.265	0.22	0.809
P*Q	1	15.87	16.764	16.764	0.32	0.579
P*R	1	6.66	6.655	6.655	0.13	0.726
Residual Error	16	837.15	837.154	52.322		
Lack-of-Fit	2	5.19	5.194	2.597	0.04	0.957
Pure Error	14	831.96	831.960	59.426		
Total	20	1204.21				

FIGURE 9.17
Revised ANOVA output after removal of QR term.

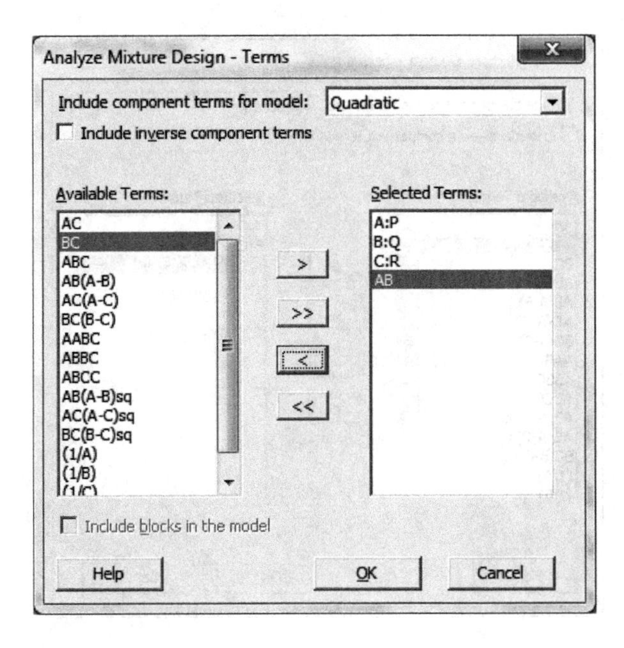

FIGURE 9.18
Removal of QR and PR terms.

"Components" and the dialog box shown in Figure 9.30 opens. Enter "100" for "Single total". Also, name the three components in the "Name" column, enter "20" for "Lower", and enter "100" for "Upper", for each of the three components. Click on "OK" and it takes you back to the dialog box shown in Figure 9.28. Click on "Options" and the dialog box shown in Figure 9.31 opens. Uncheck the "Randomize runs" box so that it is easier for you to replicate the results in this case study. Click on "OK" and it takes you back to the dialog box shown in Figure 9.28. Click on "OK" and the partial mixture design shown in Figure 9.32 is the result. In order to see the simplex

Analysis of Variance for Sulfation (component proportions)

Source	DF	Seq SS	Adj SS	Adj MS	F	P
Regression	3	360.40	360.40	120.134	2.42	0.102
Linear	2	344.53	358.76	179.379	3.61	0.049
Quadratic	1	15.87	15.87	15.874	0.32	0.579
P*Q	1	15.87	15.87	15.874	0.32	0.579
Residual Error	17	843.81	843.81	49.636		
Lack-of-Fit	3	11.85	11.85	3.950	0.07	0.977
Pure Error	14	831.96	831.96	59.426		
Total	20	1204.21				

FIGURE 9.19
Revised ANOVA output after removal of QR and PR terms.

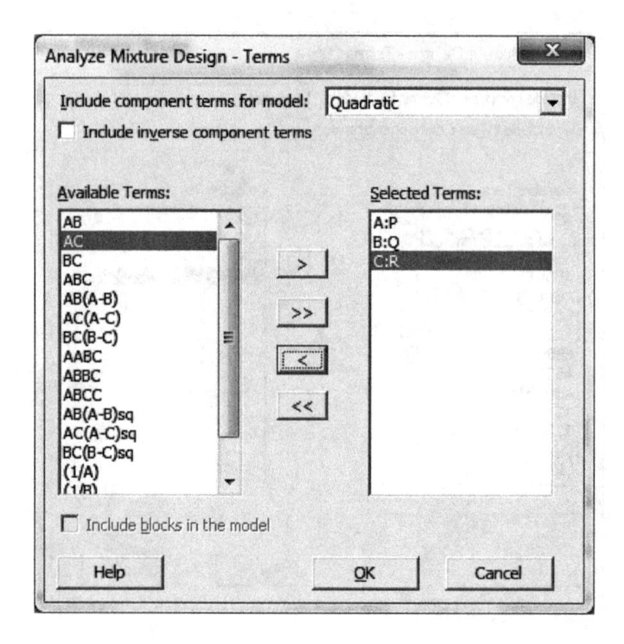

FIGURE 9.20
Removal of PQ term.

design plot, select "Simplex Design Plot" as was shown in Figure 9.8. Doing so opens the dialog box shown in Figure 9.33. Click on "OK" and the simplex design plot shown in Figure 9.34 is the result. As shown in Figure 9.35, label an empty column as "Temperature" and enter the data. (For example, in the first run, the temperature is found to be "398" when P = 60%, Y = 20%, and Z = 20%.) Open the CHAPTER_9_ZEO.MTW worksheet for the data shown in Figure 9.35 (the worksheet is available at the publisher's website; the data from the worksheet are also provided in the Appendix).

To analyze the design created for fuel Zeo, select "Analyze Mixture Design" as shown in Figure 9.36. Doing so opens the dialog box shown in Figure 9.37.

Analysis of Variance for Sulfation (component proportions)

Source	DF	Seq SS	Adj SS	Adj MS	F	P
Regression	2	344.53	344.53	172.264	3.61	0.048
Linear	2	344.53	344.53	172.264	3.61	0.048
Residual Error	18	859.68	859.68	47.760		
Lack-of-Fit	4	27.72	27.72	6.931	0.12	0.974
Pure Error	14	831.96	831.96	59.426		
Total	20	1204.21				

FIGURE 9.21
Revised ANOVA output after removal of PQ term.

File Edit Data Calc Stat Graph Editor Tools Window Help Assistant

Basic Statistics
Regression
ANOVA
DOE
Control Charts
Quality Tools
Reliability/Survival
Multivariate
Time Series
Tables
Nonparametrics
EDA
Power and Sample Size

Factorial
Response Surface
Mixture
Taguchi
Modify Design...
Display Design...

Create Mixture Design...
Define Custom Mixture Design...
Select Optimal Design...
Simplex Design Plot...
Factorial Plots...
Analyze Mixture Design...
Response Trace Plot...
Contour/Surface Plots...
Overlaid Contour Plot...
Response Optimizer...

↓	C1	C2						C8	C9	C10
	StdOrder	RunOrd						Sulfation		
1	1									
2	2									
3	3									
4	4									
5	5									
6	6				0.000	50.000	50.			
7	7				33.333	33.333	33.			
8	8				00.000	0.000	0.			
9	9				0.000	100.000	0.			
10	10	10	1	1	0.000	0.000	100.			
11	11	11	2	1	50.000	50.000	0.			
12	12	12	2	1	50.000	0.000	50.			
13	13	13	2	1	0.000	50.000	50.000	12		
14	14	14	0	1	33.333	33.333	33.333	16		
15	15	15	1	1	100.000	0.000	0.000	17		
16	16	16	1	1	0.000	100.000	0.000	12		
17	17	17	1	1	0.000	0.000	100.000	1		
18	18	18	2	1	50.000	50.000	0.000	13		

FIGURE 9.22
Selection of "Response Optimizer" for fuel Neo.

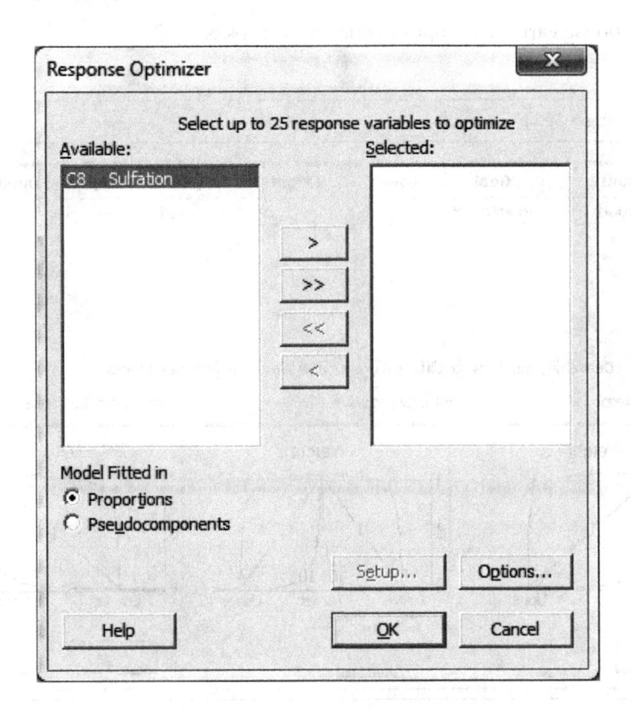

FIGURE 9.23
Available response variable for optimization for fuel Neo.

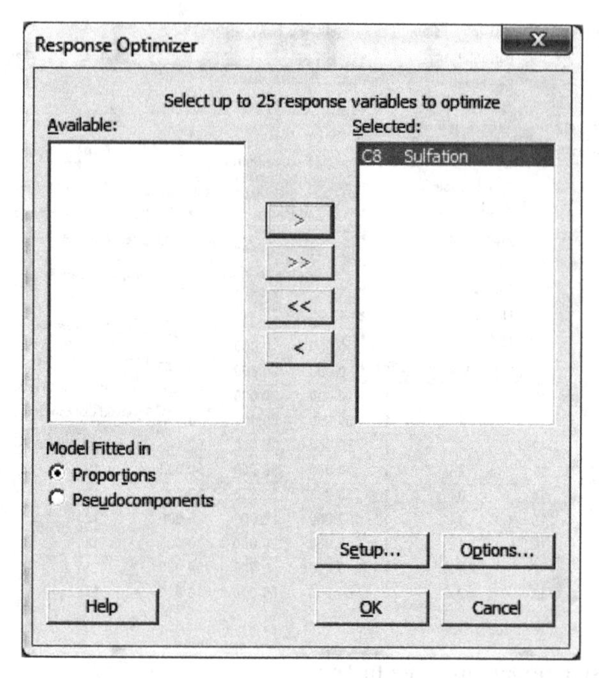

FIGURE 9.24
Selection of response variable for optimization for fuel Neo.

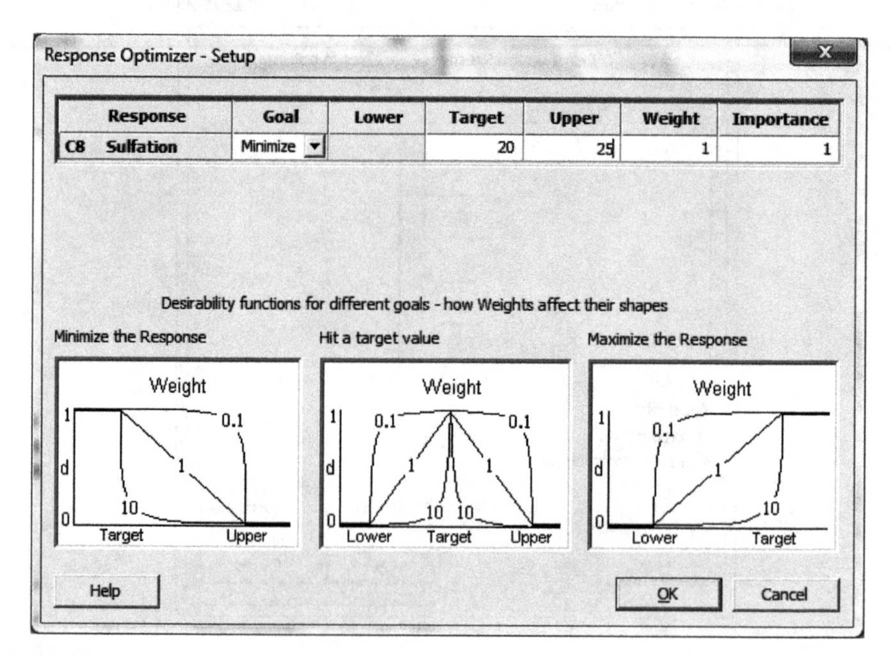

FIGURE 9.25
Entry of upper bound for sulfation.

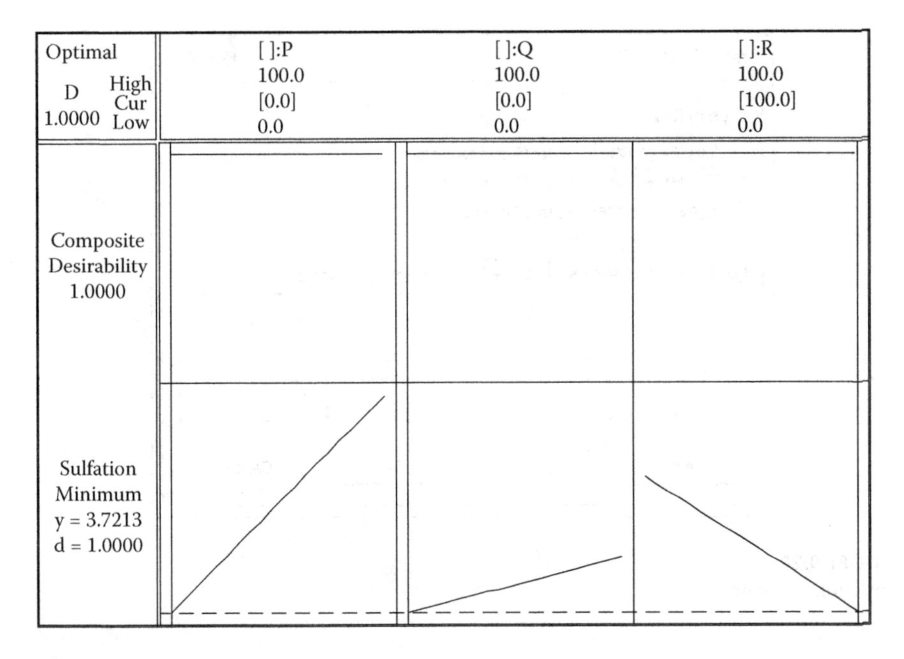

FIGURE 9.26
Optimal amounts of fuel Neo components.

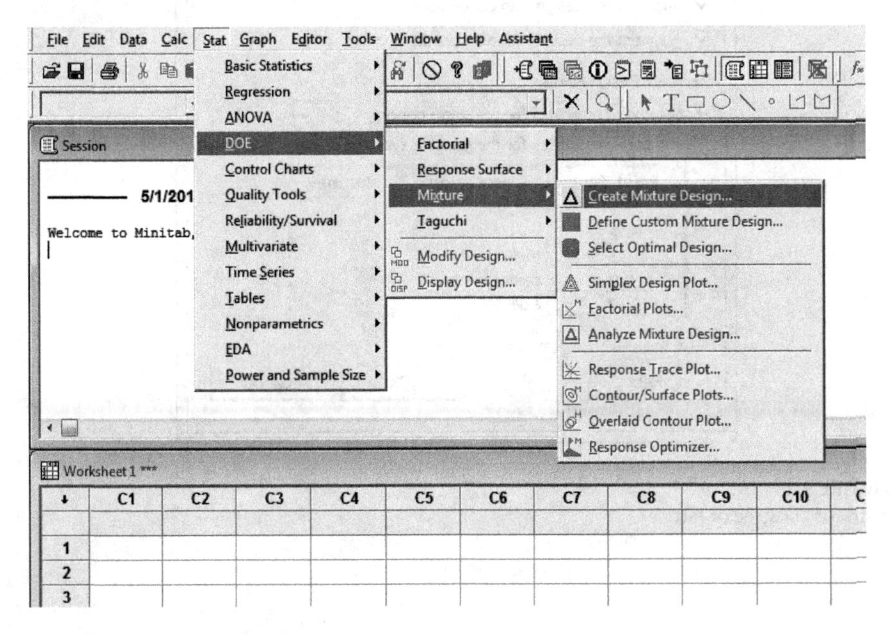

FIGURE 9.27
Selection of "Create Mixture Design" for fuel Zeo.

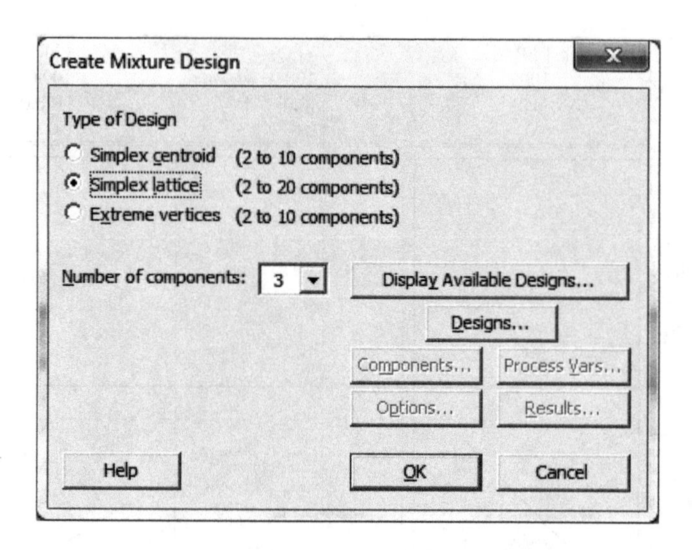

FIGURE 9.28
Selection of "Simplex lattice".

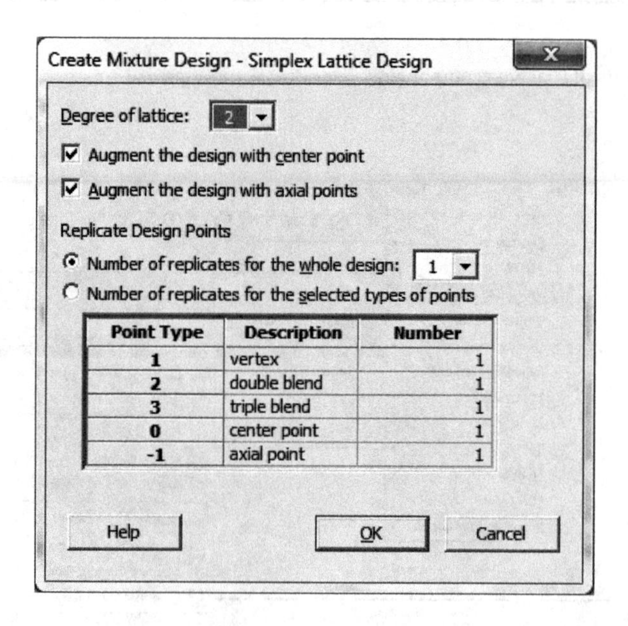

FIGURE 9.29
Entry of "Degree of lattice".

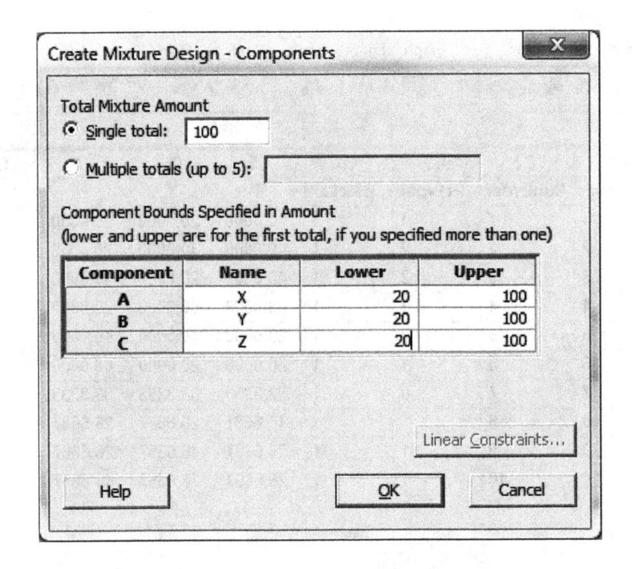

FIGURE 9.30
Entry of lower and upper bounds for components of fuel Zeo.

Select "Temperature" for "Responses" and click on "Terms". This opens the dialog box shown in Figure 9.38. Choose "Quadratic" for "Include component terms for model" and choose the "Selected Terms" as shown in Figure 9.38. Click on "OK" and it takes you back to the dialog box shown in Figure 9.37. Click on "OK" and the output shown in Figure 9.39 is the result. Notice that all the *P*-values in Figure 9.39 are less than 0.05, and hence this model in Figure 9.39 can be considered now for optimization. For optimization of temperature, select "Response Optimizer" as shown in Figure 9.40. Doing so opens the dialog box shown in Figure 9.41. Move the "Temperature" column from "Available" to "Selected", as shown in Figure 9.42. Click on "Setup" and

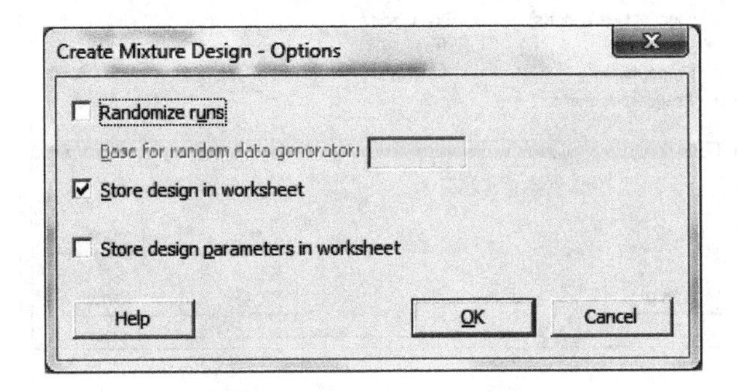

FIGURE 9.31
Unchecking "Randomize runs" box for fuel Zeo.

File Edit Data Calc Stat Graph Editor Tools Window Help Assistant

↓	C1	C2	C3	C4	C5	C6	C7	C8
	StdOrder	RunOrder	PtType	Blocks	X	Y	Z	
1	1	1	1	1	60.0000	20.0000	20.0000	
2	2	2	2	1	40.0000	40.0000	20.0000	
3	3	3	2	1	40.0000	20.0000	40.0000	
4	4	4	1	1	20.0000	60.0000	20.0000	
5	5	5	2	1	20.0000	40.0000	40.0000	
6	6	6	1	1	20.0000	20.0000	60.0000	
7	7	7	0	1	33.3333	33.3333	33.3333	
8	8	8	-1	1	46.6667	26.6667	26.6667	
9	9	9	-1	1	26.6667	46.6667	26.6667	
10	10	10	-1	1	26.6667	26.6667	46.6667	
11								
12								
13								
14								

FIGURE 9.32
Partial mixture design for fuel Zeo.

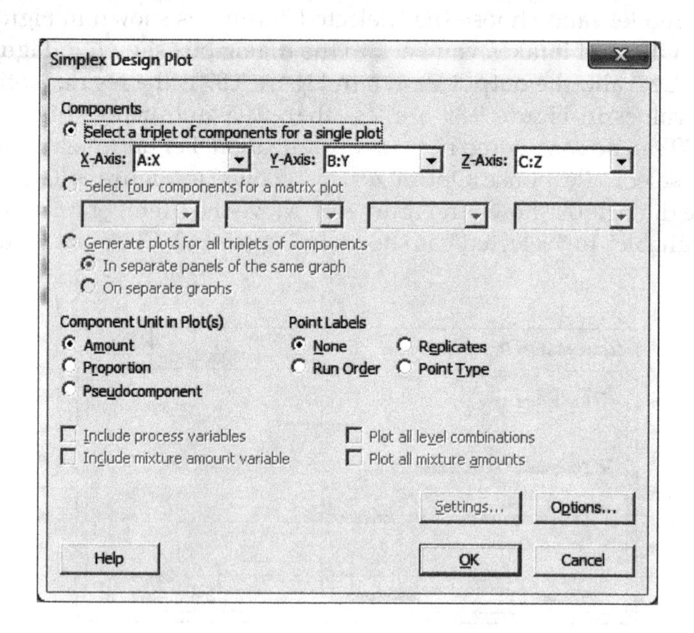

FIGURE 9.33
Options for "Simplex Design Plot" for fuel Zeo.

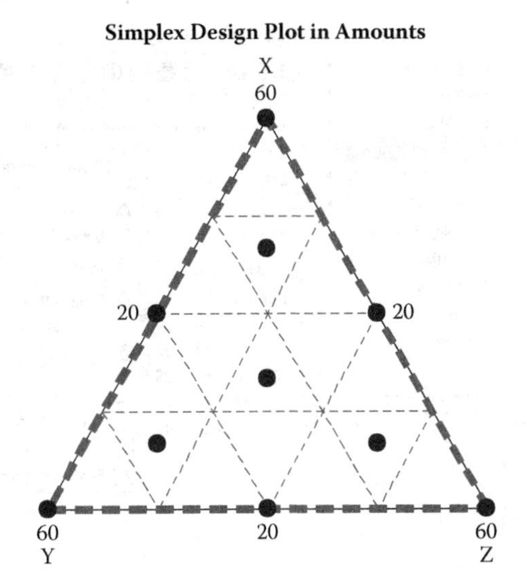

FIGURE 9.34
"Simplex Design Plot" for fuel Zeo.

File Edit Data Calc Stat Graph Editor Tools Window Help Assistant

↓	C1	C2	C3	C4	C5	C6	C7	C8
	StdOrder	RunOrder	PtType	Blocks	X	Y	Z	Temperature
1	1	1	1	1	60.0000	20.0000	20.0000	398
2	2	2	2	1	40.0000	40.0000	20.0000	484
3	3	3	2	1	40.0000	20.0000	40.0000	424
4	4	4	1	1	20.0000	60.0000	20.0000	400
5	5	5	2	1	20.0000	40.0000	40.0000	409
6	6	6	1	1	20.0000	20.0000	60.0000	342
7	7	7	0	1	33.3333	33.3333	33.3333	456
8	8	8	-1	1	46.6667	26.6667	26.6667	460
9	9	9	-1	1	26.6667	46.6667	26.6667	450
10	10	10	-1	1	26.6667	26.6667	46.6667	411
11								

FIGURE 9.35
Mixture design for fuel Zeo.

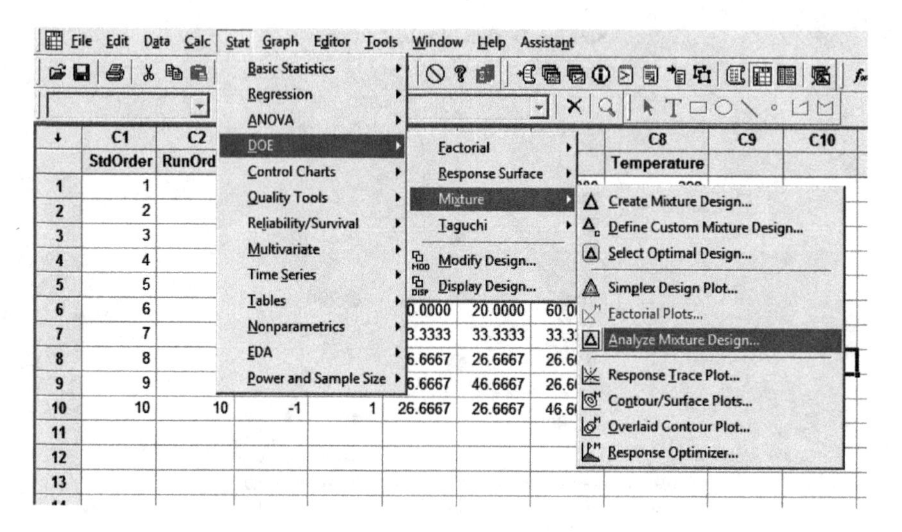

FIGURE 9.36
Selection of "Analyze Mixture Design" for fuel Zeo.

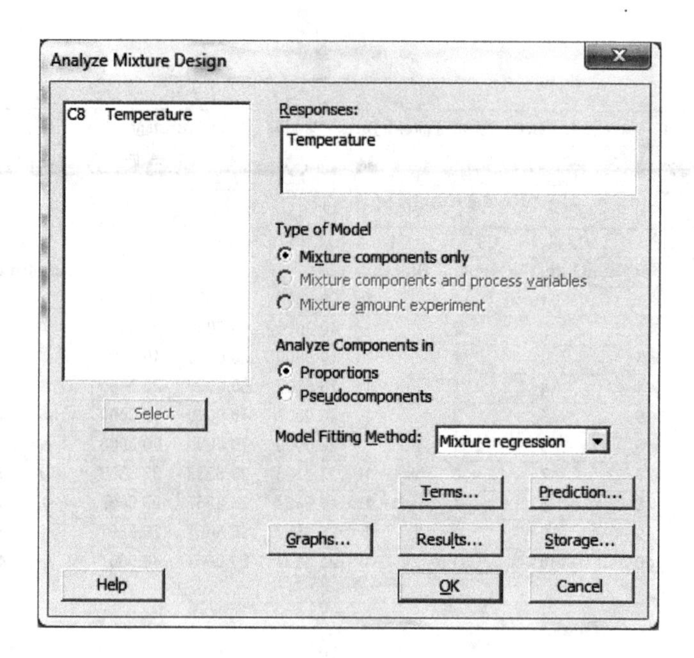

FIGURE 9.37
Selection of "Temperature" column for "Responses".

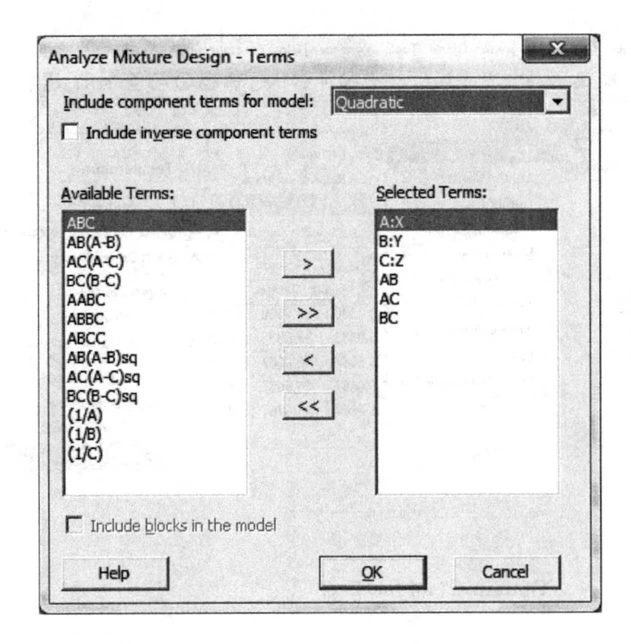

FIGURE 9.38
Quadratic terms for fuel Zeo mixture design.

the dialog box shown in Figure 9.43 opens. Select "Target" for "Goal", enter "399" for "Lower", and enter "400" for "Target" and "401" for "Upper". (The laboratory aims for a target temperature of 400 units.) Click on "OK" and the optimal solution (X: 20%, Y: 35.6443%, and Z: 44.3557%) shown in Figure 9.44 is the result.

Analysis of Variance for Temperature (component proportions)

Source	DF	Seq SS	Adj SS	Adj MS	F	P
Regression	5	14933.7	14933.71	2986.74	416.36	0.000
Linear	2	5686.8	109.00	54.50	7.60	0.043
Quadratic	3	9246.9	9246.93	3082.31	429.69	0.000
X*Y	1	5775.6	5851.62	5851.62	815.74	0.000
X*Z	1	2346.0	2365.84	2365.84	329.81	0.000
Y*Z	1	1125.3	1125.35	1125.35	156.88	0.000
Residual Error	4	28.7	28.69	7.17		
Total	9	14962.4				

FIGURE 9.39
ANOVA output for fuel Zeo.

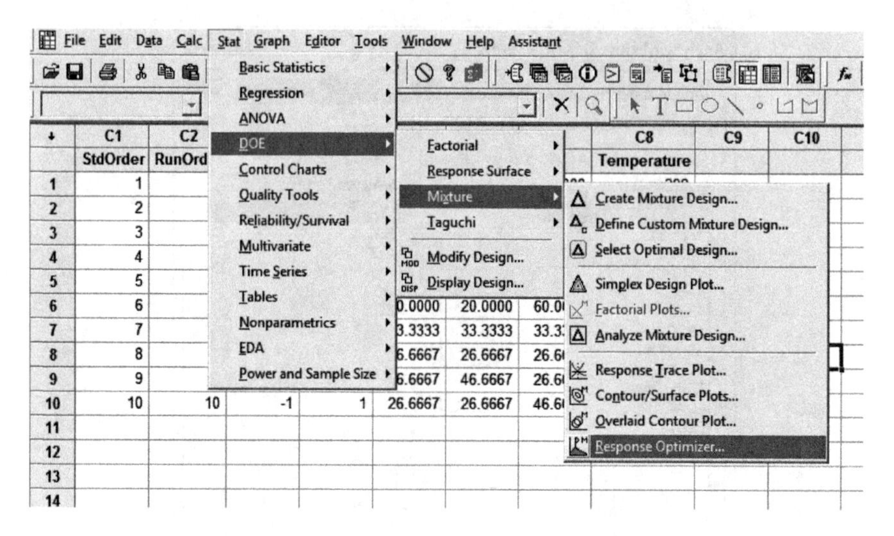

FIGURE 9.40
Selection of "Response Optimizer" for fuel Zeo.

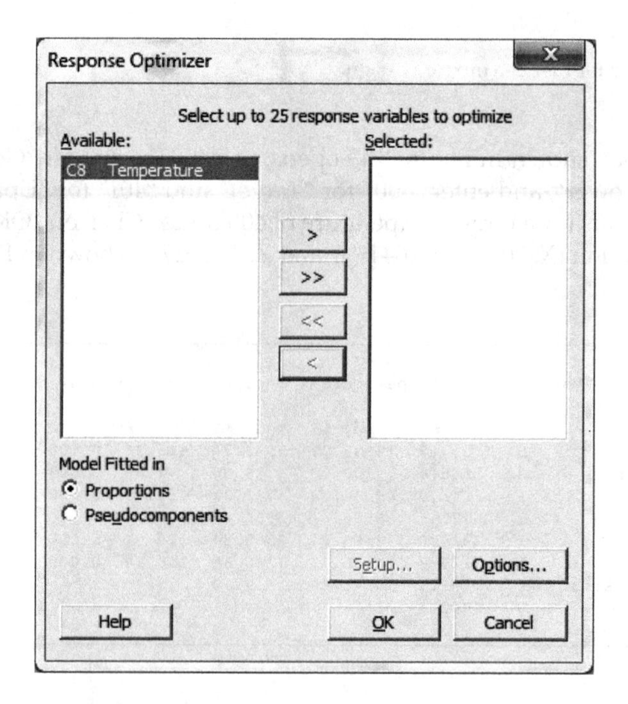

FIGURE 9.41
Available response variable for optimization for fuel Zeo.

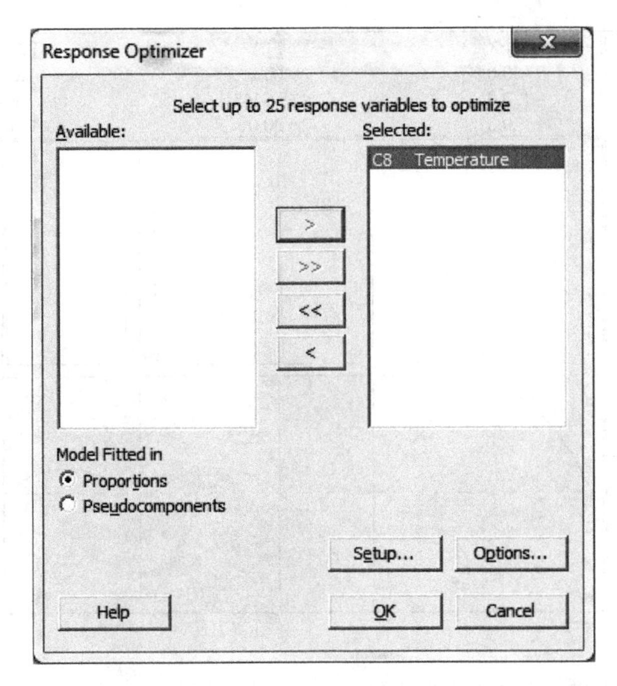

FIGURE 9.42
Selection of response variable for optimization for fuel Zeo.

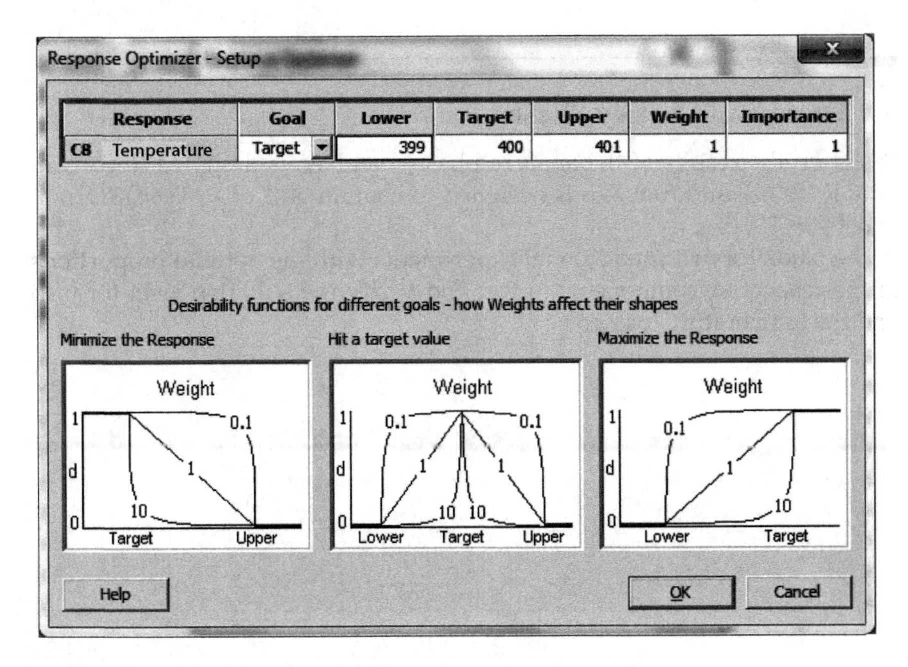

FIGURE 9.43
Entry of bounds for temperature.

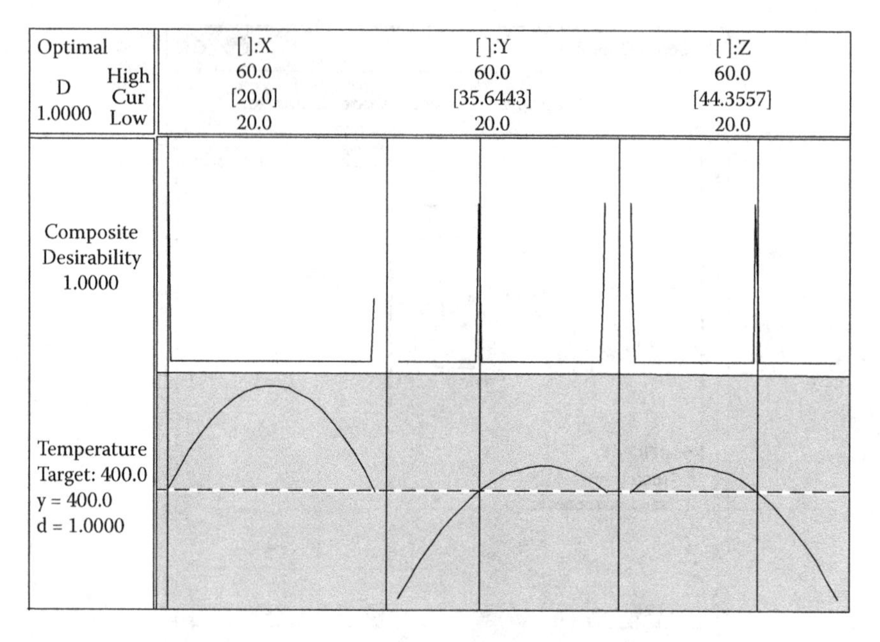

FIGURE 9.44
Optimal amounts of fuel Zeo components.

9.4 Design and Verify Phases

Based on the results of the analyze phase, fuel Neo is designed to contain only R (100%), and fuel Zeo is designed to contain 20% of X, 35.6443% of Y, and 44.3557% of Z.

The laboratory runs additional experiments with the optimal proportions of the respective components to test and verify the sulfation level for Neo and the temperature for Zeo.

10

Multivariate Analysis to Reduce Patient Waiting Time at a Medical Center

This case study is about a Six Sigma project implemented by the quality department at Quickfix Medical Center. The administration aims to reduce the patient waiting time before treatment.

Sections 10.1 and 10.2 give brief descriptions of the define phase and the measure phase, respectively. Section 10.3 illustrates the analyze phase with detailed instructions for using Minitab®. Finally, Sections 10.4 and 10.5 give brief descriptions of the improve phase and the control phase, respectively.

10.1 Define Phase

The quality department has been asked by the administration to reduce the patient waiting time. To this end, the quality department brainstorms and identifies the following two factors that they believe might have an effect on the patient waiting time:

- Location (Quickfix has two locations: Copley and Kenmore.)
- Service type (Quickfix serves in three areas: internal medicine, pediatrics, dermatology.)

The problem statement is "to reduce the patient waiting time".

10.2 Measure Phase

For a preliminary process analysis, the quality department collects data (waiting time in minutes) for a random sample of four patients for each combination of location and service type, as shown in Table 10.1.

The qualitative variables (e.g., Copley, pediatrics, etc.) are then coded as shown in Table 10.2 so that it is easier to enter the data in Minitab®.

TABLE 10.1
Sample Data

	Internal Medicine	Pediatrics	Dermatology
Copley	35, 24, 42, 21	12, 54, 10, 9	25, 32, 34, 45
Kenmore	42, 35, 21, 34	19, 18, 17, 23	10, 9, 4, 11

TABLE 10.2
Revised Sample Data (Coded Levels of Factors)

	1	2	3
1	35, 24, 42, 21	12, 54, 10, 9	25, 32, 34, 45
2	42, 35, 21, 34	19, 18, 17, 23	10, 9, 4, 11

In Table 10.2, the levels of Location are coded as 1 (for Copley) and 2 (for Kenmore). The levels of Service Type are coded as 1 (for Internal Medicine), 2 (for Pediatrics), and 3 (for Dermatology).

Figure 10.1 partially shows the waiting times of the patients as entered in the Minitab® worksheet. Refer to the CHAPTER_10_1.MTW worksheet for

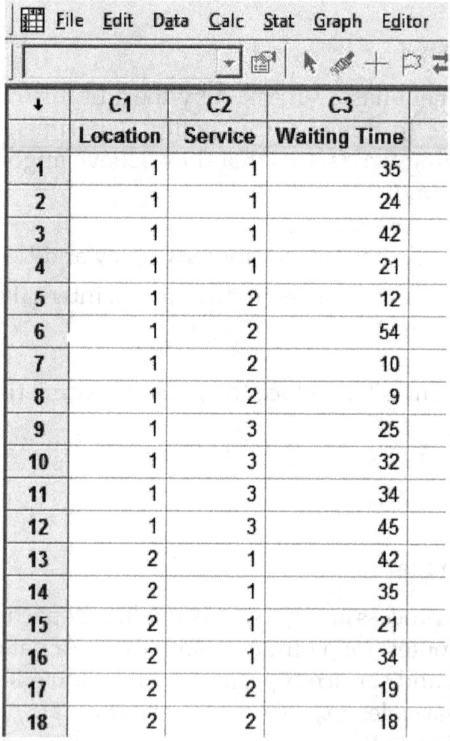

File Edit Data Calc Stat Graph Editor

↓	C1	C2	C3
	Location	Service	Waiting Time
1	1	1	35
2	1	1	24
3	1	1	42
4	1	1	21
5	1	2	12
6	1	2	54
7	1	2	10
8	1	2	9
9	1	3	25
10	1	3	32
11	1	3	34
12	1	3	45
13	2	1	42
14	2	1	35
15	2	1	21
16	2	1	34
17	2	2	19
18	2	2	18

FIGURE 10.1
Sample of patient waiting times.

the complete set of data (the worksheet is available at the publisher's website; the data from the worksheet are also provided in the Appendix).

10.3 Analyze Phase

The quality team plots multivariate charts to verify whether waiting time indeed depends on location and service type, and also whether there is any interaction between location and service type.

Open the CHAPTER_10_1.MTW worksheet, and click on "Multi-Vari Chart" as shown in Figure 10.2. As a result, the dialog box shown in Figure 10.3 opens. Using the "Select" button, select "Waiting Time" for "Response", "Location" for "Factor 1", and "Service Type" for "Factor 2". Then, click on "Options", and the dialog box shown in Figure 10.4 opens. Check the boxes for "Connect Means for Factor 1" and "Connect Means for Factor 2" if they

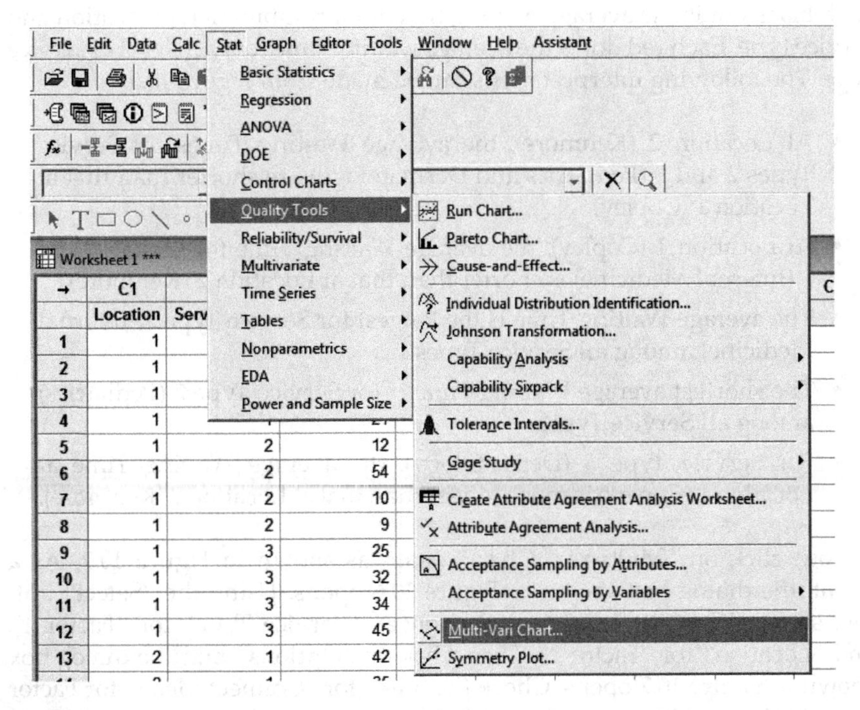

FIGURE 10.2
Selection of "Multi-Vari Chart".

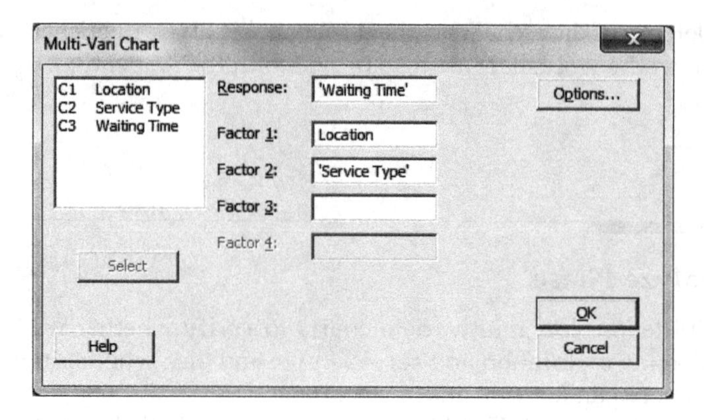

FIGURE 10.3
Selection of response and factors with "Location" as "Factor 1".

are not already checked. Enter the title in the "Title" box as shown, and click on "OK". It will take you back to the dialog box shown in Figure 10.3. Click on "OK", and the multivariate chart shown in Figure 10.5 is plotted. (Please note that the chart is shown in grayscale here, although it has colored dots.) Each black dot is the average waiting time for a combination of location and service type. Each red dot is the average waiting time for each level of service type. The following interpretations can be made from Figure 10.5:

- At Location 2 (Kenmore), the average Waiting Times for Service Types 2 and 3 (Pediatrics and Dermatology) are shorter than that at Location 1 (Copley).
- At Location 1 (Copley), the average Waiting Time for Service Type 1 (Internal Medicine) is shorter than that at Location 2 (Kenmore).
- The average Waiting Time is the longest for Service Type 1 (Internal Medicine) among all Service Types.
- The shortest average Waiting Time is for Service Type 2 (Pediatrics) among all Service Types.
- For Service Type 3 (Dermatology), the average Waiting Time at Location 1 (Copley) is much longer than that at Location 2 (Kenmore).

Now, click on "Multi-Vari Chart" again as shown in Figure 10.2. As a result, the dialog box shown in Figure 10.6 opens. Using the "Select" button, select "Waiting Time" for "Response", "Service Type" for "Factor 1", and "Location" for "Factor 2". Then, click on "Options", and the dialog box shown in Figure 10.7 opens. Check the boxes for "Connect Means for Factor 1" and "Connect Means for Factor 2" if they are not already checked. Enter the title in the "Title" box as shown, and click on "OK". It will take you back

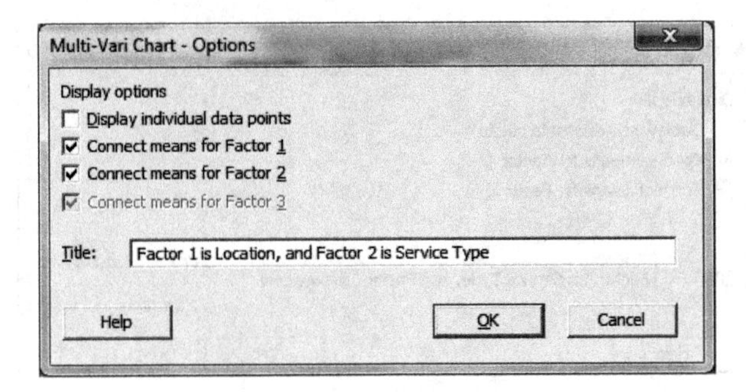

FIGURE 10.4
Connection of means with "Location" as "Factor 1".

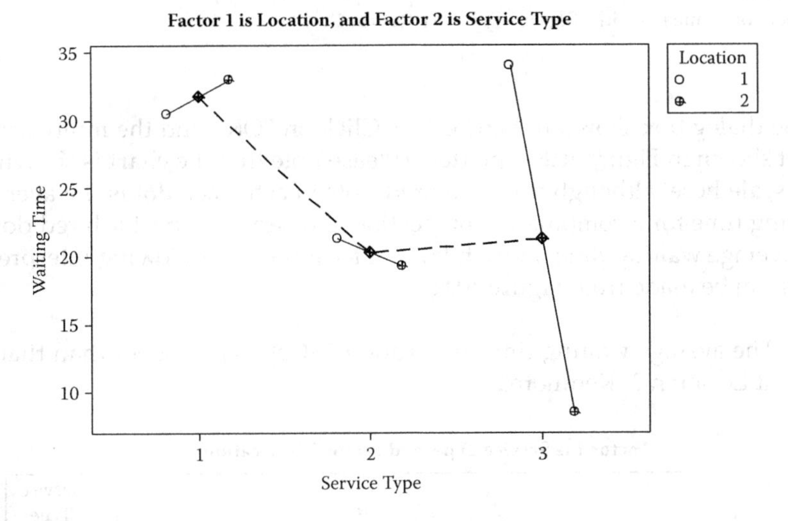

FIGURE 10.5
Multivariate chart with "Location" as "Factor 1".

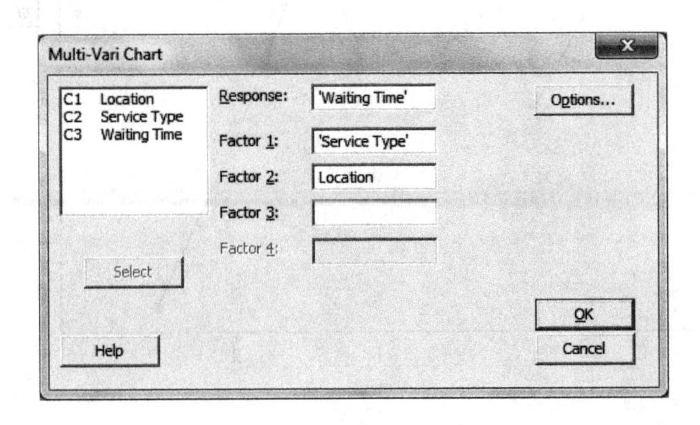

FIGURE 10.6
Multivariate chart with "Service Type" as "Factor 1".

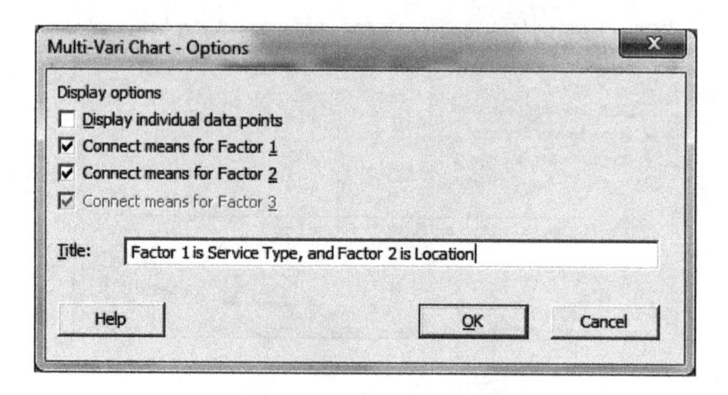

FIGURE 10.7
Connection of means with "Service Type" as "Factor 1".

to the dialog box shown in Figure 10.6. Click on "OK", and the multivariate chart shown in Figure 10.8 is plotted. (Please note that the chart is shown in grayscale here, although it has colored dots.) Each black dot is the average waiting time for a combination of location and service type. Each red dot is the average waiting time for each level of location. The following interpretations can be made from Figure 10.8:

- The average waiting time at Location 1 (Copley) is longer than that at Location 2 (Kenmore).

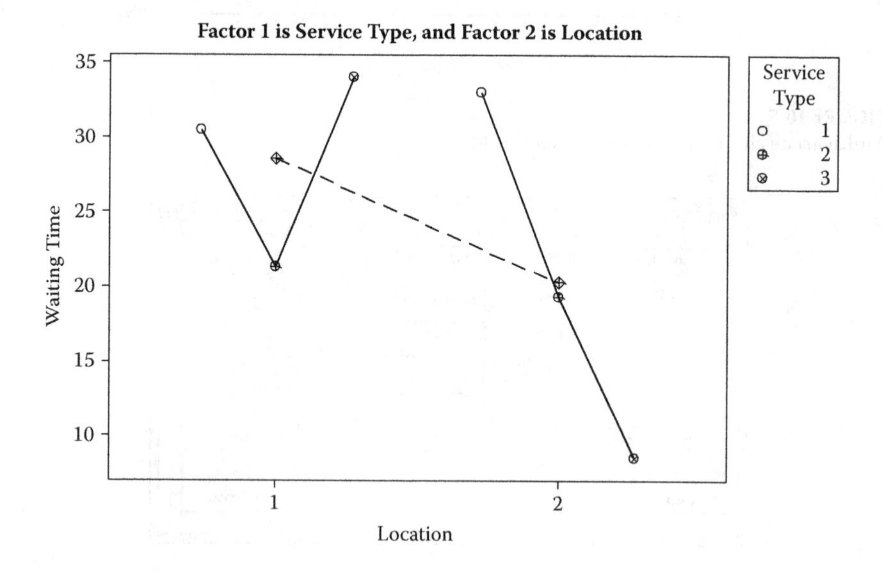

FIGURE 10.8
Multivariate chart with "Service Type" as "Factor 1".

- At Location 1 (Copley), the average waiting time for Service Type 2 (Pediatrics) is the shortest among all service types.
- At Location 2 (Kenmore), the average waiting time for Service Type 3 (Dermatology) is the shortest among all service types.

Boxplots can show the mean, median, first quartile, third quartile, and any outliers for the waiting times for the two locations and the three service types.

Click on "Boxplot" as shown in Figure 10.9, and the dialog box shown in Figure 10.10 opens. Select "With Groups" under "One Y" and the dialog box shown in Figure 10.11 opens. Select "Waiting Time" for "Graph Variables" and "Location" for "Categorical variables for grouping", and then click on "Labels". The dialog box shown in Figure 10.12 opens. Under "Data Labels", select "Means" from the drop-down menu and click on "OK". It takes you back to the dialog box shown in Figure 10.11. Click on "Data View" and the dialog box shown in Figure 3.13 opens. Check "Interquartile range box", "Outlier symbols", and "Mean symbol". Also, select "Location" for

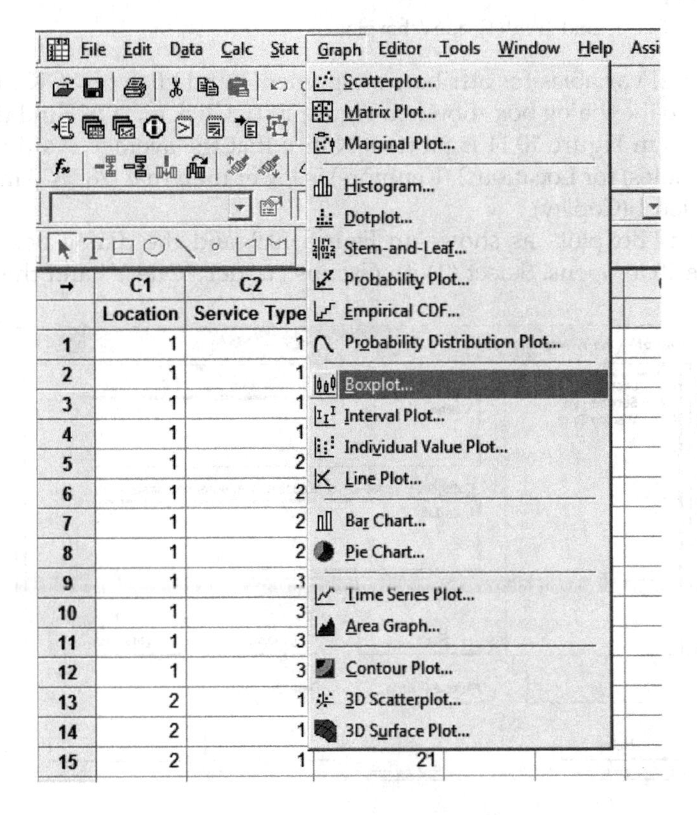

FIGURE 10.9
Selection of "Boxplot".

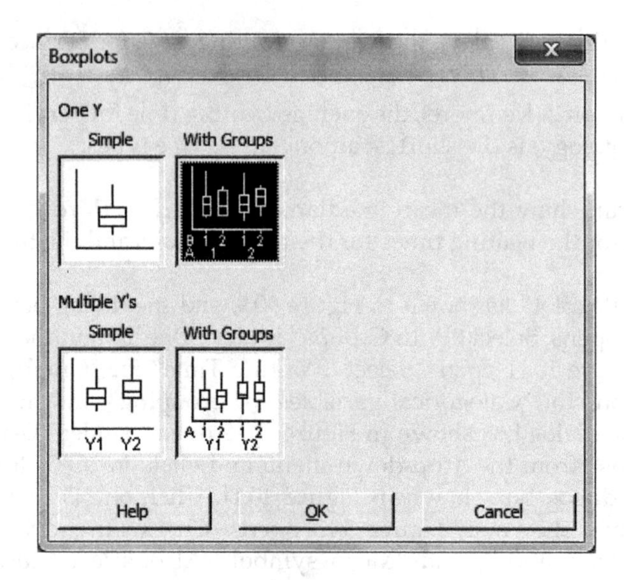

FIGURE 10.10
Selection of "One Y" and "With Groups" boxplot.

"Categorical variables for attribute assignment", and click on "OK". It takes you back to the dialog box shown in Figure 10.11. Click on "OK" and the boxplot shown in Figure 10.14 is plotted. Notice that the average Waiting Time (20.25 minutes) for Location 2 (Kenmore) is lower than that (28.5833 minutes) for Location 1 (Copley).

Click on "Boxplot" as shown in Figure 10.9, and the dialog box shown in Figure 10.10 opens. Select "With Groups" under "One Y" and the dialog

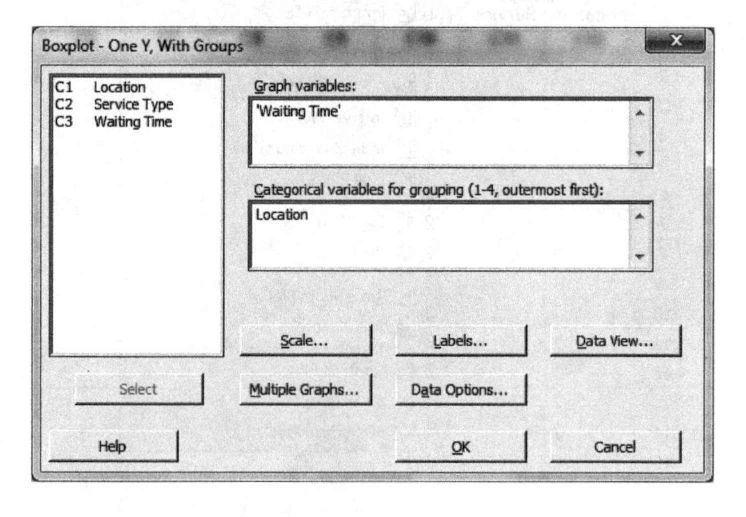

FIGURE 10.11
"Location" is "Categorical variable".

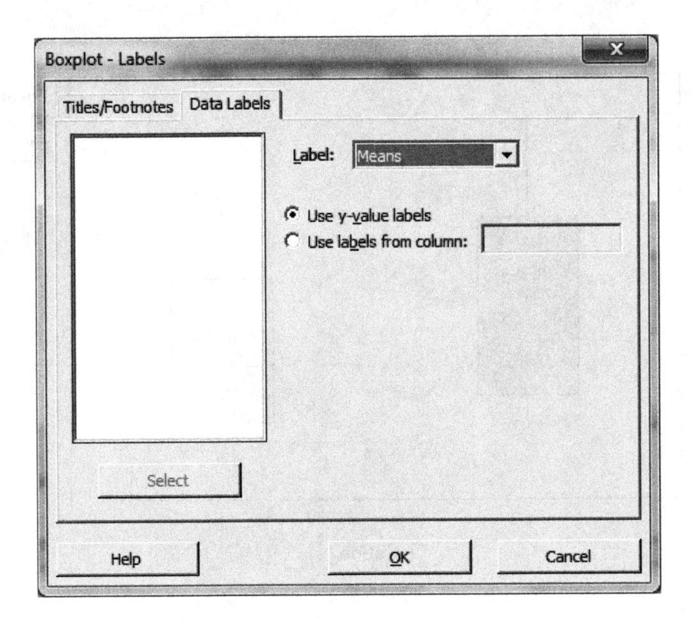

FIGURE 10.12
Labeling the means for boxplot with "Location" as "Categorical variable".

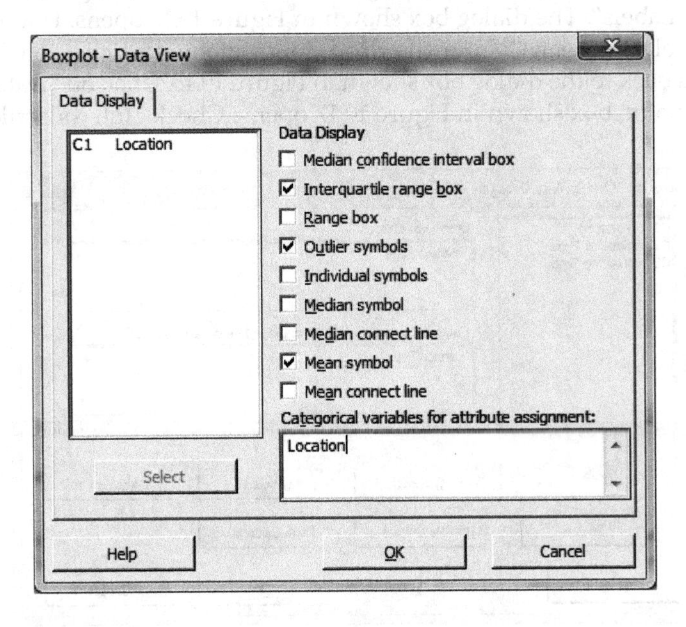

FIGURE 10.13
Data display for boxplot with "Location" as "Categorical variable".

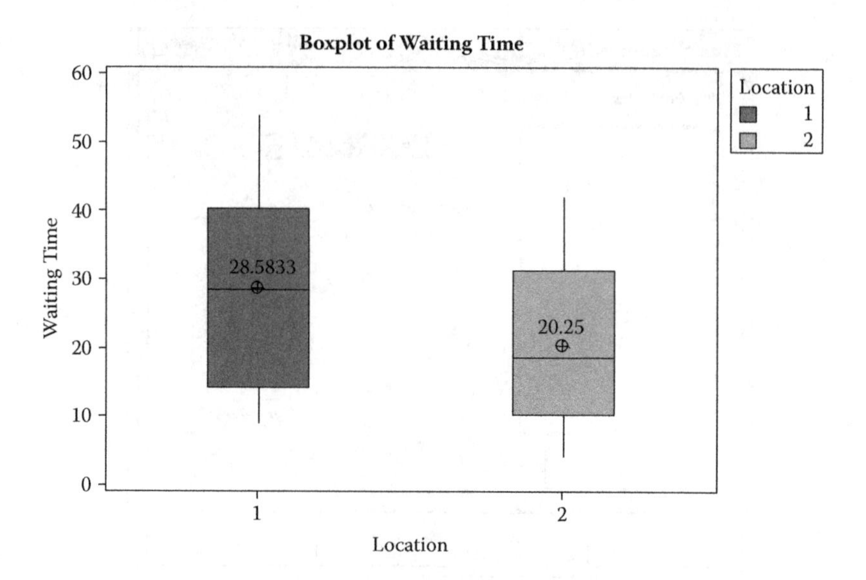

FIGURE 10.14
Boxplot with "Location" as "Categorical variable".

box shown in Figure 10.15 opens. Select "Waiting Time" for "Graph variables" and "Service Type" for "Categorical variables for grouping", and then click on "Labels". The dialog box shown in Figure 10.16 opens. Under "Data Labels", select "Means" from the drop-down menu and click on "OK". It takes you back to the dialog box shown in Figure 10.15. Click on "Data View" and the dialog box shown in Figure 10.17 opens. Check "Interquartile range

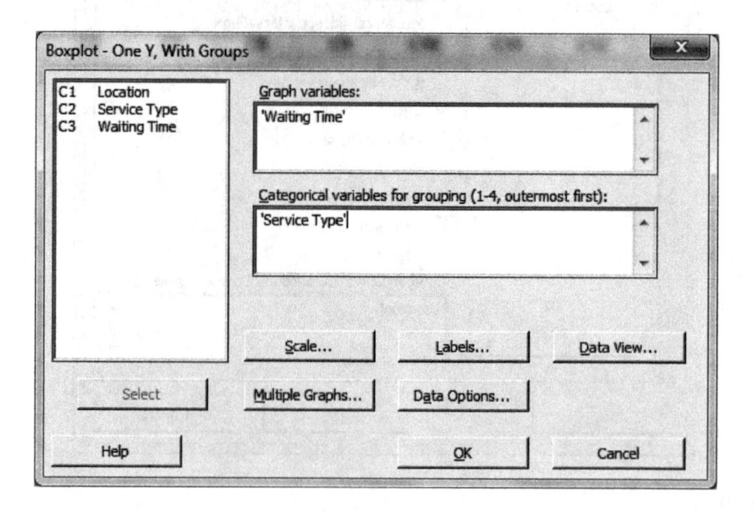

FIGURE 10.15
"Service Type" is "Categorical variable".

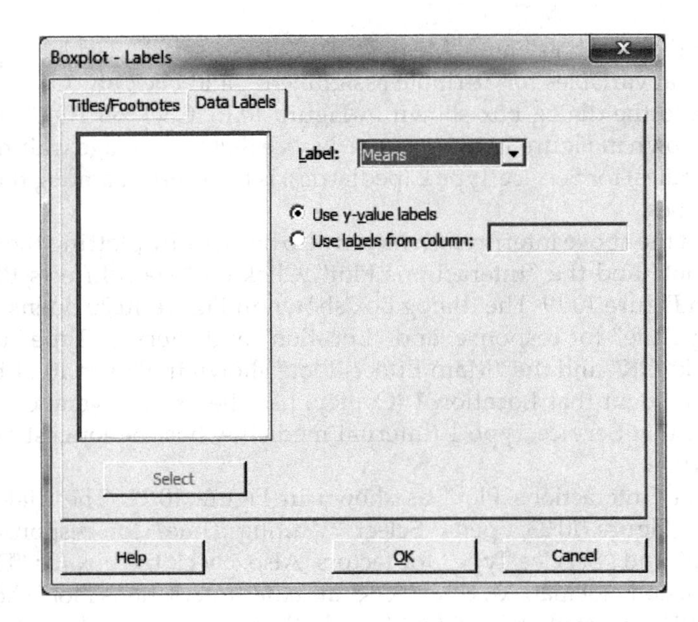

FIGURE 10.16
Labeling the means for boxplot with "Service Type" as "Categorical variable".

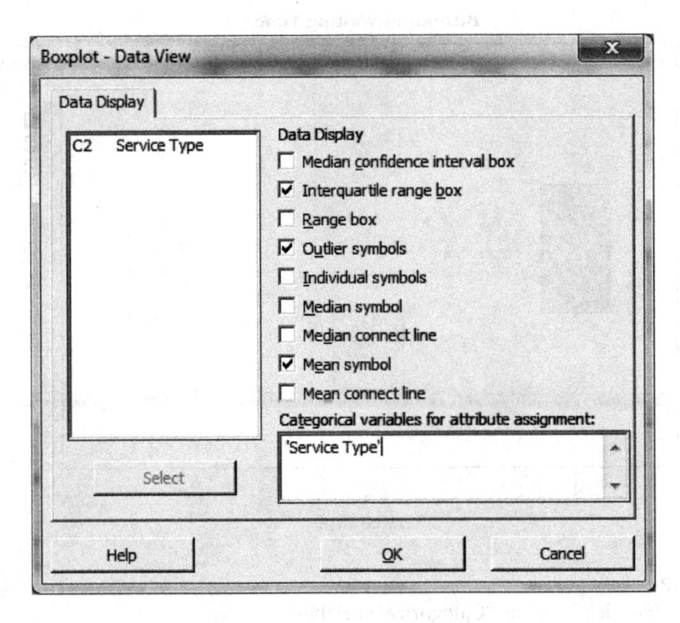

FIGURE 10.17
Data display for boxplot with "Service Type" as "Categorical variable".

box", "Outlier symbols", and "Mean symbol". Also, select "Service Type" for "Categorical variables for attribute assignment", and click on "OK". It takes you back to the dialog box shown in Figure 10.15. Click on "OK" and the boxplot shown in Figure 10.18 is plotted. Notice that the average waiting time (20.25 minutes) for Service Type 2 (pediatrics) is the shortest among the three service types.

Many of the above interpretations can also be made by plotting the "Main Effects Plot" and the "Interactions Plot". Click on "Main Effects Plot" as shown in Figure 10.19. The dialog box shown in Figure 10.20 opens. Select "Waiting Time" for response, and "Location" and "Service Type" for factors. Click "OK" and the "Main Effects Plot" shown in Figure 10.21 is plotted. It is evident that Location 1 (Copley) has the longer average waiting time, and that Service Type 1 (internal medicine) has the longest average waiting time.

Click on "Interactions Plot" as shown in Figure 10.22. The dialog box shown in Figure 10.23 opens. Select "Waiting Time" for response, and "Location" and "Service Type" for factors. Also, check the box for "Display full interaction plot matrix". Click "OK" and the "Interactions Plot" shown in Figure 10.24 is plotted. It is evident from both the plots (left side bottom plot and right side top plot) that the longest average waiting time is for Service Type 3 (dermatology) at Location 1 (Copley).

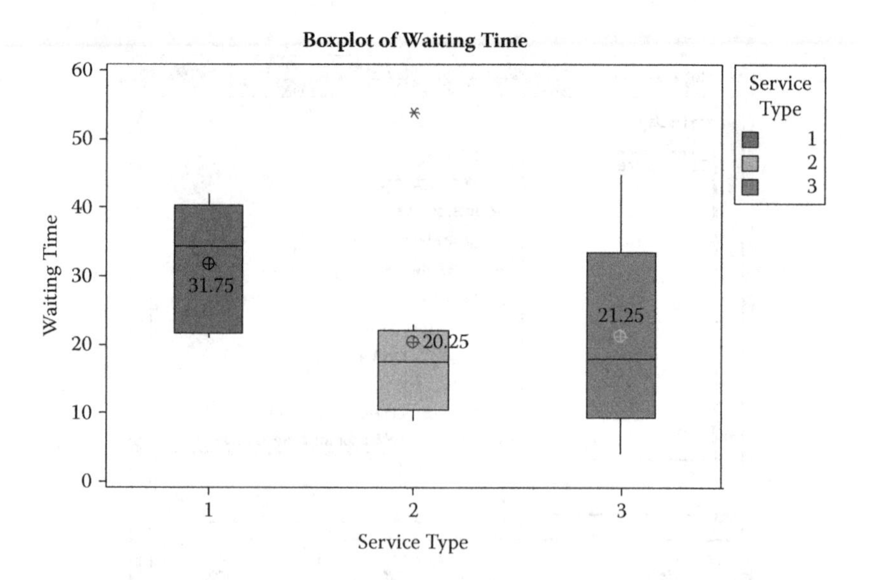

FIGURE 10.18
Boxplot with "Service Type" as "Categorical variable".

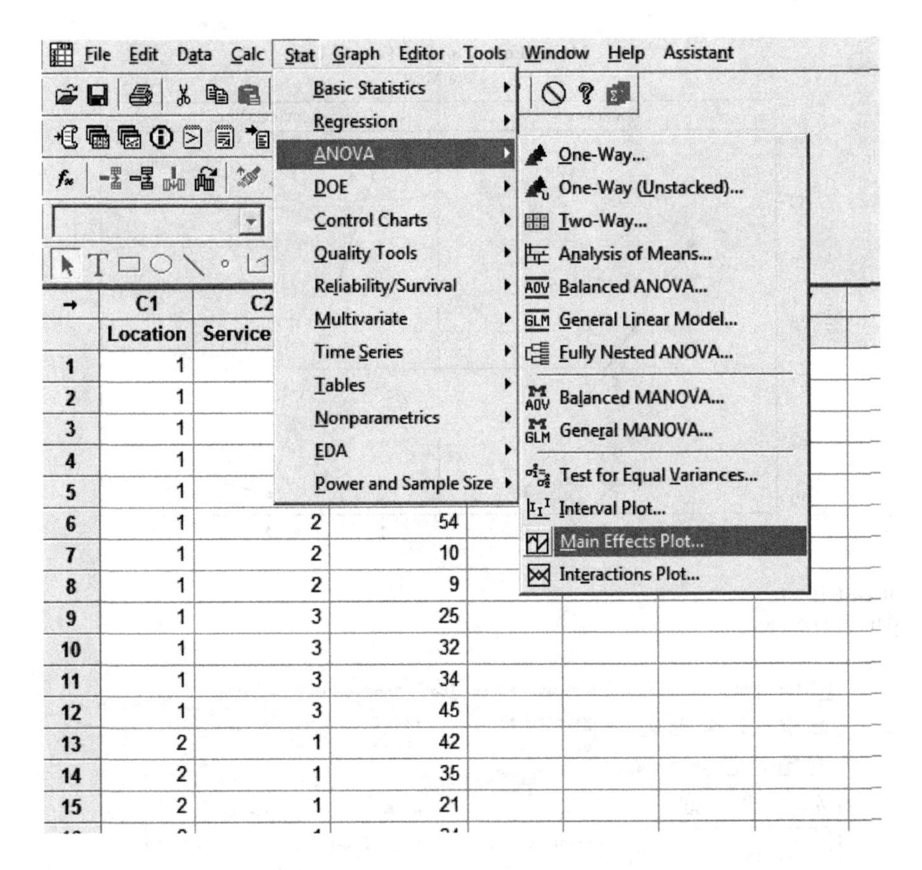

FIGURE 10.19
Selection of "Main Effects Plot".

FIGURE 10.20
Selection of responses and factors for main effects plot.

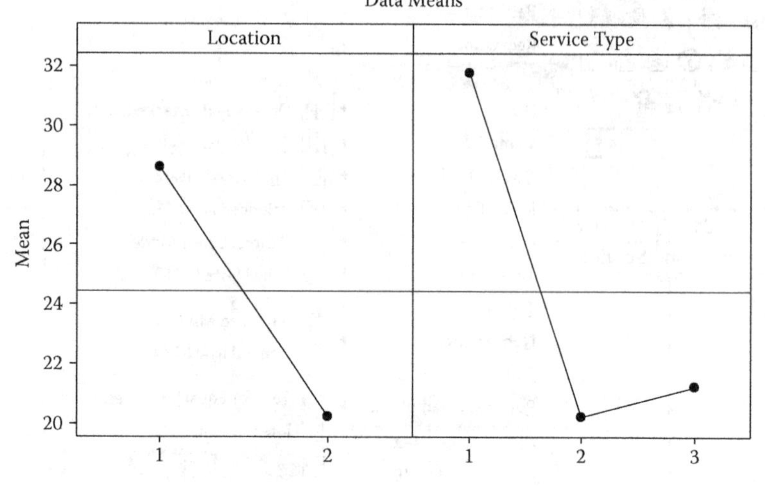

FIGURE 10.21
Main effects plot.

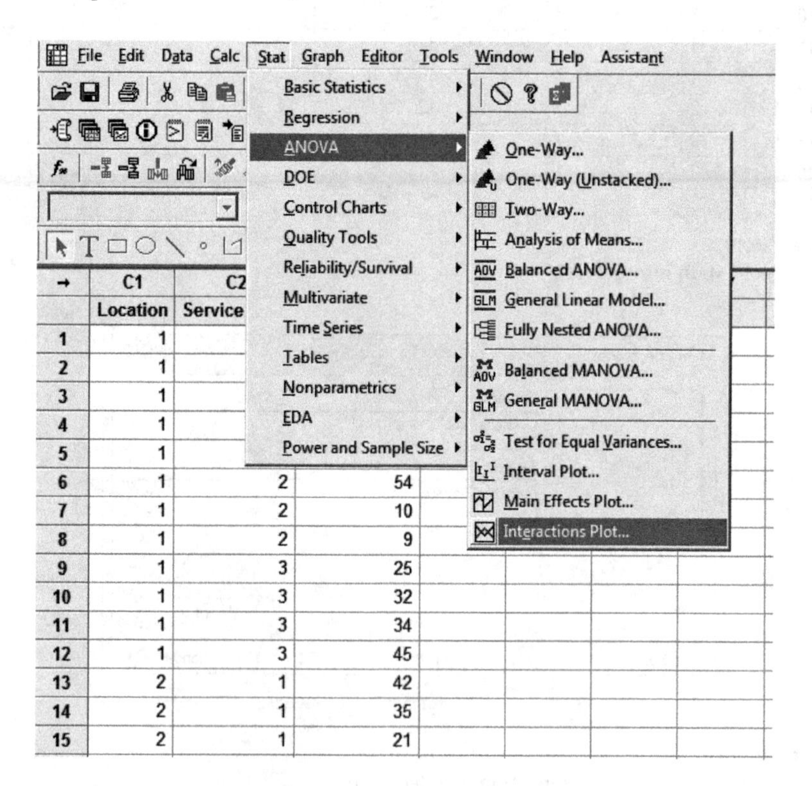

	C1 Location	C2 Service	
1	1		
2	1		
3	1		
4	1		
5	1		
6	1	2	54
7	1	2	10
8	1	2	9
9	1	3	25
10	1	3	32
11	1	3	34
12	1	3	45
13	2	1	42
14	2	1	35
15	2	1	21

FIGURE 10.22
Selection of "Interactions Plot".

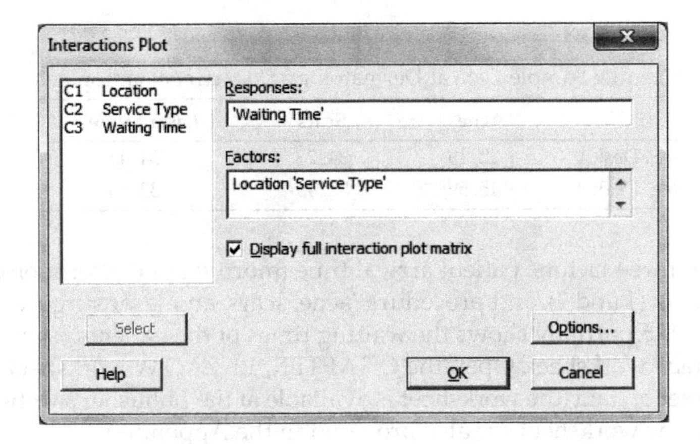

FIGURE 10.23
Selection of "Responses" and "Factors" for interaction plot.

The quality department now focuses on the longest average waiting time for Service Type 3 (dermatology) at Location 1 (Copley). The dermatology department at Copley has two registration desks (1 and 2), and they perform three types of procedures (acne, scars, and laser surgery). The department collects data (waiting time in minutes) for a random sample of two patients for each combination of registration desk and procedure, both in the morning and the afternoon. Table 10.3 shows the data for the morning, and Table 10.4 shows the data for the afternoon. Hence, the response variable is waiting

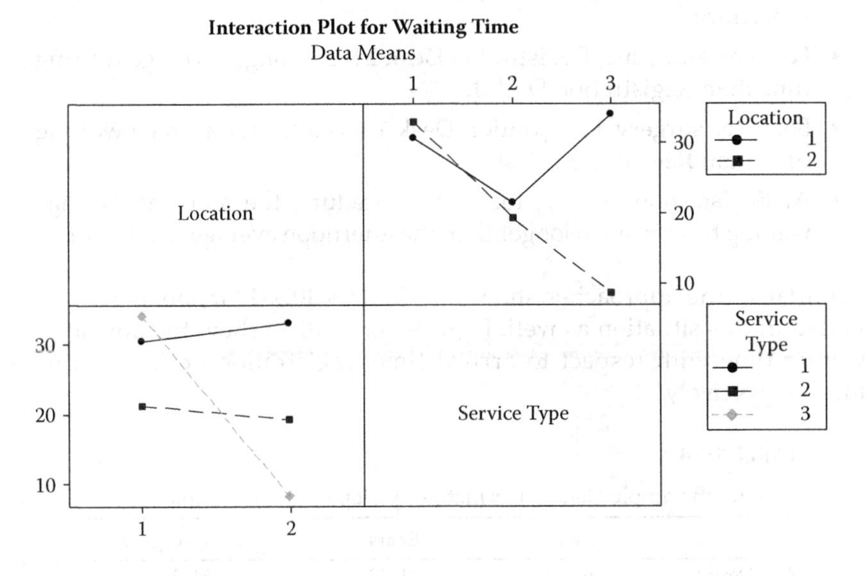

FIGURE 10.24
Interaction plot.

TABLE 10.3

Morning Sample Data at Dermatology Department at Copley

	Acne	Scars	Laser Surgery
Reg. Desk 1	22, 19	34, 23	34, 45
Reg. Desk 2	45, 59	32, 45	33, 63

time, with three factors: patient arrival time (morning and afternoon), registration desks (1 and 2), and procedure (acne, scars, and laser surgery).

Figure 10.25 partially shows the waiting times of the patients as entered in the Minitab® worksheet. Open the CHAPTER_10_2.MTW worksheet for the complete set of data (the worksheet is available at the publisher's website; the data from the worksheet are also provided in the Appendix).

Click on "Multi-Vari Chart" as was shown in Figure 10.2. As a result, the dialog box shown in Figure 10.26 opens. Using the "Select" button, select "Waiting Time" for "Response", "Arrival Time" for "Factor 1", "Registration Desk" for "Factor 2", and "Procedure" for "Factor 3". Click on "OK", and the multivariate chart shown in Figure 10.27 is plotted. (Please note that the chart is shown in grayscale here, although it has colored dots.) Each black dot is the average waiting time for a combination of registration desk and arrival time, for each procedure. Each red dot is the average waiting time for each registration desk and procedure. Each green dot is the average waiting time for each procedure. The following interpretations can be made from Figure 10.27:

- The longest average waiting time is for laser surgery, among all procedures.
- For acne and scars, Registration Desk 2 has a longer average waiting time than Registration Desk 1.
- For laser surgery, Registration Desk 1 has a longer average waiting time than Registration Desk 2.
- At Registration Desk 2, for each procedure, the morning average waiting time is much longer than the afternoon average waiting time.

Similar to the approaches shown in Figures 10.9–10.18, boxplots can be plotted in this situation as well. Figures 10.28–10.30 show the boxplots for waiting time with respect to arrival time, registration desk, and procedure, respectively.

TABLE 10.4

Afternoon Sample Data at Dermatology Department at Copley

	Acne	Scars	Laser Surgery
Reg. Desk 1	10, 13	11, 12	34, 23
Reg. Desk 2	10, 9	15, 9	11, 10

→	C1-T	C2	C3-T	C4
	Arrival Time	Registration Desk	Procedure	Waiting Time
1	M	1	Acne	22
2	M	1	Acne	19
3	M	1	Scars	34
4	M	1	Scars	23
5	M	1	Laser Surgery	34
6	M	1	Laser Surgery	45
7	M	2	Acne	45
8	M	2	Acne	59
9	M	2	Scars	32
10	M	2	Scars	45
11	M	2	Laser Surgery	33
12	M	2	Laser Surgery	63
13	A	1	Acne	10
14	A	1	Acne	13
15	A	1	Scars	11

FIGURE 10.25
Sample of patient waiting times in dermatology department at Copley.

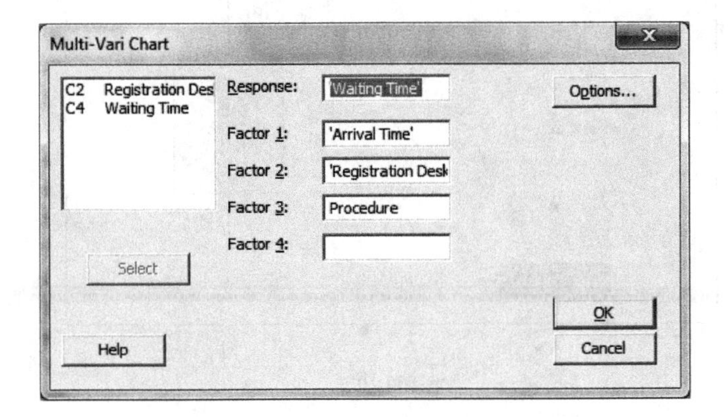

FIGURE 10.26
Selection of "Responses" and "Factors" for dermatology department at Copley.

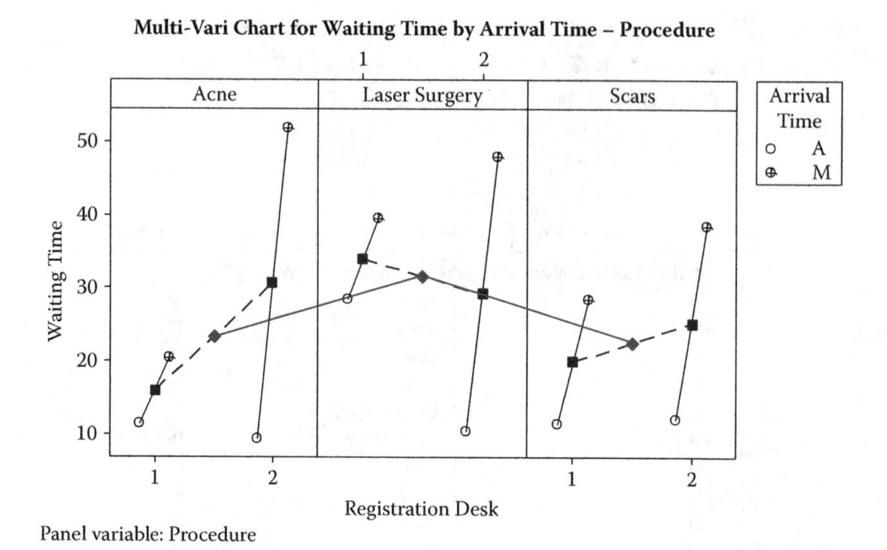

FIGURE 10.27
Multivariate chart for dermatology department at Copley.

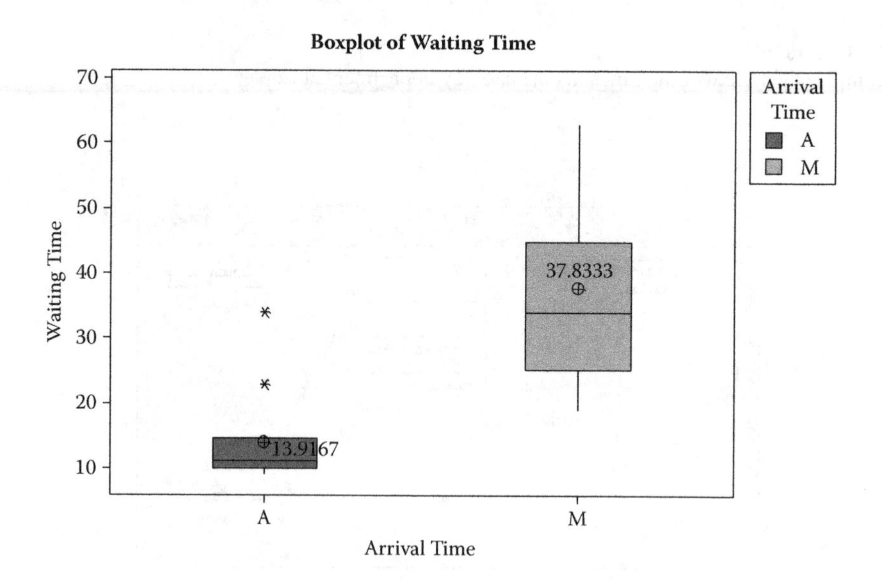

FIGURE 10.28
Boxplot for arrival time in Copley dermatology.

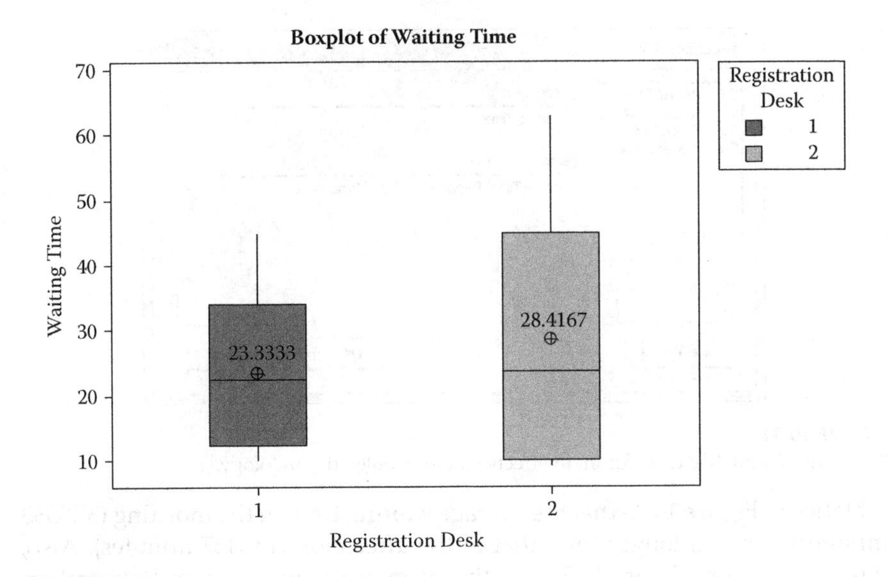

FIGURE 10.29
Boxplot for registration desk in Copley dermatology.

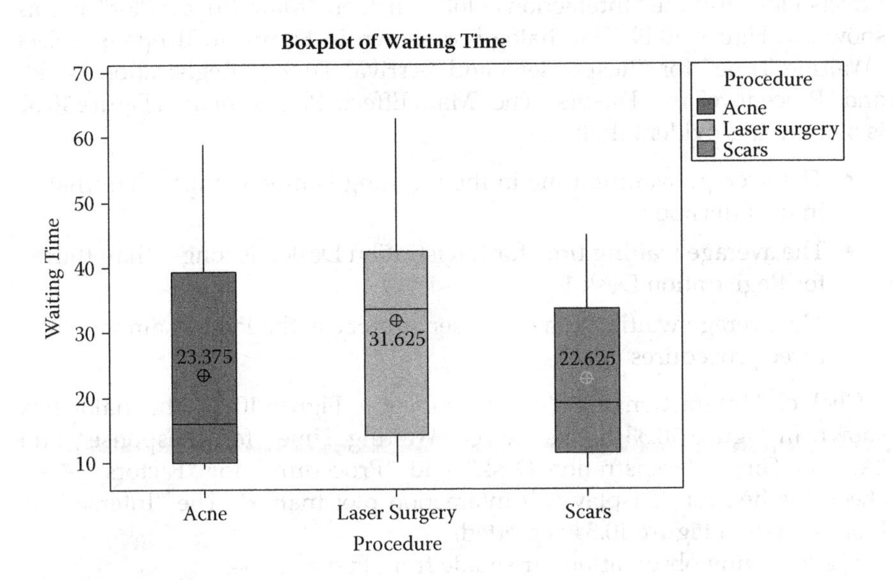

FIGURE 10.30
Boxplot for procedure in Copley dermatology.

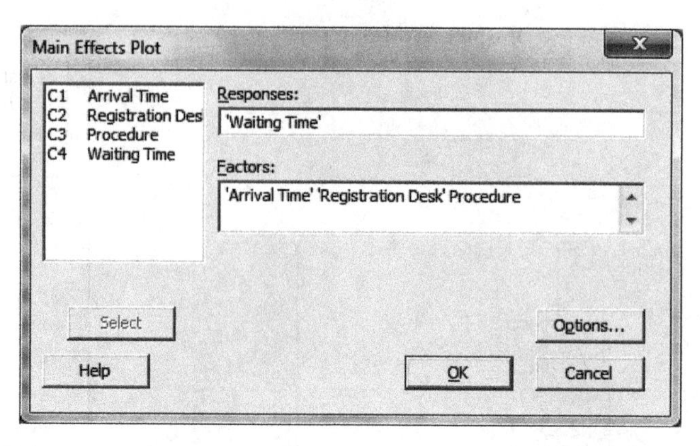

FIGURE 10.31
"Responses" and "Factors" for main effects plot for Copley dermatology.

Notice in Figure 10.28 that the average waiting time in the morning (37.8333 minutes) is much longer than that in the afternoon (13.9167 minutes). Also, it is evident from Figure 10.29 that the average waiting time for Registration Desk 2 is longer than for Registration Desk 1. Furthermore, it is evident from Figure 10.30 that the average waiting time for laser surgery (31.625 minutes) is the longest among all of the three procedures.

Many of the above interpretations can also be made by plotting the "Main Effects Plot" and the "Interactions Plot". Click on "Main Effects Plot" as was shown in Figure 10.19. The dialog box shown in Figure 10.31 opens. Select "Waiting Time" for "Responses", and "Arrival Time", "Registration Desk", and "Procedure" for "Factors". The "Main Effects Plot" shown in Figure 10.32 is plotted. It is evident that:

- The average waiting time in the morning is much longer than that in the afternoon.
- The average waiting time for Registration Desk 2 is longer than that for Registration Desk 1.
- The average waiting time for laser surgery is the longest among all three procedures.

Click on "Interactions Plot" as was shown in Figure 10.22. The dialog box shown in Figure 10.33 opens. Select "Waiting Time" for "Response", and "Arrival Time", "Registration Desk", and "Procedure" for "Factors". Also, check the box for "Display full interaction plot matrix". The "Interactions Plot" shown in Figure 10.34 is plotted.

The following observations are made from Figure 10.34:

- Top-right corner: Regardless of the procedure, the average waiting time is longer for the morning patients than that for the afternoon patients.
- Middle-right: For acne and scars, the average waiting times at Registration Desk 2 are longer than those at Registration Desk 1.

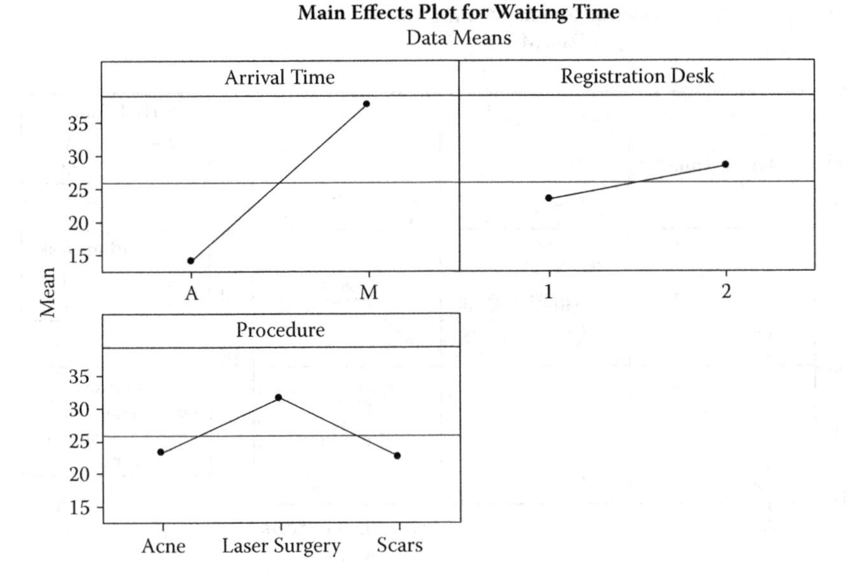

FIGURE 10.32
Main effects plot for Copley dermatology.

- Middle-bottom: The differences in the average waiting times among the three procedures at Registration Desk 1 are higher than the differences at Registration Desk 2.
- Middle-top: The difference in the average waiting times between the two arrival times is higher at Registration Desk 2 than that at Registration Desk 1.

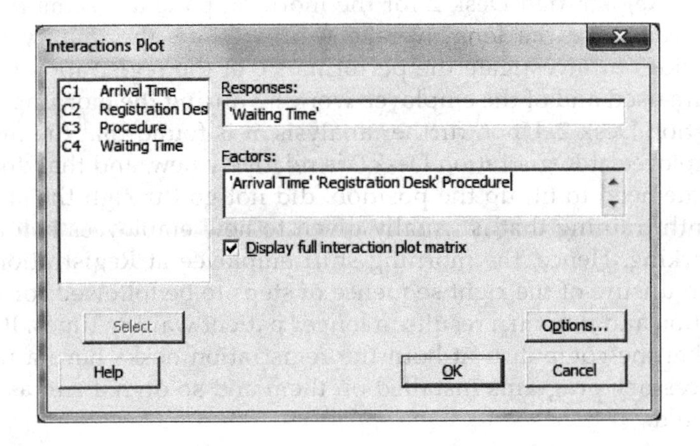

FIGURE 10.33
"Responses" and "Factors" for interactions plot for Copley dermatology.

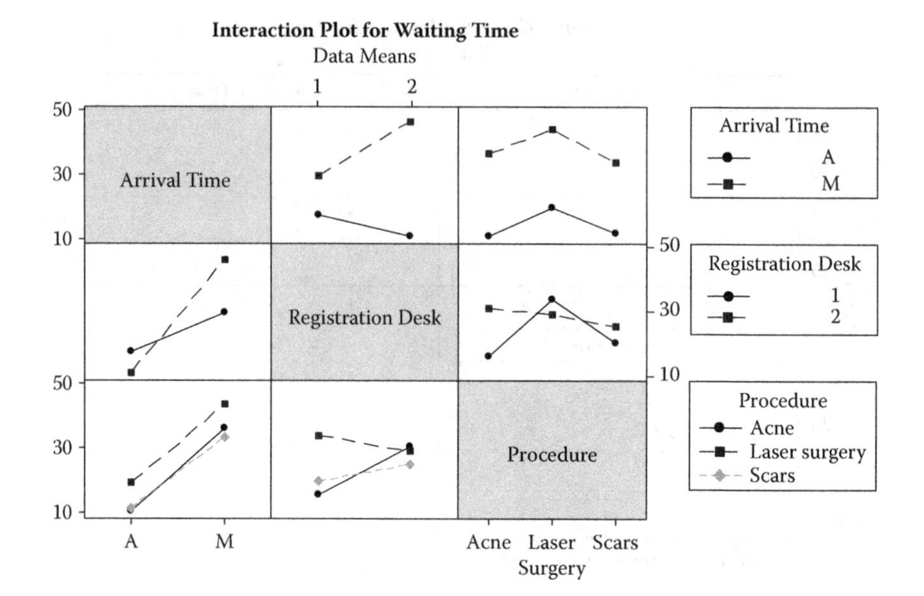

FIGURE 10.34
Interactions plot for Copley dermatology.

- Bottom-left corner: Regardless of the procedure, the average waiting time is longer for the morning patients than that for the afternoon patients.
- Middle-left: The average waiting time for the morning patients is longer at Registration Desk 2, and the average waiting time for the afternoon patients is longer at Registration Desk 1.

Because Registration Desk 2 for the morning patients seems to be the primary cause for the long average waiting time, the quality department decides to investigate the performance of the registration system/procedure used and of the employee working during the morning shift at Registration Desk 2. Upon further analysis, it is found that the morning shift employee at Registration Desk 2 is relatively new, and that due to an immediate need to fill up the position, did not go through the standard one-month training that is usually given to new employees before they start working. Hence, the morning shift employee at Registration Desk 2 is often unsure of the right sequence of steps to be followed for patient registration and it in turn results in longer patient waiting times. It is also found that the computers at both the registration desks have a number of unnecessary programs installed on them and so do not run as fast as they should.

10.4 Improve Phase

The following improvements are made in response to the analyses made in the analyze phase:

- The morning shift employee at Registration Desk 2 in the dermatology department at Copley begins undergoing training every day in the afternoon, that is, after his shift is over. This training lasts for a month.
- The programs that are no longer necessary for patient registration are uninstalled from the computers at both of the registration desks in the dermatology department at Copley.

10.5 Control Phase

The quality team decides to immediately implement compulsory training for all new employees, and regular maintenance of computers to uninstall unnecessary programs.

11

Pareto Chart and Fishbone Diagram to Minimize Recyclable Waste Disposal in a Town

Town XYZ collects the following recyclable trash from its residents: glass, plastic, paper, aluminum, yard waste, and iron. One of its primary aims for next year is to encourage residents to reuse the recyclables as much as possible before disposing of them.

Section 11.1 briefly describes the define and measure phases. Section 11.2 illustrates the analyze phase with detailed instructions for using Minitab®. The improve and control phases are briefly discussed in Section 11.3.

11.1 Define and Measure Phases

Of late, the town has been receiving excessive recyclable trash from its residents, and hence, the town hall aims to communicate to the residents different ways to reuse the recyclables before the residents dispose of them. To this end, the town hall collects data regarding how much waste of each category was collected and recycled during the previous year. These data are in the CHAPTER_11_1.MTW worksheet (the worksheet is available at the publisher's website; the data from the worksheet are also provided in the Appendix). Figure 11.1 is a screenshot of the worksheet.

The town hall first wishes to identify the categories of waste that contribute the most to the total waste. A Pareto chart can be used for this purpose, and Figure 11.2 shows how to select a Pareto chart. Doing so will open the dialog box shown in Figure 11.3. Select the "Recycled Trash" column for "Defects or attribute data in", and "Number of Tons (in hundreds)" for "Frequencies in". Select the option of "Do not combine" so that the categories of waste are not combined. Clicking on "Options" will open the dialog box shown in Figure 11.4. Enter the *X*-axis label as "Recycled Trash" and the *Y*-axis label as "Number of Tons (in hundreds)". Also, enter the title of the Pareto chart as shown. Click on "OK" and it takes you back to the dialog box shown in Figure 11.3. Click on "OK" and the Pareto chart shown in Figure 11.5 is the

	C1-T	C2
	Recycled Trash	Number of Tons (in hundreds)
1	Glass	200
2	Plastic	50
3	Paper	600
4	Aluminum	260
5	Yard waste	200
6	Iron	210

FIGURE 11.1
Recycled trash.

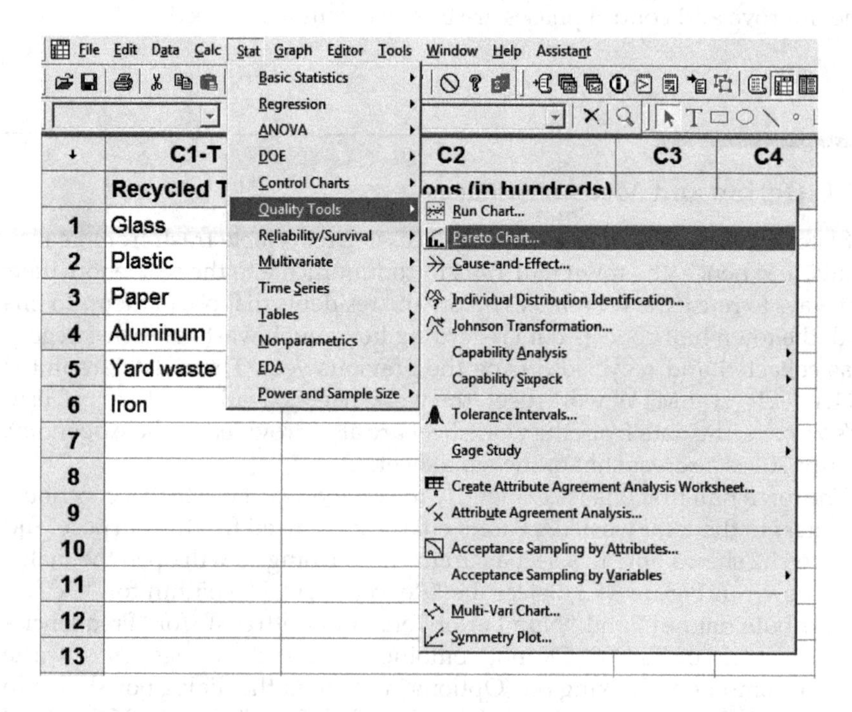

FIGURE 11.2
Selection of "Pareto Chart".

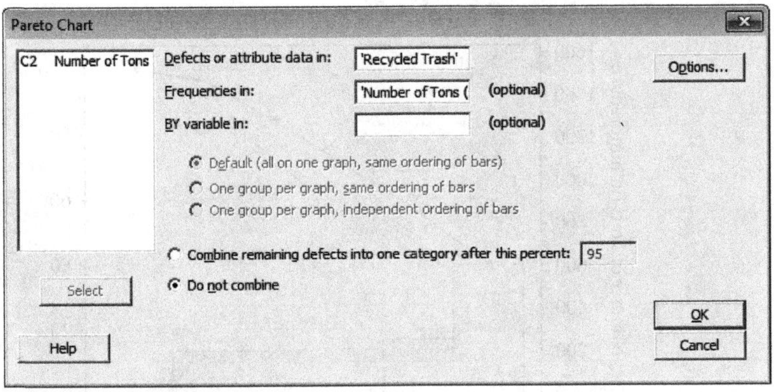

FIGURE 11.3
Entry of defects and frequencies for Pareto chart.

result. It is clear from the Pareto chart that almost 40% of the recycled waste is paper. Hence, the town hall decides to focus on minimizing the paper waste.

11.2 Analyze Phase

The employees of the town hall brainstorm to identify the different ways to minimize the paper waste. They classify paper waste into three types: office, kitchen, and other. The examples of these three types are in the C4, C5, and C6 columns, respectively, in the CHAPTER_11_2.MTW worksheet (the worksheet is available at the publisher's website; the data from the worksheet are also provided in the Appendix). Figure 11.6 is a screenshot of the worksheet. Notice that, in addition to the examples in the C4, C5, and C6 columns, a

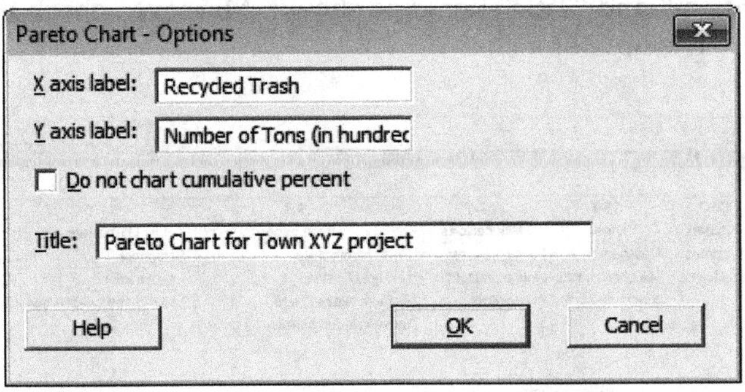

FIGURE 11.4
Entry of labels for Pareto chart.

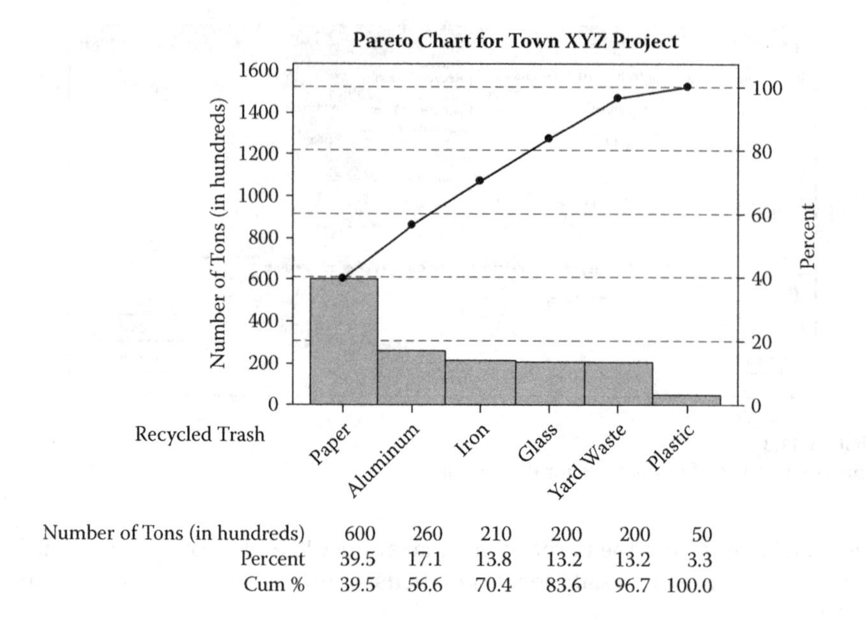

FIGURE 11.5
Pareto chart.

separate column is created for each example, where ways to minimize the waste for that example have been listed. For example, for the "Office paper" example in the C4 column, a separate column (C8) is used to list various ways ("print on both sides", "use email instead", etc.) to minimize waste for this example.

To facilitate understanding of the above by all the employees, a fishbone diagram can be created. Figure 11.7 shows how to select a fishbone diagram (also called a cause-and-effect diagram). Doing so will open the dialog box shown in Figure 11.8. Select the three categories of paper waste for "Causes" and type in the same names for "Label", as shown in Figure 11.9. Also, complete the "Effect" and "Title" boxes as shown. Moreover, check the box for "Do not display empty branches".

Data Calc Stat Graph Editor Tools Window Help Assistant

C4-T	C5-T	C6-T	C7-T	C8-T	C9-T	C10-T
Office	Kitchen	Other	Milk cartons	Office paper	Magazines	Junk mail
Office paper	Milk cartons	Newspapers	Paint containers	Print on both sides	Scrapbook	Envelopes
Magazines	Paper towels	Telephone books	Floor protectors	Use email instead	Room borders	Wrapping paper
		Junk mail	Bird feeders	Fit more text per page	Use pictures to teach kids	
				Use one-sided printouts for drafts		

FIGURE 11.6
Data for paper waste.

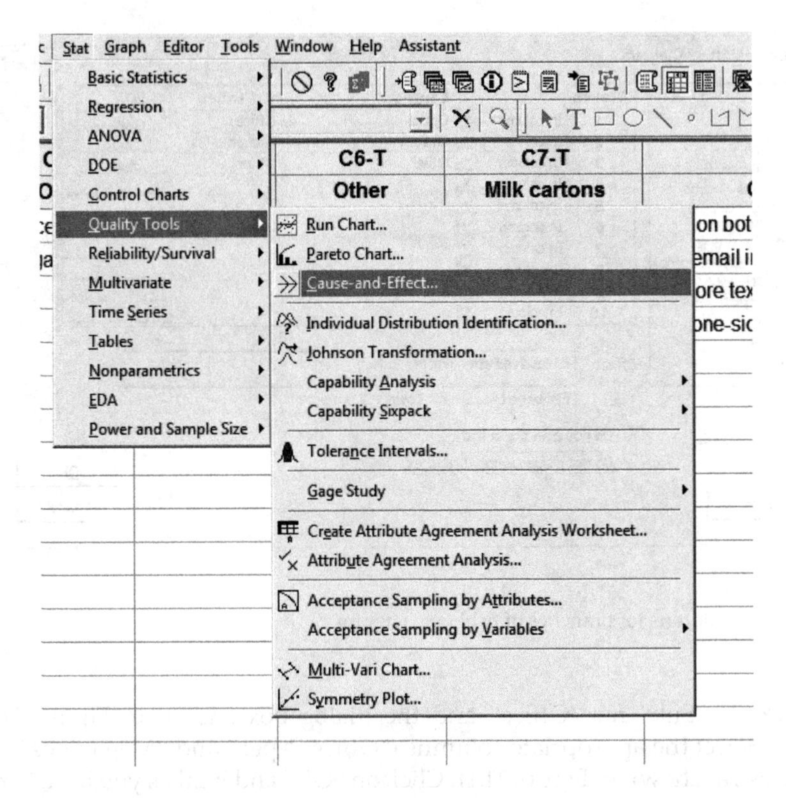

FIGURE 11.7
Selection of fishbone diagram ("Cause-and-Effect").

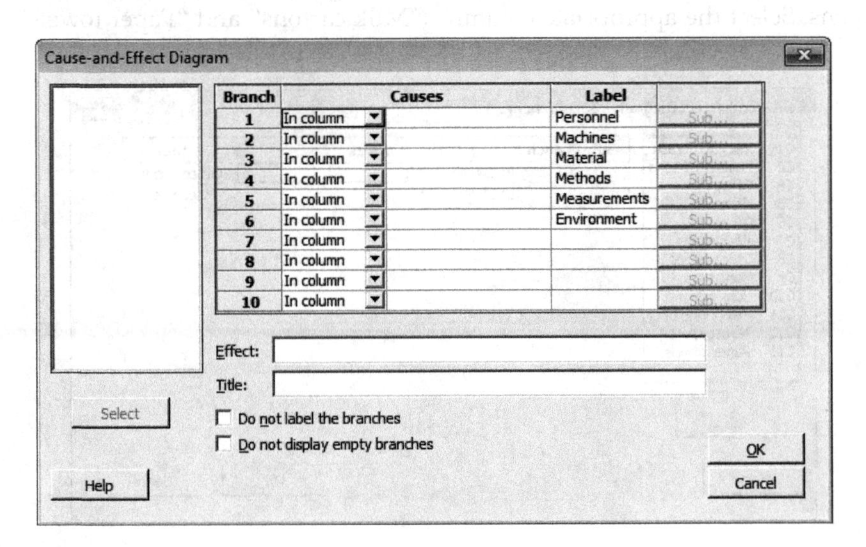

FIGURE 11.8
Categories for branches of fishbone diagram.

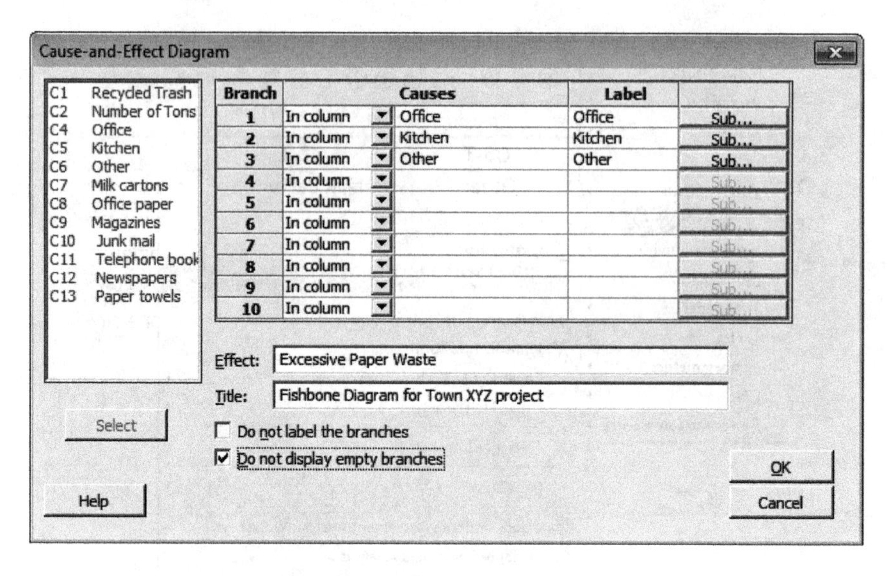

FIGURE 11.9
Selection of columns for branches of fishbone diagram.

Click on "Sub" for "Office" and the dialog box shown in Figure 11.10 opens. Select the appropriate columns ("Office Paper" and "Magazines") for "Causes", as shown in Figure 11.11. Click on "OK" and it takes you back to the dialog box shown in Figure 11.9.

Click on "Sub" for "Kitchen" and the dialog box shown in Figure 11.12 opens. Select the appropriate columns ("Milk cartons" and "Paper towels")

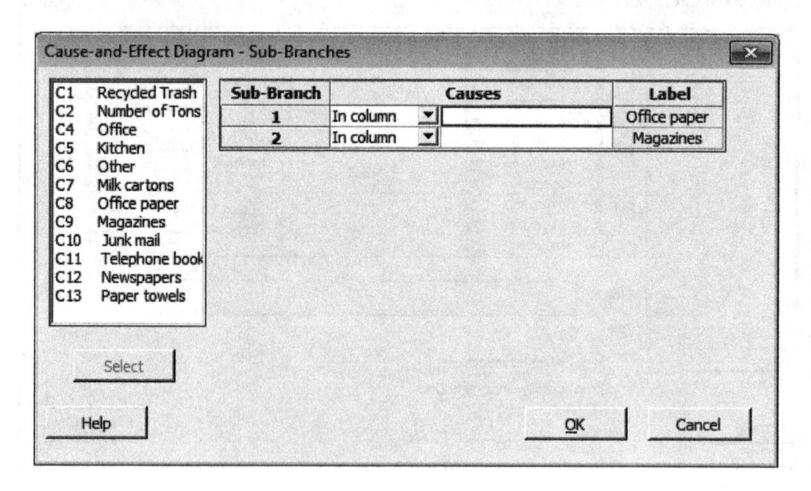

FIGURE 11.10
Subcategories for "Office" branch.

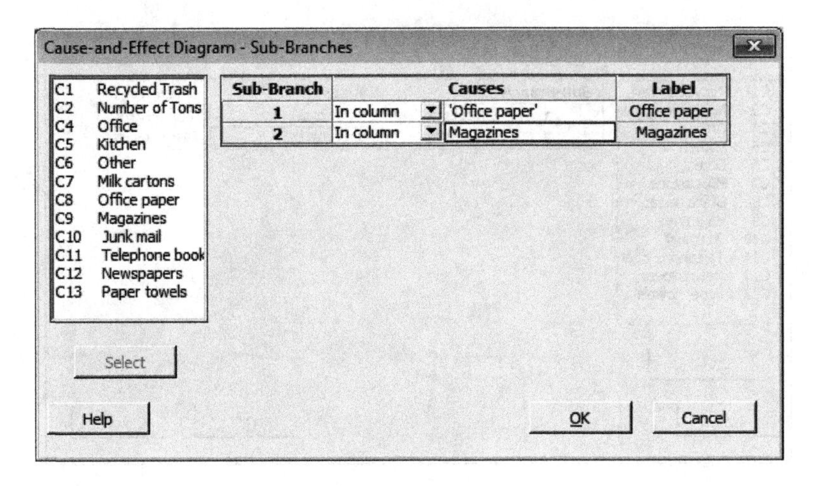

FIGURE 11.11
Selection of subcategories for "Office" branch.

for "Causes", as shown in Figure 11.13. Click on "OK" and it takes you back to the dialog box shown in Figure 11.9.

Click on "Sub" for "Other" and the dialog box shown in Figure 11.14 opens. Select the appropriate columns ("Newspapers", "Telephone books", and "Junk mail") for "Causes", as shown in Figure 11.15. Click on "OK" and it takes you back to the dialog box shown in Figure 11.9. Click on "OK" and the fishbone diagram shown in Figure 11.16 is the result.

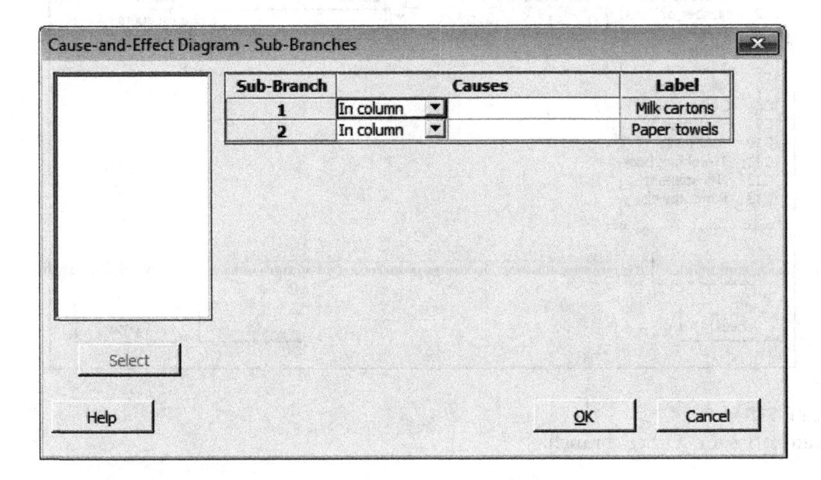

FIGURE 11.12
Subcategories for "Kitchen" branch.

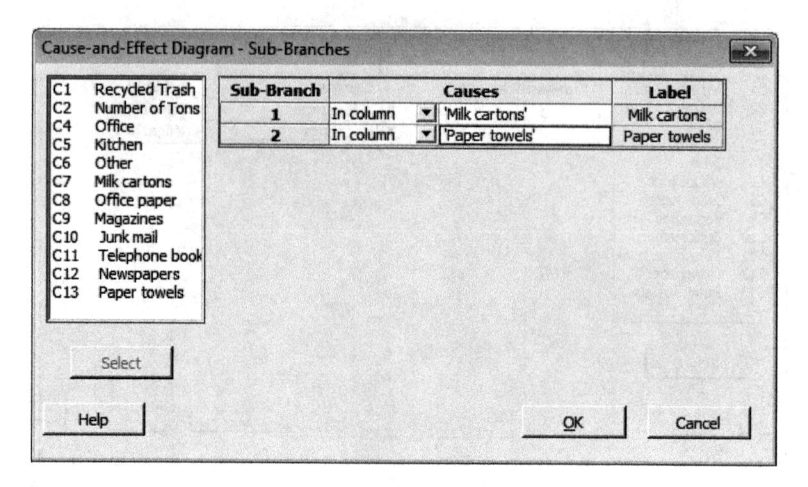

FIGURE 11.13
Selection of subcategories for "Kitchen" branch.

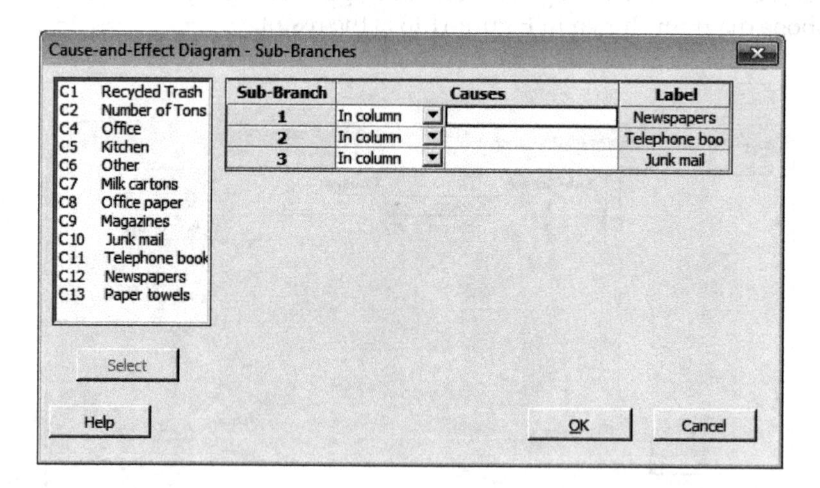

FIGURE 11.14
Subcategories for "Other" branch.

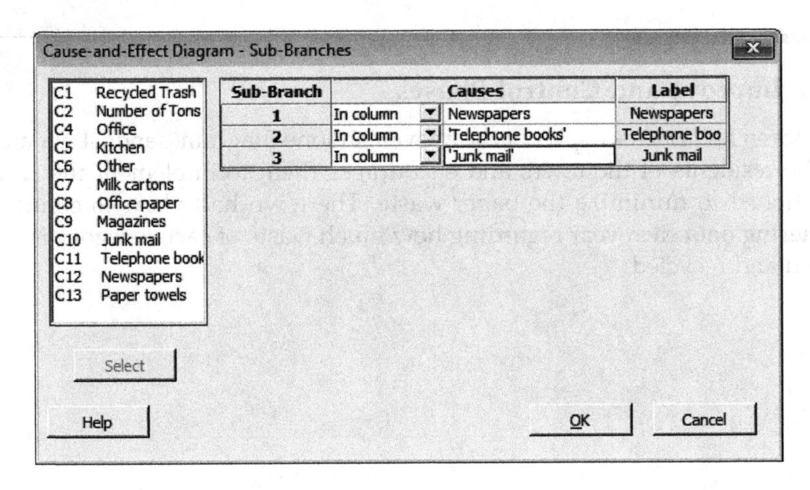

FIGURE 11.15
Selection of subcategories for "Other" branch.

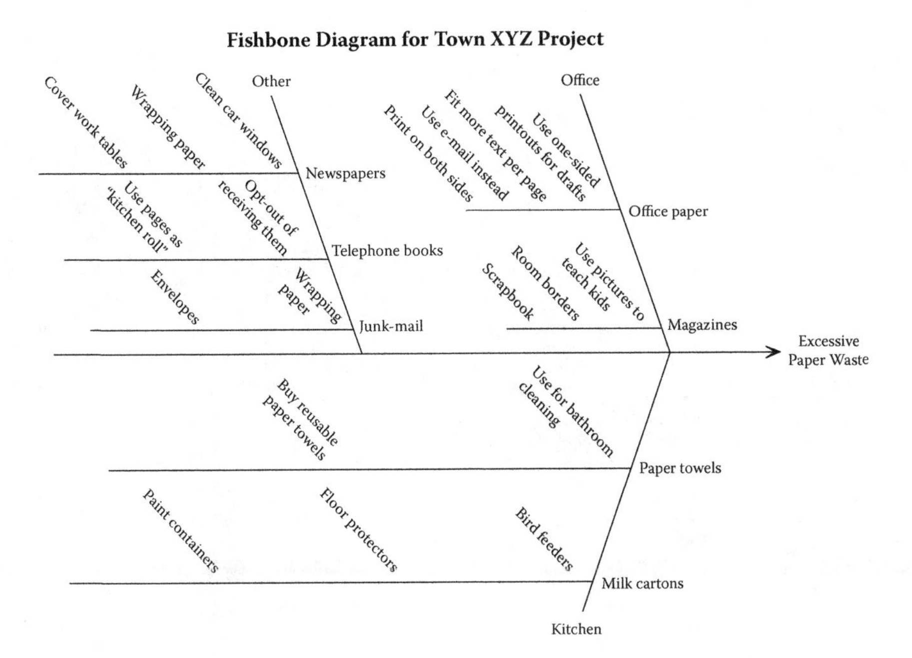

FIGURE 11.16
Fishbone diagram.

11.3 Improve and Control Phases

The town hall makes copies of the above fishbone diagram, sends it via mail to the residents of the town, and encourages them to implement the ways mentioned to minimize the paper waste. The town hall plans to continue gathering data each year regarding how much waste of each category is collected and recycled.

12

Measurement System Analysis at a Medical Equipment Manufacturer

This case study is about a Six Sigma project implemented by a supplier of medical equipment parts. The aim is to reduce the percentage of defective parts.

Section 12.1 gives a brief description of the define phase. Section 12.2 illustrates the measure phase with detailed instructions for using Minitab®. Finally, Section 12.3 gives brief descriptions of the analyze, improve, and control phases.

12.1 Define Phase

A number of units of a certain part that is used in popular medical equipment sold to major hospitals are being scrapped due to inaccurate diameters or imperfect shapes. The objective is to reduce the percentage of these defective medical equipment parts.

12.2 Measure Phase

Before measuring the current performance of the process that makes the parts, the quality head wishes to ensure that the system used to measure and inspect the parts is efficient.

In order to check the measurement system performance for the part diameters (variable data), a random sample of 9 parts is taken, and two operators measure each of the 9 parts twice (2 trials). See Table 12.1 for the collected data.

Open the CHAPTER_12_VARIABLE.MTW worksheet (the worksheet is available at the publisher's website; the data from the worksheet are also provided in the Appendix). Figure 12.1 is a screenshot of the partial worksheet. Figure 12.2 shows to how to select "Gage R&R Study (Crossed)". (The "crossed" option is chosen because the two operators measured the same parts.) Doing so opens the dialog box shown in Figure 12.3. As shown, select "Part" for "Part Numbers", "Operator" for "Operators", and "Diameter" for

TABLE 12.1

Variable Data

Part	Operator	Diameter
8	1	9.013
2	1	9.012
1	1	9.014
4	1	9.013
3	1	9.012
9	1	9.012
5	1	9.013
6	1	9.010
7	1	9.013
9	2	9.012
4	2	9.013
7	2	9.013
1	2	9.014
6	2	9.010
5	2	9.012
3	2	9.012
2	2	9.011
8	2	9.013
6	1	9.010
1	1	9.014
2	1	9.011
5	1	9.012
7	1	9.013
8	1	9.013
4	1	9.013
3	1	9.012
9	1	9.012
4	2	9.013
6	2	9.010
7	2	9.013
9	2	9.012
2	2	9.012
1	2	9.014
3	2	9.012
8	2	9.013
5	2	9.013

"Measurement Data". Then click on "Gage Info" in the dialog box shown in Figure 12.3, and the dialog box shown in Figure 12.4 opens. Enter the info as shown in Figure 12.4, and click on "OK". It takes you back to the dialog box shown in Figure 12.3. Click on "Options", and the dialog box shown in Figure 12.5 opens. Enter the info as shown in Figure 12.5 (leave the default values of "6" for "Study Variation" and "0.25" for "Alpha to remove interaction term"), and

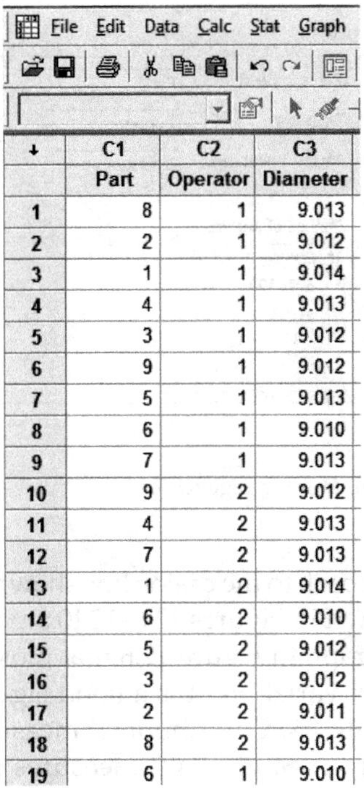

↓	C1 Part	C2 Operator	C3 Diameter
1	8	1	9.013
2	2	1	9.012
3	1	1	9.014
4	4	1	9.013
5	3	1	9.012
6	9	1	9.012
7	5	1	9.013
8	6	1	9.010
9	7	1	9.013
10	9	2	9.012
11	4	2	9.013
12	7	2	9.013
13	1	2	9.014
14	6	2	9.010
15	5	2	9.012
16	3	2	9.012
17	2	2	9.011
18	8	2	9.013
19	6	1	9.010

FIGURE 12.1
Variable data for gage repeatability and reproducibility.

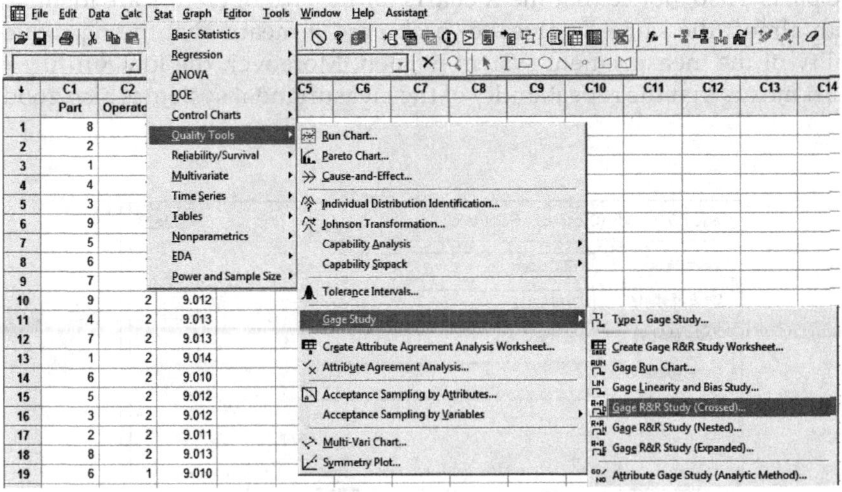

FIGURE 12.2
Selection of "Gage R&R Study (Crossed)".

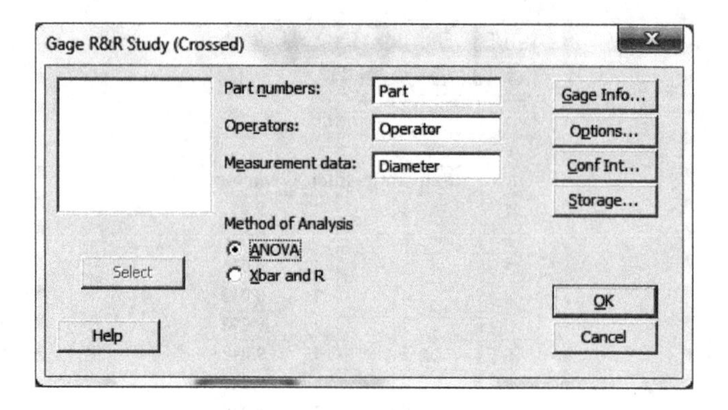

FIGURE 12.3
Entry of part, operator, and diameter for gage R&R.

click on "OK". It takes you back to the dialog box shown in Figure 12.3. Click on "OK", and the graphs shown in Figures 12.6–12.10 are produced.

It is evident from the graph in Figure 12.6 that most of the variation in the data is due to part-to-part variation. It is a good sign that the variation due to Gage R&R is very small. "% Contribution" measures the contribution to variance of the data and "% Study Var" measures the contribution to the standard deviation of the data.

Figure 12.7 shows the $\overline{X}$ (means) and R (ranges) charts. It is a good sign that the $\overline{X}$ chart for each of the two operators is out of statistical control, because it shows that there is a lot of variation due to the difference among the parts. Also, notice that the $\overline{X}$ charts for the two operators are identical with almost the same $\overline{X}$ values for each part. It means that the reproducibility of the measurement system is good. Moreover, the low $\overline{R}$ in the R chart means that the repeatability of the measurement system is also good.

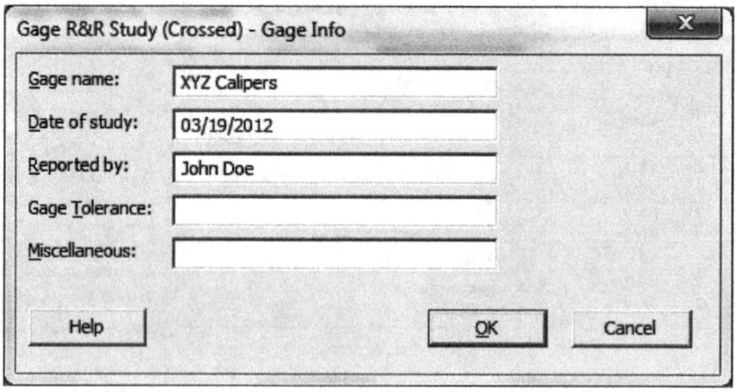

Gage R&R Study (Crossed) - Gage Info
Gage name: XYZ Calipers
Date of study: 03/19/2012
Reported by: John Doe
Gage Tolerance:
Miscellaneous:
Help
OK
Cancel

FIGURE 12.4
Entry of gage info for gage R&R.

Gage R&R Study (Crossed) - ANOVA Options

Study variation: 6 (number of standard deviations)

Process tolerance

- (•) Enter at least one specification limit
  - Lower spec:
  - Upper spec:
- ( ) Upper spec - Lower spec:

Historical standard deviation:

Alpha to remove interaction term: 0.25

- [ ] Display probabilities of misclassification
- [ ] Do not display percent contribution
- [ ] Do not display percent study variation
- [x] Draw graphs on separate graphs, one graph per page

Title: ABC Project

Help | OK | Cancel

FIGURE 12.5
Entry of title for gage R&R.

ABC Project

		Reported by: John Doe
Gage name:	XYZ Calipers	Tolerance:
Date of study:	03/19/2012	Misc:

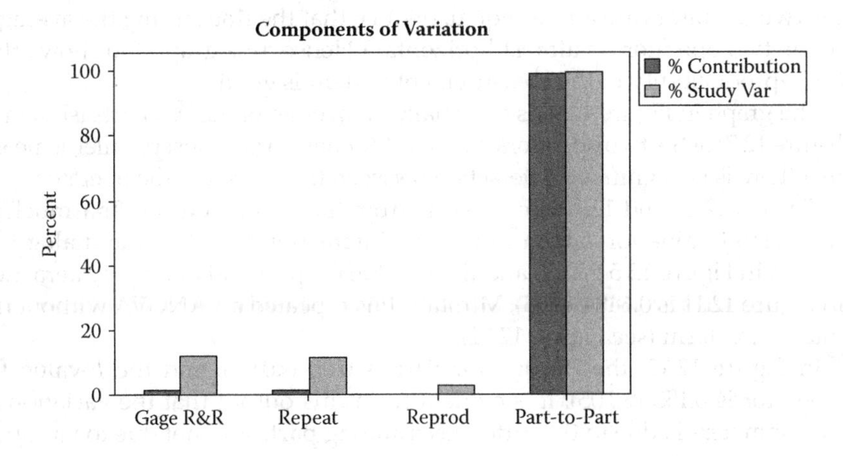

FIGURE 12.6
Components of variation.

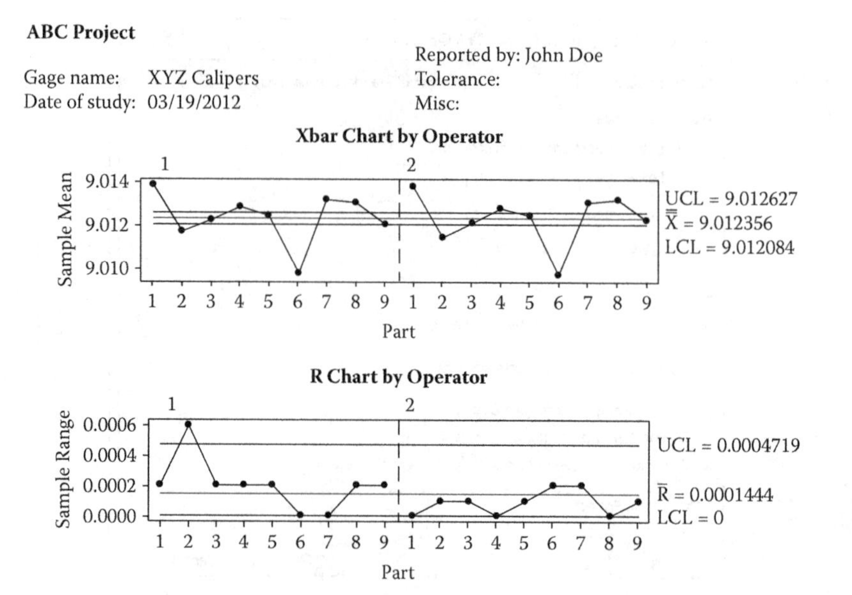

FIGURE 12.7
Xbar and R charts.

The graph in Figure 12.8 shows the $\overline{X}$ values of the two operators. The averages of the $\overline{X}$ values are joined with straight lines. Notice that, for each part, the $\overline{X}$ values of the two operators are very close. Hence, this graph also means that the reproducibility of the measurement system is good. The graph in Figure 12.9 too shows that the average diameters of all the parts for the two operators are almost equal. (Notice that the line joining the averages of the two operators is almost horizontal.) Hence, this graph too shows that the reproducibility of the measurement system is good.

The graph in Figure 12.10 is essentially an overlap of the $\overline{X}$ charts (shown in Figure 12.7) of the two operators. Because the charts are almost parallel, it means that there is no significant interaction between the parts and the operators.

Figures 12.11 and 12.12 are excerpts from the session output. Inasmuch as the default value for "alpha to remove interaction term" in the dialog box shown in Figure 12.5 is 0.25 and the *P*-value for part and operator interaction in Figure 12.11 is 0.649 (>0.25), Minitab® has repeated an ANOVA without the interaction term (see Figure 12.12).

In Figure 12.12, the *P*-value for Part is 0.00 (<0.05), and the *P*-value for Operator is 0.153 (>0.05). It is evident from this output that the variation in the diameters is due to the differences among parts and not due to variation between operators.

Figures 12.13 and 12.14 are also excerpts from the session output. It is clear from the "%Contribution" values (variance values) in Figure 12.13 that just 1.39% of the variation in the diameters is due to variation in gage

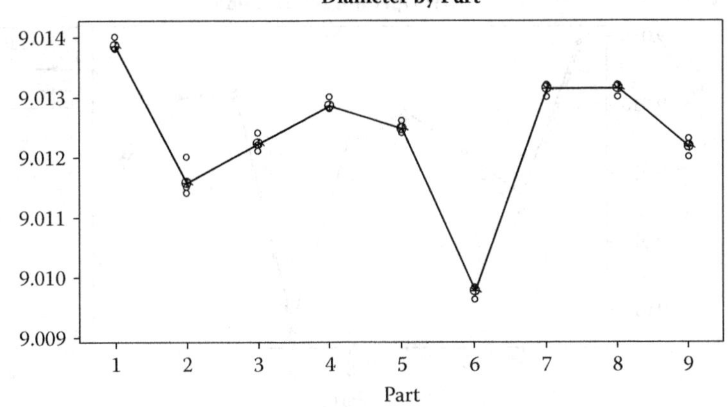

FIGURE 12.8
Diameter by part.

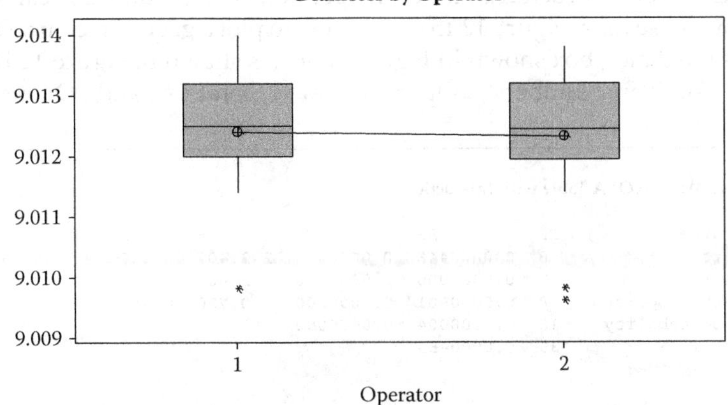

FIGURE 12.9
Diameter by operator.

ABC Project

Gage name: XYZ Calipers
Date of study: 03/19/2012

Reported by: John Doe
Tolerance:
Misc:

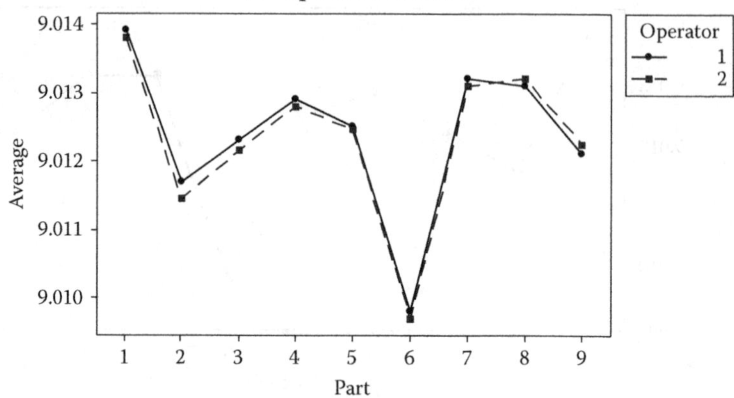

FIGURE 12.10
Interaction between part and operator.

repeatability and gage reproducibility. Note that 1.30% for repeatability and 0.08% for reproducibility are not adding up to 1.39% due to rounding. Whereas "% Contribution" in Figure 12.13 measures contribution to variance of the data, "% Study Var" in Figure 12.14 measures contribution to standard deviation of the data.

Notice that none of the graphs in Figures 12.6–12.10 shows the individual part diameters for each trial for each operator. A gage run chart can be used to show those data. Figure 12.15 shows how to plot a gage run chart. Doing so opens the dialog box shown in Figure 12.16. As shown in Figure 12.16, select "Part" for "Part Numbers", "Operator" for "Operators", and "Diameter" for

Two-Way ANOVA Table with Interaction

Source	DF	SS	MS	F	P
Part	8	0.0000448	0.0000056	373.407	0.000
Operator	1	0.0000000	0.0000000	2.667	0.141
Part * Operator	8	0.0000001	0.0000000	0.750	0.649
Repeatability	18	0.0000004	0.0000000		
Total	35	0.0000453			

Alpha to remove interaction term = 0.25

FIGURE 12.11
ANOVA with interaction.

Two-Way ANOVA Table without Interaction

Source	DF	SS	MS	F	P
Part	8	0.0000448	0.0000056	303.394	0.000
Operator	1	0.0000000	0.0000000	2.167	0.153
Repeatability	26	0.0000005	0.0000000		
Total	35	0.0000453			

FIGURE 12.12
ANOVA without interaction.

Source	VarComp	%Contribution (of VarComp)
Total Gage R&R	0.0000000	1.39
Repeatability	0.0000000	1.30
Reproducibility	0.0000000	0.08
Operator	0.0000000	0.08
Part-To-Part	0.0000014	98.61
Total Variation	0.0000014	100.00

FIGURE 12.13
Distribution of variances.

"Measurement Data". Click on the "Gage Info" button, and the dialog box shown in Figure 12.17 opens. Enter the information as shown in Figure 12.17 and click on "OK". It takes you back to the dialog box shown in Figure 12.16. Click on the "Options" button and the dialog box shown in Figure 12.18 opens. Enter the title as shown in Figure 12.18 and click on "OK". It takes you back to the dialog box shown in Figure 12.16. Click on "OK" and the gage run chart shown in Figure 12.19 results. It is clear from the gage run chart that the repeatability of each operator is good and also the reproducibility between the two operators is good.

Source	StdDev (SD)	Study Var (6 * SD)	%Study Var (%SV)
Total Gage R&R	0.0001402	0.0008412	11.79
Repeatability	0.0001359	0.0008152	11.42
Reproducibility	0.0000346	0.0002075	2.91
Operator	0.0000346	0.0002075	2.91
Part-To-Part	0.0011814	0.0070883	99.30
Total Variation	0.0011897	0.0071380	100.00

FIGURE 12.14
Distribution of standard deviations.

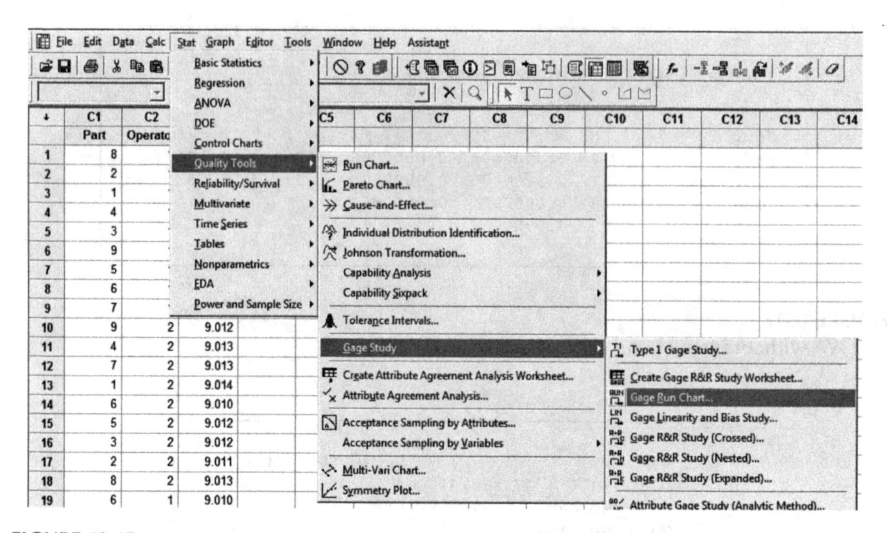

FIGURE 12.15
Selection of "Gage Run Chart".

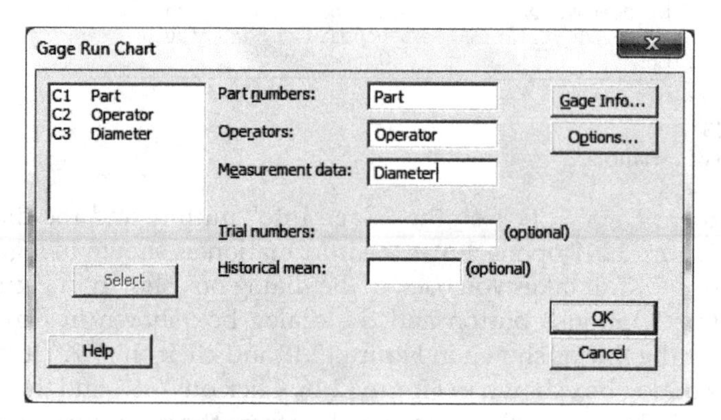

FIGURE 12.16
Entry of part, operator, and diameter for gage run chart.

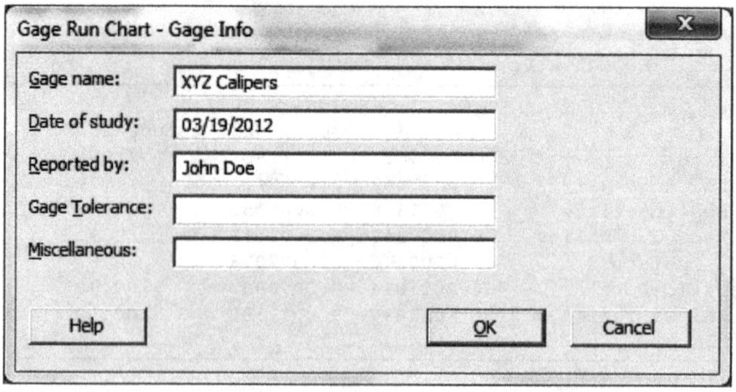

Gage Run Chart - Gage Info

Gage name: XYZ Calipers
Date of study: 03/19/2012
Reported by: John Doe
Gage Tolerance:
Miscellaneous:

Help OK Cancel

FIGURE 12.17
Entry of gage info for gage run chart.

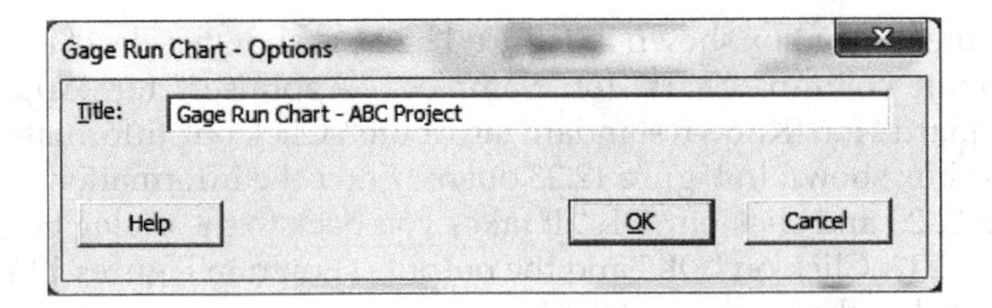

FIGURE 12.18
Entry of title for gage run chart.

The quality head then wishes to ensure that the system used to inspect the parts for "pass/fail" is efficient. Because this is attribute data, the procedure to evaluate the measurement system is different. Ten parts are randomly selected and each of the 3 operators (John, Mary, and Buddy) is asked to evaluate whether each part is "pass" or "fail". This exercise is repeated for each operator (2 trials). Table 12.2 shows the data collected. The last column ("Standard") of Table 12.2 is the response by a "master inspector" who is assumed to be correct.

Open the CHAPTER_12_ATTRIBUTE.MTW worksheet (the worksheet is available at the publisher's website; the data from the worksheet are also provided in the Appendix). Figure 12.20 is a screenshot of the partial worksheet. Figure 12.21 shows to how to select "Attribute Agreement Analysis". Doing

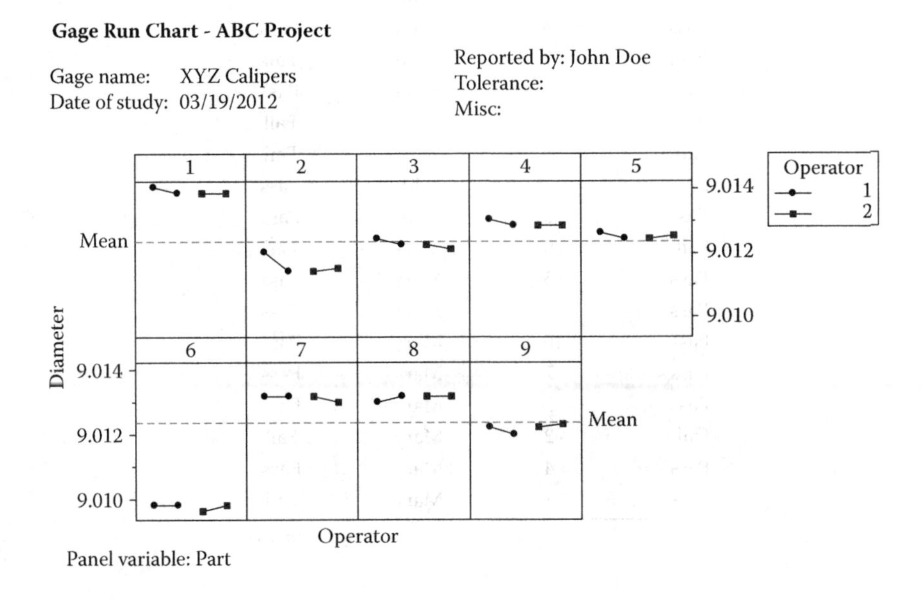

FIGURE 12.19
Gage run chart.

so opens the dialog box shown in Figure 12.22. As shown, select "Response" for "Attribute column", "Part" for "Samples", "Appraiser" for "Appraisers", and "Standard" for "Known standard/attribute". Click on "Information" and the dialog box shown in Figure 12.23 opens. Enter the information as shown in Figure 12.23 and click on "OK". It takes you back to the dialog box shown in Figure 12.22. Click on "OK" and the outputs shown in Figures 12.24–12.28 are produced in the session output file.

It is evident from Figure 12.24 that Buddy and John are perfect with respect to repeatability, and Mary's repeatability is 90%. Typically, 90% or over is considered acceptable for repeatability.

Figure 12.25 shows that the reproducibility of the system is 80%. Typically, 80%–90% is considered marginally acceptable for reproducibility.

It is evident from Figure 12.26 that Buddy's accuracy (percentage of times Buddy's responses matched with the standard responses) is 100%, John's accuracy is 90%, and Mary's accuracy is 90%. Typically, 90% or over is considered acceptable for accuracy.

TABLE 12.2

Attribute Data

Response	Part	Appraiser	Standard
Fail	6	John	Fail
Pass	7	John	Pass
Pass	10	John	Pass
Pass	8	John	Pass
Pass	2	John	Fail
Pass	4	John	Pass
Fail	1	John	Fail
Fail	9	John	Fail
Pass	3	John	Pass
Pass	5	John	Pass
Fail	9	Mary	Fail
Pass	5	Mary	Pass
Pass	8	Mary	Pass
Pass	10	Mary	Pass
Pass	7	Mary	Pass
Pass	3	Mary	Pass
Fail	2	Mary	Fail
Pass	4	Mary	Pass
Pass	1	Mary	Fail

(Continued)

TABLE 12.2 (CONTINUED)

Attribute Data

Response	Part	Appraiser	Standard
Fail	6	Mary	Fail
Pass	3	Buddy	Pass
Pass	4	Buddy	Pass
Fail	2	Buddy	Fail
Fail	6	Buddy	Fail
Pass	5	Buddy	Pass
Fail	9	Buddy	Fail
Pass	8	Buddy	Pass
Pass	10	Buddy	Pass
Pass	7	Buddy	Pass
Fail	1	Buddy	Fail
Pass	3	John	Pass
Fail	6	John	Fail
Pass	5	John	Pass
Pass	8	John	Pass
Pass	4	John	Pass
Pass	7	John	Pass
Fail	1	John	Fail
Pass	10	John	Pass
Fail	9	John	Fail
Pass	2	John	Fail
Fail	6	Mary	Fail
Pass	3	Mary	Pass
Pass	4	Mary	Pass
Pass	10	Mary	Pass
Pass	5	Mary	Pass
Pass	7	Mary	Pass
Fail	2	Mary	Fail
Fail	1	Mary	Fail
Fail	9	Mary	Fail
Pass	8	Mary	Pass
Fail	1	Buddy	Fail
Fail	9	Buddy	Fail
Pass	10	Buddy	Pass
Fail	6	Buddy	Fail
Pass	4	Buddy	Pass
Pass	5	Buddy	Pass
Pass	8	Buddy	Pass
Pass	7	Buddy	Pass
Pass	3	Buddy	Pass
Fail	2	Buddy	Fail

File Edit Data Calc Stat Graph Editor Tools

↓	C1-T	C2	C3-T	C4-T
	Response	Part	Appraiser	Standard
1	Fail	6	John	Fail
2	Pass	7	John	Pass
3	Pass	10	John	Pass
4	Pass	8	John	Pass
5	Pass	2	John	Fail
6	Pass	4	John	Pass
7	Fail	1	John	Fail
8	Fail	9	John	Fail
9	Pass	3	John	Pass
10	Pass	5	John	Pass
11	Fail	9	Mary	Fail
12	Pass	5	Mary	Pass
13	Pass	8	Mary	Pass
14	Pass	10	Mary	Pass
15	Pass	7	Mary	Pass
16	Pass	3	Mary	Pass
17	Fail	2	Mary	Fail
18	Pass	4	Mary	Pass
19	Pass	1	Mary	Fail

FIGURE 12.20
Attribute data for "Attribute Agreement Analysis".

Figure 12.27 shows that the overall accuracy (percentage of times the responses of all of the inspectors matched with the standard response) is 80%. Typically, 80%–90% is considered marginally acceptable for reproducibility.

In Figures 12.24–12.27, notice that the Kappa coefficients are 1.0 for performances (repeatability or reproducibility or individual accuracy or overall accuracy) of 100%. The better the performances are, the greater the Kappa coefficients are, with a maximum possible value of 1.0.

Figure 12.28 shows the confidence intervals for repeatability and individual accuracy of each inspector. The intervals are not so precise because of the small sample size of parts.

Because the measurement system is at least marginally acceptable with respect to repeatability, reproducibility, individual accuracy, and overall accuracy, the quality head proceeds to get further samples to measure the current defects per million opportunities (DPMO) of the manufacturing process.

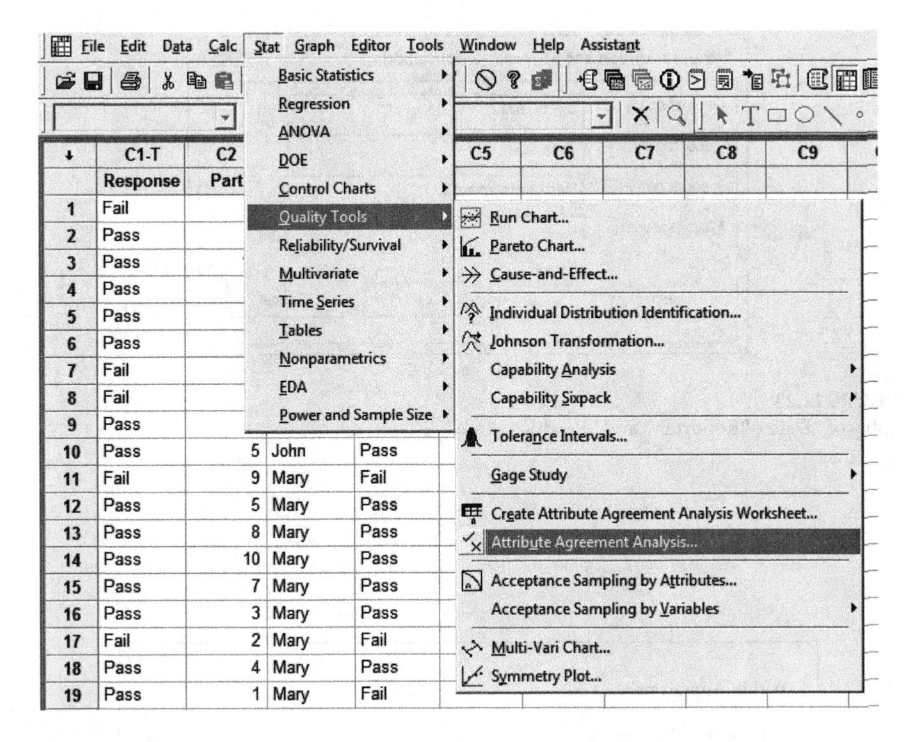

FIGURE 12.21
Selection of "Attribute Agreement Analysis".

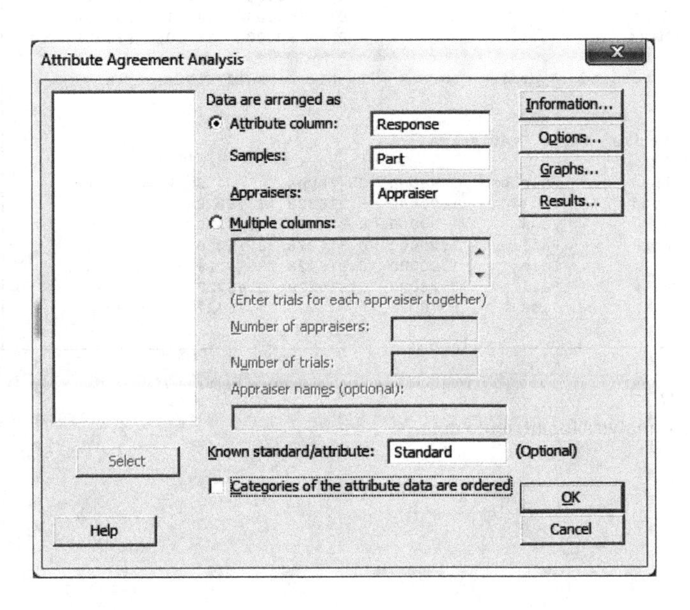

FIGURE 12.22
Selection of "Response," "Part," and "Appraiser".

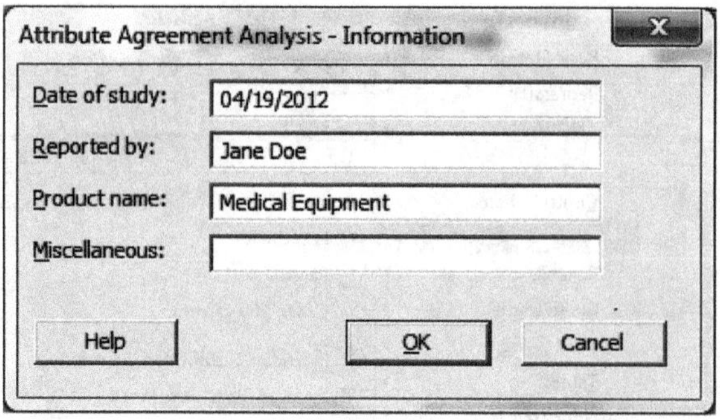

Attribute Agreement Analysis - Information

Date of study:	04/19/2012
Reported by:	Jane Doe
Product name:	Medical Equipment
Miscellaneous:	

Help OK Cancel

FIGURE 12.23
Entry of "Date," "Reporter," and "Product name."

Within Appraisers

Assessment Agreement

Appraiser	# Inspected	# Matched	Percent	95% CI
Buddy	10	10	100.00	(74.11, 100.00)
John	10	10	100.00	(74.11, 100.00)
Mary	10	9	90.00	(55.50, 99.75)

Matched: Appraiser agrees with him/herself across trials.

Fleiss' Kappa Statistics

Appraiser	Response	Kappa	SE Kappa	Z	P(vs > 0)
Buddy	Fail	1.00000	0.316228	3.16228	0.0008
	Pass	1.00000	0.316228	3.16228	0.0008
John	Fail	1.00000	0.316228	3.16228	0.0008
	Pass	1.00000	0.316228	3.16228	0.0008
Mary	Fail	0.78022	0.316228	2.46727	0.0068
	Pass	0.78022	0.316228	2.46727	0.0068

FIGURE 12.24
Repeatability analysis for attribute data.

Between Appraisers

Assessment Agreement

# Inspected	# Matched	Percent	95% CI
10	8	80.00	(44.39, 97.48)

Matched: All appraisers' assessments agree with each other.

Fleiss' Kappa Statistics

Response	Kappa	SE Kappa	Z	P(vs > 0)
Fail	0.809524	0.0816497	9.91460	0.0000
Pass	0.809524	0.0816497	9.91460	0.0000

FIGURE 12.25
Reproducibility analysis for attribute data.

Each Appraiser vs Standard

Assessment Agreement

Appraiser	# Inspected	# Matched	Percent	95% CI
Buddy	10	10	100.00	(74.11, 100.00)
John	10	9	90.00	(55.50, 99.75)
Mary	10	9	90.00	(55.50, 99.75)

Matched: Appraiser's assessment across trials agrees with the known standard.

Assessment Disagreement

Appraiser	# Pass / Fail	Percent	# Fail / Pass	Percent	# Mixed	Percent
Buddy	0	0.00	0	0.00	0	0.00
John	1	25.00	0	0.00	0	0.00
Mary	0	0.00	0	0.00	1	10.00

Pass / Fail: Assessments across trials = Pass / standard = Fail.
Fail / Pass: Assessments across trials = Fail / standard = Pass.
Mixed: Assessments across trials are not identical.

Fleiss' Kappa Statistics

Appraiser	Response	Kappa	SE Kappa	Z	P(vs > 0)
Buddy	Fail	1.00000	0.223607	4.47214	0.0000
	Pass	1.00000	0.223607	4.47214	0.0000
John	Fail	0.78022	0.223607	3.48925	0.0002
	Pass	0.78022	0.223607	3.48925	0.0002
Mary	Fail	0.89011	0.223607	3.98069	0.0000
	Pass	0.89011	0.223607	3.98069	0.0000

FIGURE 12.26
Accuracy analysis for attribute data.

```
All Appraisers vs Standard

Assessment Agreement

# Inspected  # Matched  Percent       95% CI
         10          8    80.00  (44.39, 97.48)

# Matched: All appraisers' assessments agree with the known
standard.


Fleiss' Kappa Statistics

Response      Kappa  SE Kappa        Z  P(vs > 0)
Fail       0.890110  0.129099  6.89476     0.0000
Pass       0.890110  0.129099  6.89476     0.0000
```

FIGURE 12.27
Overall accuracy analysis for attribute data.

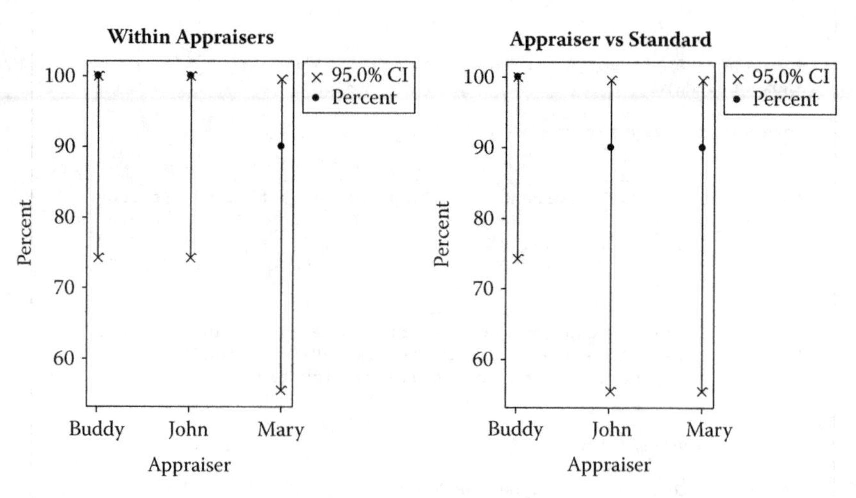

FIGURE 12.28
Confidence intervals for repeatability and accuracy for each appraiser.

12.3 Analyze, Improve, and Control Phases

Upon brainstorming with the technicians, it becomes clear that one of the five machines in the assembly line has a faulty motor coupler. Hence, the faulty motor coupler is replaced, and the process is improved. Confirmation of the improvement is made by recalculating the DPMO value. It is recommended that regular maintenance of the five machines in the assembly line be performed.

13

Taguchi Design to Improve Customer Satisfaction of an Airline Company

This case study is about a Six Sigma project implemented by an airline company to reduce the customer waiting time at a major airport.

Sections 13.1 and 13.2 explain the define phase and the measure phase, respectively. Section 13.3 illustrates the analyze phase with detailed instructions for using Minitab®. Finally, Section 13.4 gives a brief description of the improve and control phases.

13.1 Define Phase

The management decides to use Taguchi design of experiments for this project because there are both controllable factors ("control factors") and uncontrollable factors ("noise factors"). Table 13.1 shows each of the control factors with its corresponding levels (1 and 2). Table 13.2 shows each of the noise factors with its corresponding levels (1 and 2).

Because there are three control factors with 2 levels in each, a total of 8 "runs" is possible, as shown in Table 13.3.

Performing the experiment with all possible runs is time-consuming, therefore the management decides to consider only a fractional set of runs as shown in Table 13.4. The specific values of the levels of Table 13.4 are shown in Table 13.5.

There are three noise factors with 2 levels in each, therefore a total of 8 "runs" is possible, as shown in Table 13.6.

Performing the experiment with all possible runs is time-consuming, therefore the management decides to consider only a fractional set of runs shown in Table 13.7. The specific values of the levels of Table 13.7 are shown in Table 13.8.

The Taguchi design is ready as shown in Table 13.9. Notice that each of the "Noise" columns in Table 13.9 is essentially each run shown in Table 13.8.

Table 13.9 can be easily created in Minitab®. Figure 13.1 shows how to select "Create Taguchi Design". Doing so opens the dialog box shown in Figure 13.2. Select "3" from the drop-down menu for "Number of factors" and click on "Designs". This opens the dialog box shown in Figure 3.3. Select

TABLE 13.1

Control Factors and Their Levels

Control Factors	Level 1	Level 2
Assigned seating	Yes (Y)	No (N)
Number of check-in bags allowed	1	2
Food and beverage supply on flight	Beverage only (B)	Food and Beverage (FB)

TABLE 13.2

Noise Factors and Their Levels

Noise Factors	Level 1	Level 2
Engine	P	Q
Flight size	Small (S)	Large (L)
Terminal	A	B

TABLE 13.3

Possible Runs for Control Factors

Runs	Assigned Seating	Number of Check-In Bags Allowed	Food and Beverage Supply on Flight
1	Level 1	Level 1	Level 1
2	Level 1	Level 2	Level 2
3	Level 2	Level 1	Level 2
4	Level 2	Level 2	Level 1
5	Level 1	Level 1	Level 2
6	Level 1	Level 2	Level 1
7	Level 2	Level 1	Level 1
8	Level 2	Level 2	Level 2

TABLE 13.4

Fractional Set of Runs for Control Factors

Runs	Assigned Seating	Number of Check-In Bags Allowed	Food and Beverage Supply on Flight
1	Level 1	Level 1	Level 1
2	Level 1	Level 2	Level 2
3	Level 2	Level 1	Level 2
4	Level 2	Level 2	Level 1

TABLE 13.5

Fractional Set of Runs for Control Factors with Specific Values for Levels

Runs	Assigned Seating	Number of Check-In Bags Allowed	Food and Beverage Supply on Flight
1	Y	1	B
2	Y	2	FB
3	N	1	FB
4	N	2	B

TABLE 13.6

Possible Runs for Noise Factors

Runs	1	2	3	4	5	6	7	8
Engine	Level 1	Level 1	Level 2	Level 2	Level 1	Level 1	Level 2	Level 2
Flight size	Level 1	Level 2	Level 1	Level 2	Level 1	Level 2	Level 1	Level 2
Terminal	Level 1	Level 2	Level 2	Level 1	Level 2	Level 1	Level 1	Level 2

the "L4" option and click on "OK". It takes you back to the dialog box shown in Figure 13.2. Click on "Factors" and the dialog box shown in Figure 13.4 opens. Enter the factor names and their levels as shown in Figure 13.4, and click on "OK". It takes you back to the dialog box shown in Figure 13.2. Click on "OK" and the Taguchi design summary shown in Figure 13.5 and the worksheet shown in Figure 13.6 are the results. As shown in Figure 13.7, label four empty columns as "PSA", "PLB", "QSB", and "QLA" just as in Table 13.9.

TABLE 13.7

Fractional Set of Runs for Noise Factors

Runs	1	2	3	4
Engine	Level 1	Level 1	Level 2	Level 2
Flight size	Level 1	Level 2	Level 1	Level 2
Terminal	Level 1	Level 2	Level 2	Level 1

TABLE 13.8

Fractional Set of Runs for Noise Factors with Specific Values for Levels

Runs	1	2	3	4
Engine	P	P	Q	Q
Flight size	S	L	S	L
Terminal	A	B	B	A

TABLE 13.9
Taguchi Design

Runs	Assigned Seating	Number of Check-In Bags Allowed	Food and Beverage Supply on Flight	Noise PSA	Noise PLB	Noise QSB	Noise QLA
1	Y	1	B				
2	Y	2	FB				
3	N	1	FB				
4	N	2	B				

13.2 Measure Phase

The experiment shown in Figure 13.7 is run, and the waiting times are obtained for each pair of control factor combination and noise factor combination. The results of the experiment are shown in Figure 13.8. For example, when Assigned = "Y", Bags = "1", "Food/Bev" = "B", Engine = "P", Flight Size = "S", and Terminal = "A", the waiting time is 31 minutes.

Enter the data in the worksheet as shown in Figure 13.8.

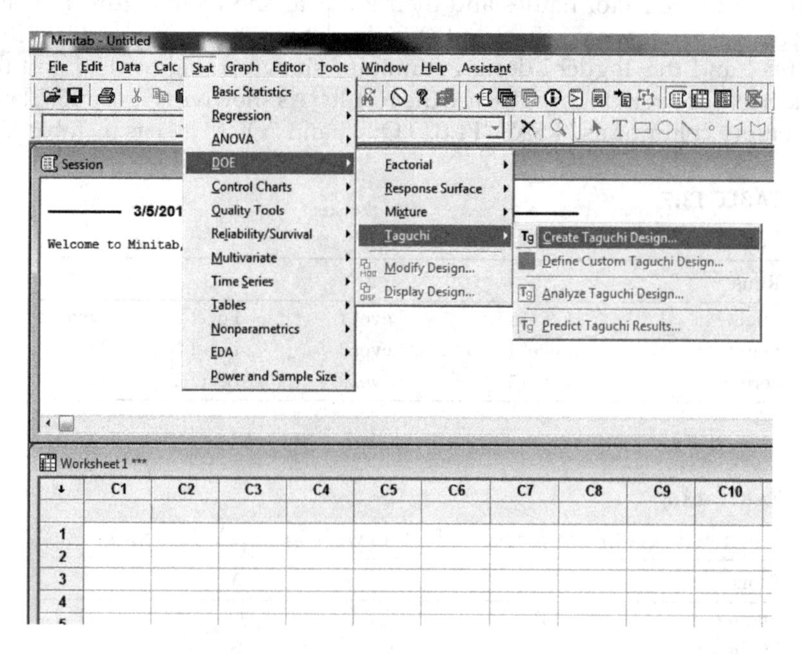

FIGURE 13.1
Selection of "Create Taguchi Design".

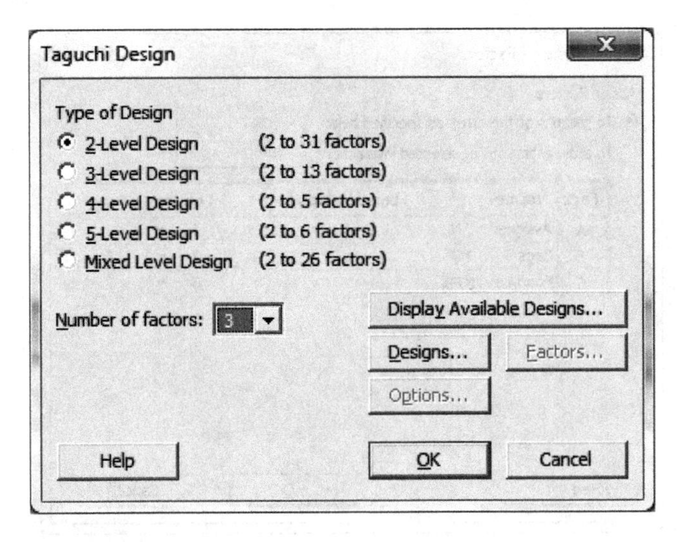

FIGURE 13.2
Selection of "Type of Design" and "Number of factors".

13.3 Analyze Phase

Select "Analyze Taguchi Design" as shown in Figure 13.9. Doing so opens the dialog box shown in Figure 13.10. Select "PSA-QLA" columns as shown in Figure 13.11, and click on "Options". This opens the dialog box shown in Figure 13.12. Because small waiting times are better, select the "smaller is

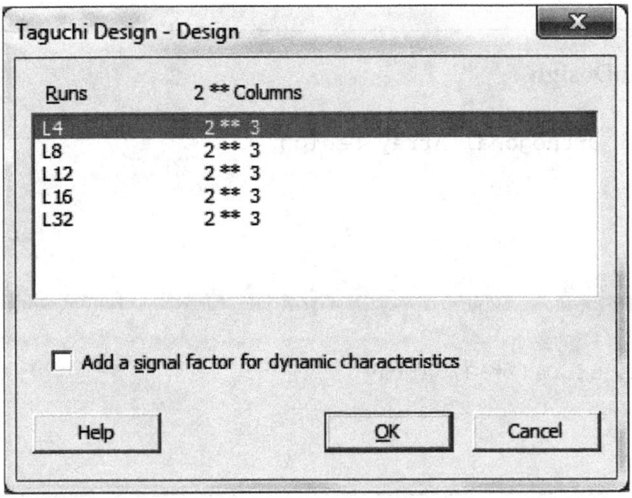

FIGURE 13.3
Selection of "Runs".

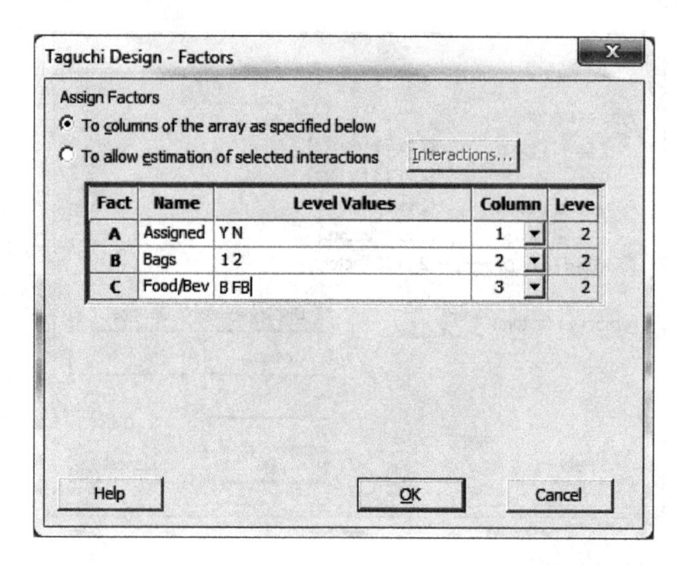

FIGURE 13.4
Entry of factors and their levels.

better" option and click on "OK". This takes you back to the dialog box shown in Figure 13.11. Click on "Graphs" and the dialog box shown in Figure 13.13 opens. Ensure that "Signal to Noise ratios" and "Means" plots are selected. Click on "OK" and it takes you back to the dialog box shown in Figure 13.11. Click on "OK" and the graphs shown in Figures 13.14 and 13.15 are plotted.

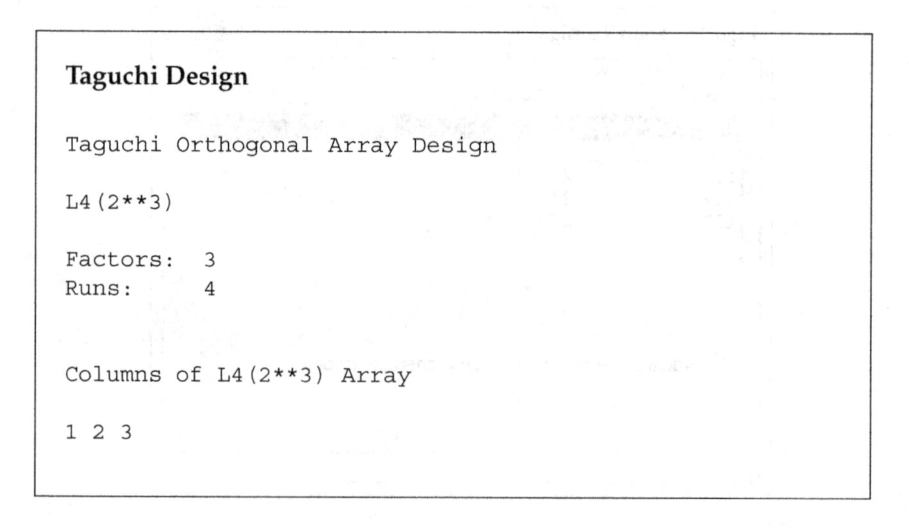

Taguchi Design

```
Taguchi Orthogonal Array Design

L4(2**3)

Factors:  3
Runs:     4


Columns of L4(2**3) Array

1 2 3
```

FIGURE 13.5
Summary of Taguchi design.

↓	C1-T	C2	C3-T	C4
	Assigned	Bags	Food/Bev	
1	Y	1	B	
2	Y	2	FB	
3	N	1	FB	
4	N	2	B	

FIGURE 13.6
Partial Taguchi design on worksheet.

→	C1-T	C2	C3-T	C4	C5	C6	C7
	Assigned	Bags	Food/Bev	PSA	PLB	QSB	QLA
1	Y	1	B				
2	Y	2	FB				
3	N	1	FB				
4	N	2	B				

FIGURE 13.7
Creation of columns for noise factors.

→	C1-T	C2	C3-T	C4	C5	C6	C7
	Assigned	Bags	Food/Bev	PSA	PLB	QSB	QLA
1	Y	1	B	31	27	29	32
2	Y	2	FB	46	40	38	35
3	N	1	FB	31	29	28	37
4	N	2	B	45	43	42	41

FIGURE 13.8
Entry of data from experiment.

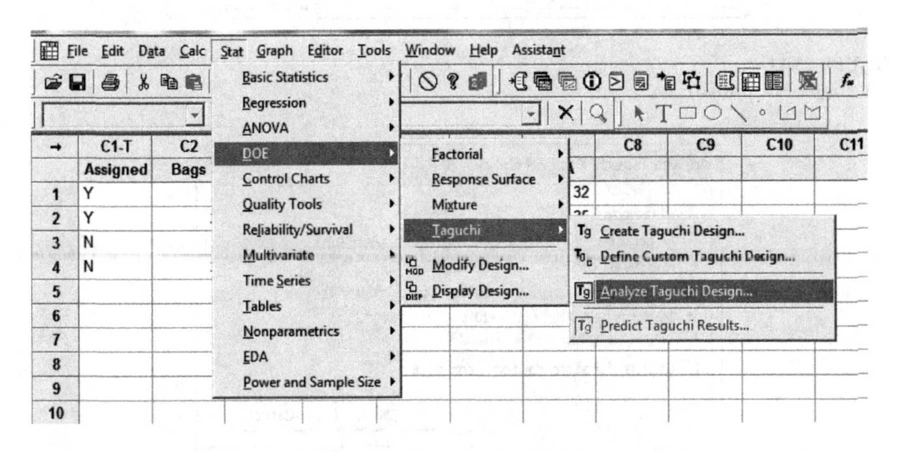

FIGURE 13.9
Selection of “Analyze Taguchi Design”.

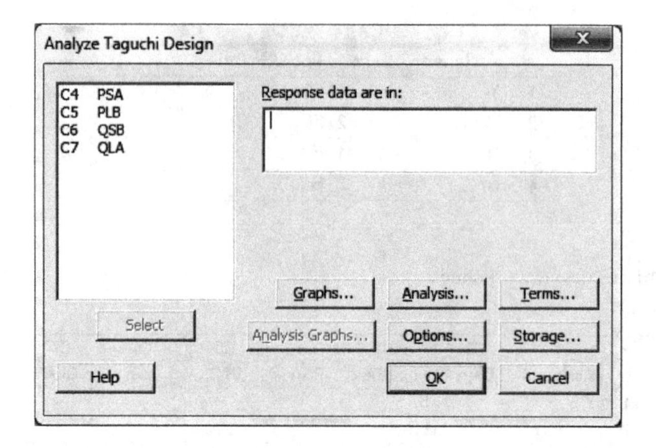

FIGURE 13.10
Dialog box for entry of response data.

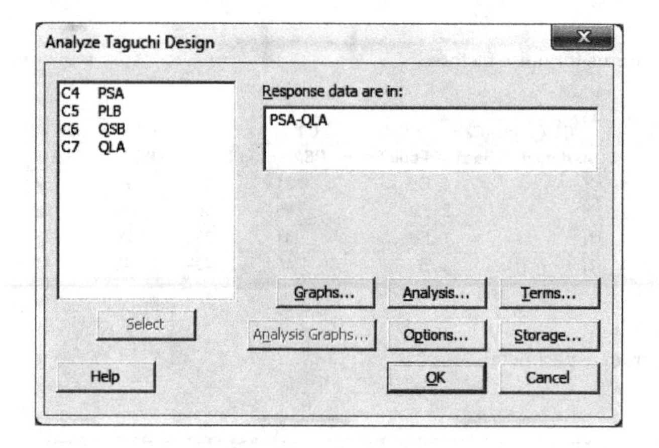

FIGURE 13.11
Selection of response data.

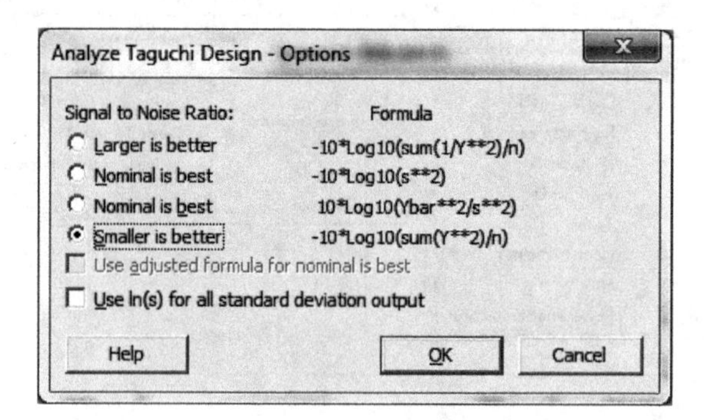

FIGURE 13.12
Selection of "Smaller is better" signal-to-noise ratio.

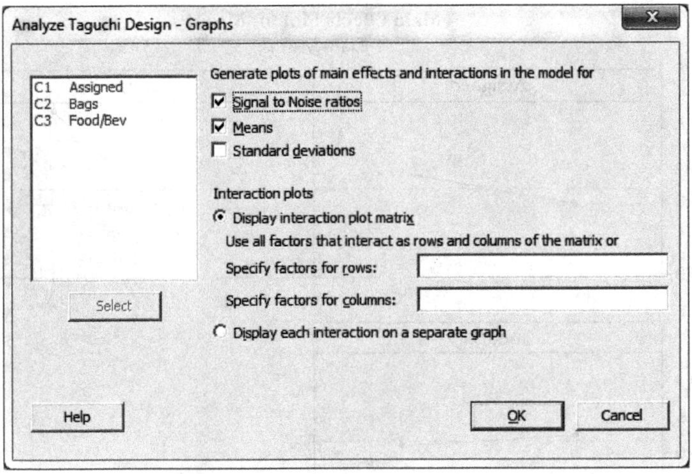

Analyze Taguchi Design - Graphs

C1 Assigned
C2 Bags
C3 Food/Bev

Select

Generate plots of main effects and interactions in the model for
☑ Signal to Noise ratios
☑ Means
☐ Standard deviations

Interaction plots
◉ Display interaction plot matrix
Use all factors that interact as rows and columns of the matrix or
Specify factors for rows:
Specify factors for columns:
○ Display each interaction on a separate graph

Help OK Cancel

FIGURE 13.13
Selection of "Signal to Noise ratios" and "Means" graphs.

Also, the summary of these plots is produced in the session output file, as shown in Figure 13.16.

For the "Y" level of the "Assigned" factor, Figure 13.14 shows a higher signal-to-noise ratio and Figure 13.15 shows a lower average waiting time. This means that we can safely recommend the "Y" level for the "Assigned"

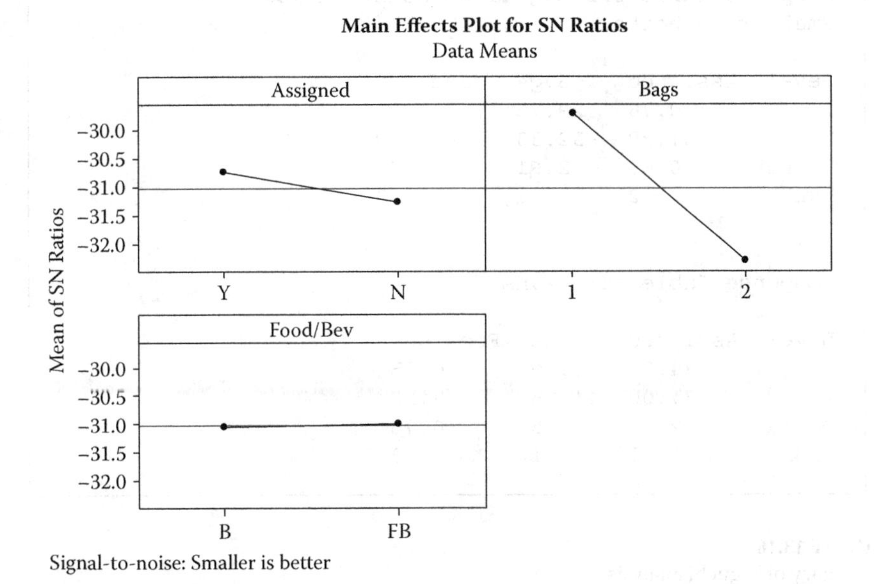

FIGURE 13.14
Signal-to-noise ratio graphs.

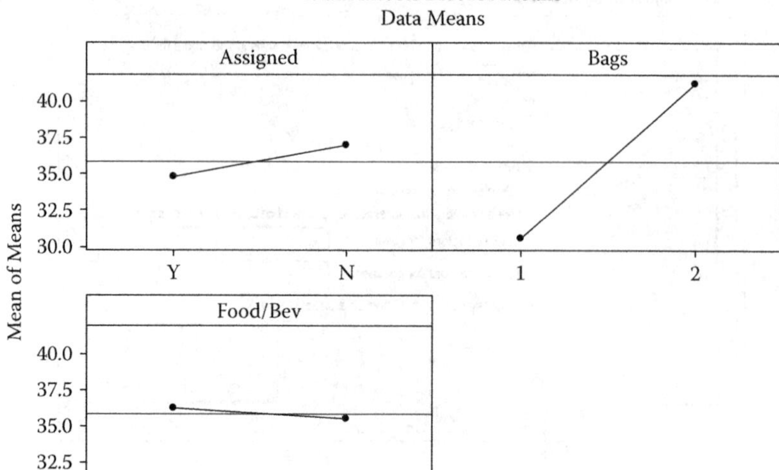

FIGURE 13.15
Main effects plot for means.

```
Response Table for Signal to Noise Ratios
Smaller is better

Level  Assigned    Bags  Food/Bev
1        -30.76  -29.72    -31.06
2        -31.29  -32.33    -30.99
Delta      0.53    2.61      0.06
Rank          2       1         3


Response Table for Means

Level  Assigned   Bags  Food/Bev
1         34.75  30.50     36.25
2         37.00  41.25     35.50
Delta      2.25  10.75      0.75
Rank          2      1         3
```

FIGURE 13.16
Summary of Taguchi analysis.

factor with the assurance that the noise factors do not have a significant impact on the waiting time.

For the "1" level of the "Bags" factor, Figure 13.14 shows a higher signal-to-noise ratio and Figure 13.15 shows a lower average waiting time. This means that we can safely recommend the "1" level for the "Bags" factor with the assurance that the noise factors do not have a significant impact on the waiting time.

For the "FB" level of the "Food/Bev" factor, Figure 13.14 shows a higher signal-to-noise ratio and Figure 13.15 shows a lower average waiting time. This means that we can safely recommend the "FB" level for the "Food/Bev" factor with the assurance that the noise factors do not have a significant impact on the waiting time.

In order to predict the waiting time for a combination of the control factors (say, Y, 1, FB, for Assigned, Bags, and Food/Bev, respectively), select "Predict Taguchi Results" as shown in Figure 13.17. Doing so opens the dialog box shown in Figure 13.18. Select the box for "Mean" and click on "Terms". This opens the dialog box shown in Figure 13.19. For the "Selected Terms" box, select the three control factors as shown in Figure 13.19 and click on "OK". This opens the dialog box shown in Figure 13.20. Select the levels of the three factors as shown in Figure 13.20, and click on "OK". The output (prediction of 29 minutes) shown in Figure 13.21 is the result. Figures 13.22 and 13.23 show the approach and the prediction for a different combination of the control factors (Y, 1, B).

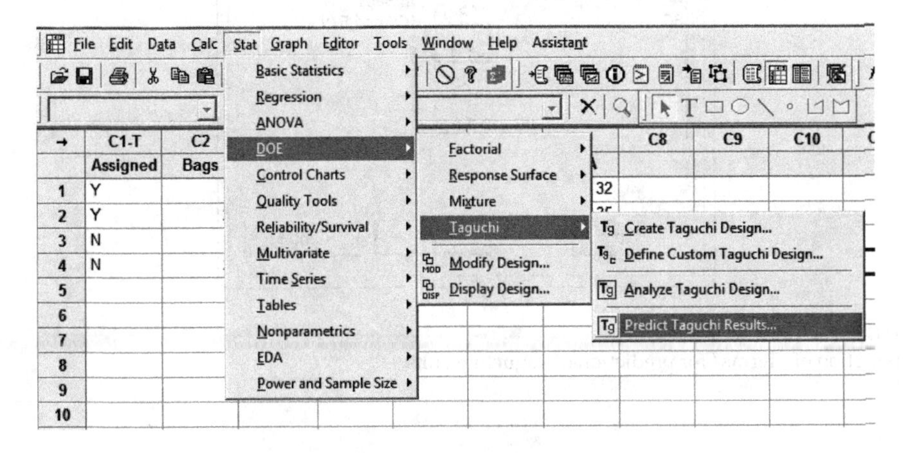

FIGURE 13.17
Selection of "Predict Taguchi Results."

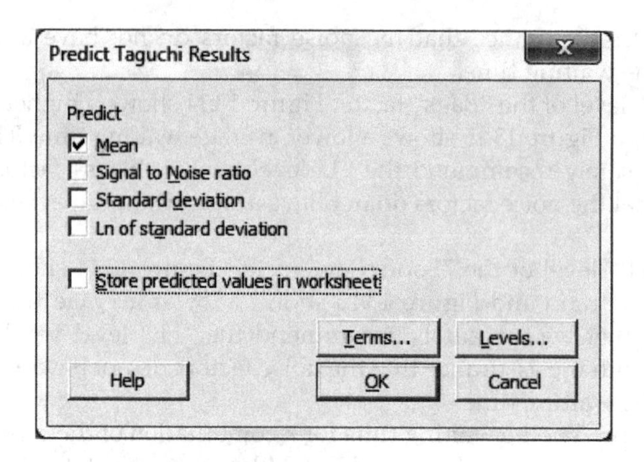

FIGURE 13.18
Selection of "Mean" for prediction of Taguchi results.

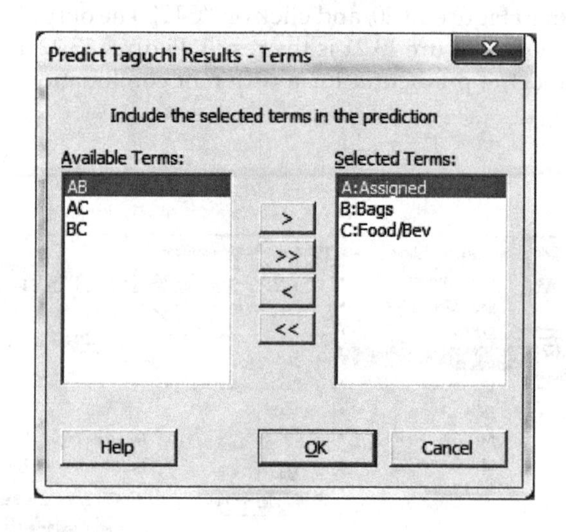

FIGURE 13.19
Selection of "Terms" for prediction of Taguchi results.

Predict Taguchi Results - Levels

C1 Assigned
C2 Bags
C3 Food/Bev
C4 PSA
C5 PLB
C6 QSB
C7 QLA

Specify new levels of factors in
Uncoded units
Coded units

Method of specifying new factor levels
Select variables stored in worksheet
Select levels from a list

Factor	Levels
Assigned	Y
Bags	1
Food/Bev	FB

Select

Help OK Cancel

FIGURE 13.20
Selection of factor levels (Y, 1, FB) for prediction of Taguchi results.

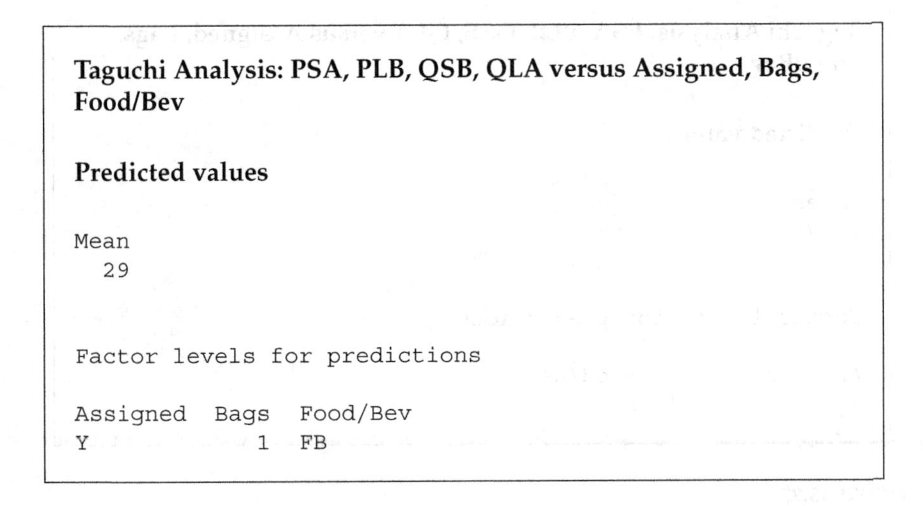

Taguchi Analysis: PSA, PLB, QSB, QLA versus Assigned, Bags, Food/Bev

Predicted values

```
Mean
  29


Factor levels for predictions

Assigned  Bags  Food/Bev
Y            1  FB
```

FIGURE 13.21
Predicted mean for factor levels (Y, 1, FB).

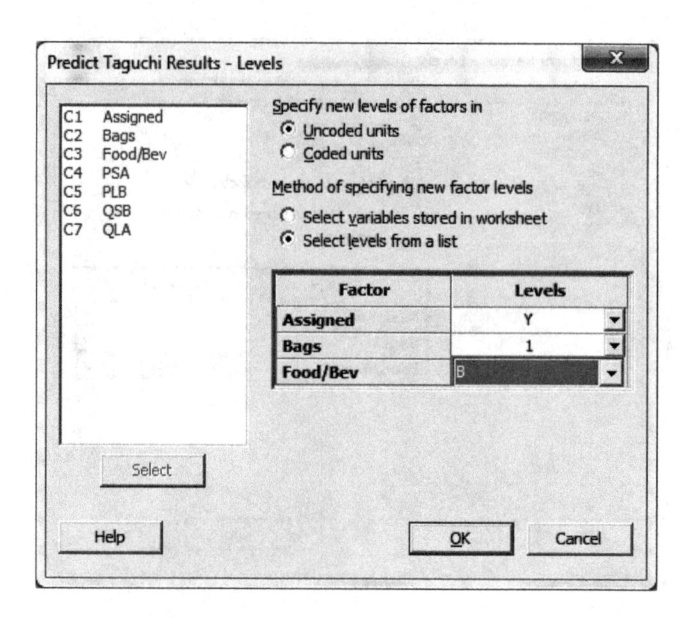

FIGURE 13.22
Selection of factor levels (Y, 1, B) for prediction of Taguchi results.

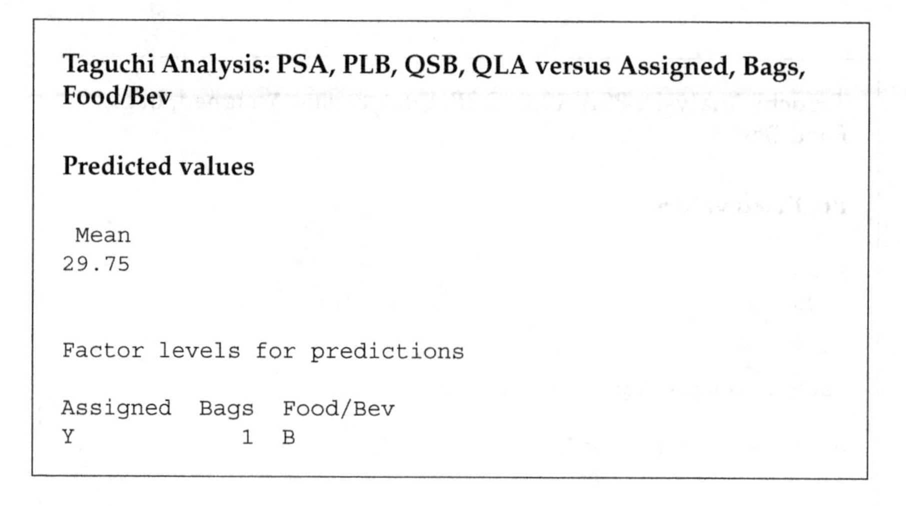

Taguchi Analysis: PSA, PLB, QSB, QLA versus Assigned, Bags, Food/Bev

Predicted values

```
 Mean
29.75


Factor levels for predictions

Assigned  Bags  Food/Bev
Y            1  B
```

FIGURE 13.23
Predicted mean for factor levels (Y, 1, B).

13.4 Improve and Control Phases

The waiting time is reduced by choosing the "Y" level for the "Assigned" factor, "1" level for the "Bags" factor, and the "FB" level for the "Food/Bev" factor. The management plans to continue taking samples of customer waiting times to verify that these levels are indeed the best levels for the three control factors.

14

Factorial Design of Experiments to Optimize a Chemical Process

This case study is about a process that produces a chemical whose yield (weight in grams) needs to be maximized while minimizing the cost ($) of production.

Sections 14.1 and 14.2 describe the define phase and the measure phase, respectively. Section 14.3 illustrates the analyze phase with detailed instructions for using Minitab®. Finally, Section 14.4 gives a brief description of the improve and control phases.

14.1 Define Phase

The objective is to maximize the yield and minimize the cost. The members of the Six Sigma project team brainstorm to identify the following factors that may be affecting the yield and cost:

- Processing time
- Processing temperature
- Catalyst

14.2 Measure Phase

Because it is impractical to design an experiment with all possible processing times, processing temperatures, and catalysts, the project team shortlisted the levels of the above three factors to what is shown in Table 14.1 for the experiment. Inasmuch as there are 3 factors and 2 levels of each factor, there are 8 (2^3) runs possible for a full factorial replication of the experiment.

Although it is possible that the time (morning shift or afternoon shift) of performing the experiment too has an effect on the yield or cost, the team knows that one replication of the experiment can be conveniently performed within the duration of a shift. In other words, the shift type is the same for all 8 runs in a replication. Hence, the shift type is not considered a factor within

TABLE 14.1
Factors and Their Levels

Factor	"Low" Level	"High" Level
Processing Time	20 min	50 min
Processing Temperature	150 Fahrenheit	200 Fahrenheit
Catalyst	Ajuba	Tapori

each replication. However, the team wants to perform two replications of the experiment (2 * 8 runs = 16 runs) to gather additional data for more reliable analysis, and this means that not all 16 runs will be from the same shift. This in turn means that the potential effect of the shift type must somehow be studied in the experiment. Hence, the team decides to perform a 2-block experiment, where one replication is performed in the morning shift (block) and the other in the afternoon shift (block). (Note: If an entire replication cannot be performed within the duration of a shift, the shift type must be considered the fourth factor with two levels: morning and afternoon. In that case, 16 (2^4) runs are possible for a full factorial replication of the experiment.)

In order to perform the experiment, the team first creates the factorial design of the experiment. The following is the approach to creating a factorial design in Minitab®:

Click on "Create Factorial Design" as shown in Figure 14.1. As a result, the dialog box shown in Figure 14.2 opens. Because there are three factors in this

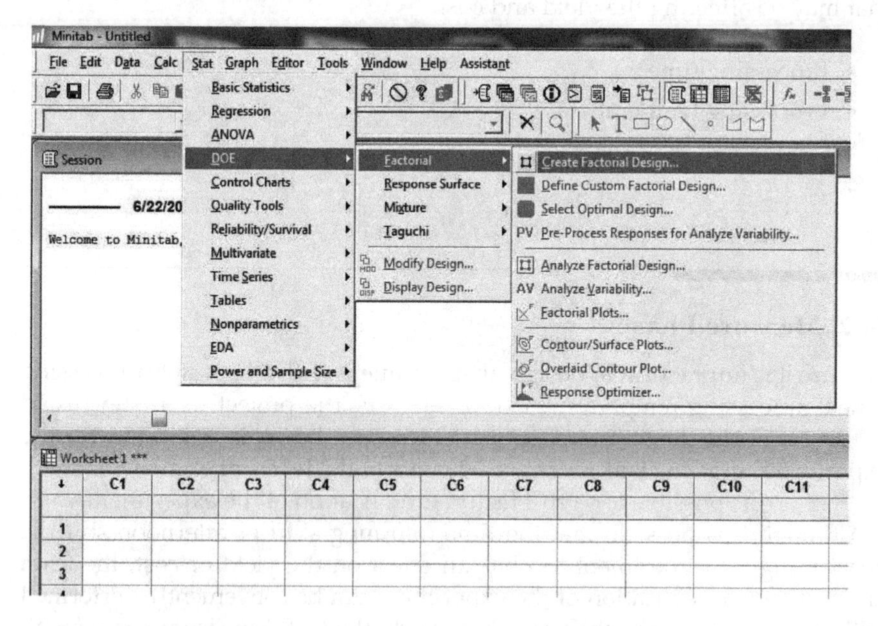

FIGURE 14.1
Selection of "Create Factorial Design".

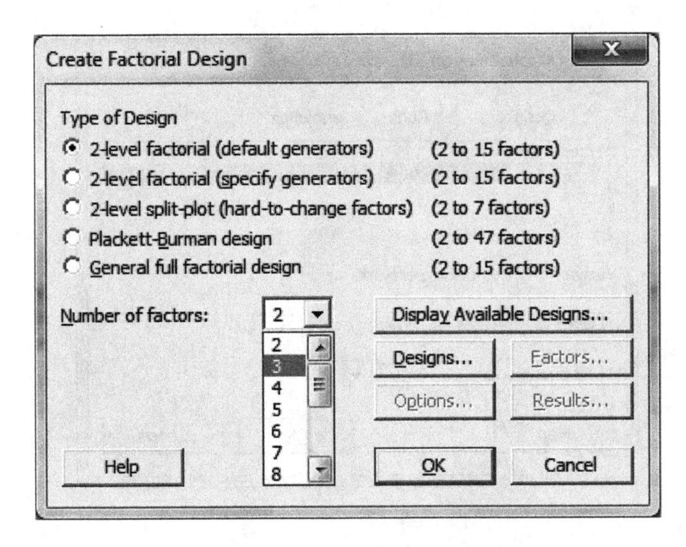

FIGURE 14.2
Selection of "Number of factors".

experiment, select "3" for "Number of factors" as shown in Figure 14.2. The "Type of Design" is "2-level factorial" (default) because each factor in this experiment has two levels. Click on "Display Available Designs" and the dialog box shown in Figure 14.3 opens. Notice that the column with 3 "Factors" has 8 "Run" in "Full" "Resolution". Click on "OK" and the dialog box shown

Create Factorial Design - Display Available Designs

Available Factorial Designs (with Resolution)

	Factors													
Run	**2**	**3**	**4**	**5**	**6**	**7**	**8**	**9**	**10**	**11**	**12**	**13**	**14**	**15**
4	Full	III												
8		Full	IV	III	III	III								
16			Full	V	IV	IV	IV	III	III	III	III	III	III	III
32				Full	VI	IV	IV	IV	IV	IV	IV	IV	IV	IV
64					Full	VII	V	IV	IV	IV	IV	IV	IV	IV
128						Full	VIII	VI	V	V	IV	IV	IV	IV

Available Resolution III Plackett-Burman Designs

Factors	Runs	Factors	Runs	Factors	Runs
2-7	12,20,24,28,...,48	20-23	24,28,32,36,...,48	36-39	40,44,48
8-11	12,20,24,28,...,48	24-27	28,32,36,40,44,48	40-43	44,48
12-15	20,24,28,36,...,48	28-31	32,36,40,44,48	44-47	48
16-19	20,24,28,32,...,48	32-35	36,40,44,48		

Help OK

FIGURE 14.3
Available designs.

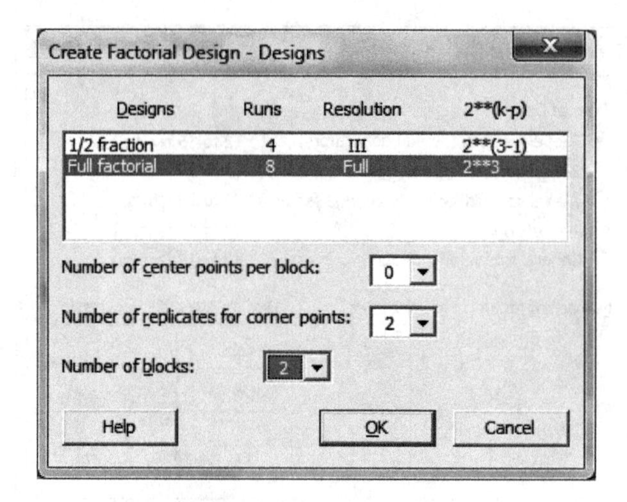

FIGURE 14.4
Selection of "Full Factorial" and entry of "Replicates" and "Blocks".

in Figure 14.2 reappears. Click on "Designs" and the dialog box shown in Figure 14.4 opens. Select "Full Factorial" design and change "Number of replicates for corner points" to "2" and "Number of blocks" to "2". Click on "OK" and the dialog box shown in Figure 14.2 reappears. Click on "Factors" and the dialog box shown in Figure 14.5 opens. Enter the "Low" and "High" levels for each factor as shown in Figure 14.5. Click on "OK" and the dialog box shown in Figure 14.2 reappears. Click on "OK" in the dialog box shown in Figure 14.2 and the factorial design is created as shown in Figure 14.6.

Create Factorial Design - Factors

Factor	Name	Type	Low	High
A	Time	Numeric	20	50
B	Temp	Numeric	150	200
C	Catalyst	Text	Ajuba	Tapori

Help OK Cancel

FIGURE 14.5
Entry of factors and their levels.

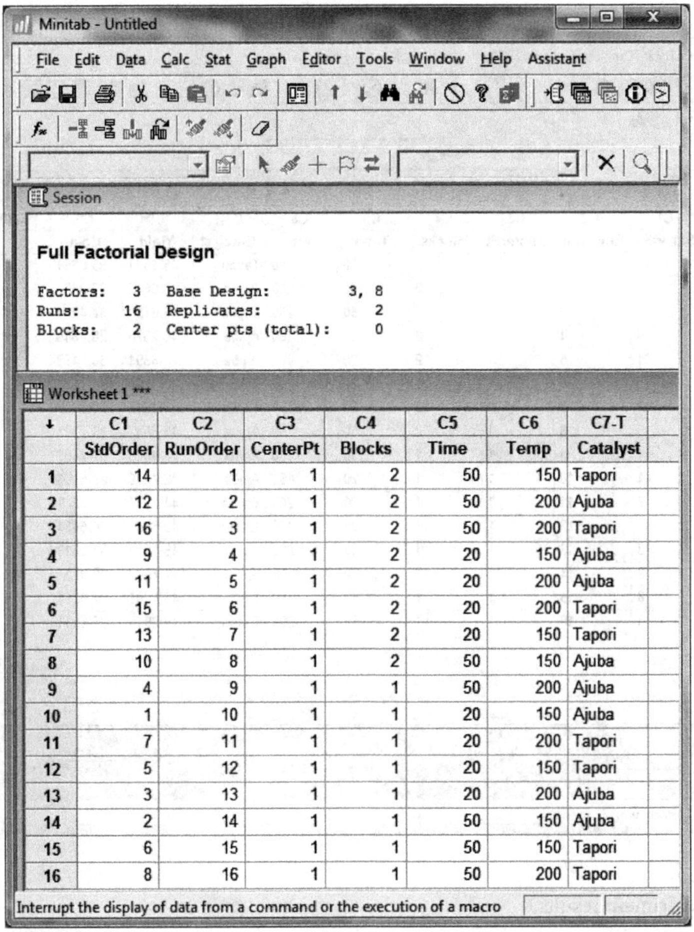

Minitab - Untitled

File Edit Data Calc Stat Graph Editor Tools Window Help Assistant

Session

Full Factorial Design

```
Factors:   3   Base Design:         3, 8
Runs:     16   Replicates:             2
Blocks:    2   Center pts (total):     0
```

Worksheet 1 ***

↓	C1	C2	C3	C4	C5	C6	C7-T
	StdOrder	RunOrder	CenterPt	Blocks	Time	Temp	Catalyst
1	14	1	1	2	50	150	Tapori
2	12	2	1	2	50	200	Ajuba
3	16	3	1	2	50	200	Tapori
4	9	4	1	2	20	150	Ajuba
5	11	5	1	2	20	200	Ajuba
6	15	6	1	2	20	200	Tapori
7	13	7	1	2	20	150	Tapori
8	10	8	1	2	50	150	Ajuba
9	4	9	1	1	50	200	Ajuba
10	1	10	1	1	20	150	Ajuba
11	7	11	1	1	20	200	Tapori
12	5	12	1	1	20	150	Tapori
13	3	13	1	1	20	200	Ajuba
14	2	14	1	1	50	150	Ajuba
15	6	15	1	1	50	150	Tapori
16	8	16	1	1	50	200	Tapori

Interrupt the display of data from a command or the execution of a macro

FIGURE 14.6
Full factorial design.

The "RunOrder" is the order of running the various factor combinations (called treatments) and this is randomized. In other words, if you repeat the above steps, the "RunOrder" number for a particular treatment (e.g., "20" Time, "150" Temp, and "Ajuba" Catalyst) may be different. The "StdOrder" is the order of all possible treatments created in a systematic fashion and it is always the same even if you repeat the above steps. For example, in the above treatment example, each of the factors is set at its low level in one particular block, and the "StdOrder" number is and always will be 1 for that treatment. The treatment where the first factor (Time) is changed to its high level and the other two factors (Temp and Catalyst) remain at their respective low

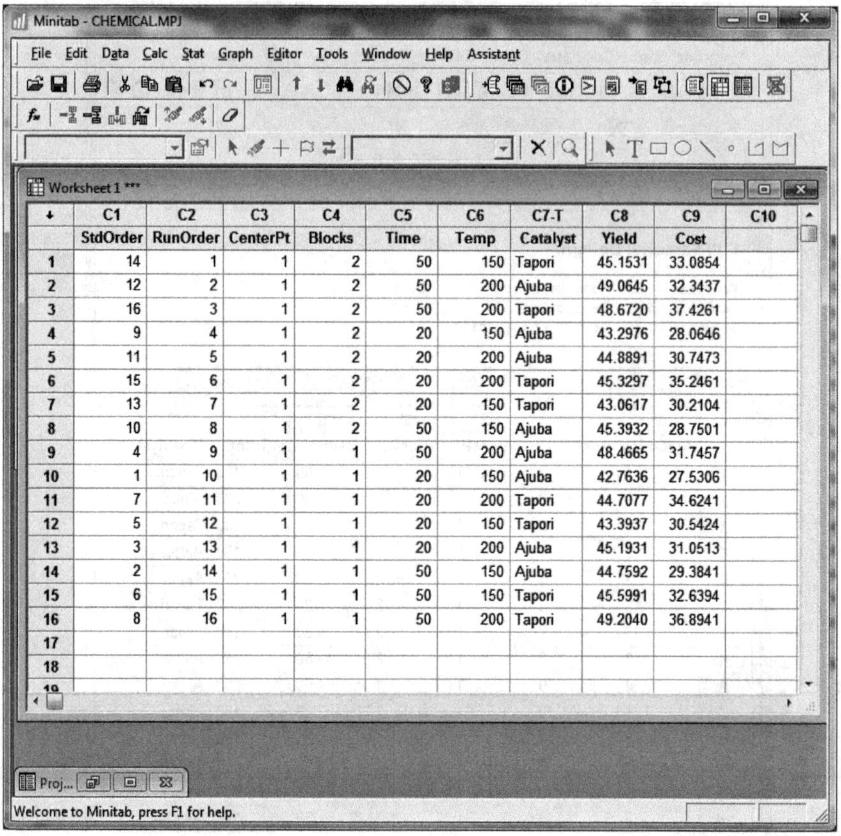

Minitab - CHEMICAL.MPJ

Worksheet 1 ***

↓	C1	C2	C3	C4	C5	C6	C7-T	C8	C9	C10
	StdOrder	RunOrder	CenterPt	Blocks	Time	Temp	Catalyst	Yield	Cost	
1	14	1	1	2	50	150	Tapori	45.1531	33.0854	
2	12	2	1	2	50	200	Ajuba	49.0645	32.3437	
3	16	3	1	2	50	200	Tapori	48.6720	37.4261	
4	9	4	1	2	20	150	Ajuba	43.2976	28.0646	
5	11	5	1	2	20	200	Ajuba	44.8891	30.7473	
6	15	6	1	2	20	200	Tapori	45.3297	35.2461	
7	13	7	1	2	20	150	Tapori	43.0617	30.2104	
8	10	8	1	2	50	150	Ajuba	45.3932	28.7501	
9	4	9	1	1	50	200	Ajuba	48.4665	31.7457	
10	1	10	1	1	20	150	Ajuba	42.7636	27.5306	
11	7	11	1	1	20	200	Tapori	44.7077	34.6241	
12	5	12	1	1	20	150	Tapori	43.3937	30.5424	
13	3	13	1	1	20	200	Ajuba	45.1931	31.0513	
14	2	14	1	1	50	150	Ajuba	44.7592	29.3841	
15	6	15	1	1	50	150	Tapori	45.5991	32.6394	
16	8	16	1	1	50	200	Tapori	49.2040	36.8941	
17										
18										

Welcome to Minitab, press F1 for help.

FIGURE 14.7
Entry of experiment results.

levels in the same block as above, has and always will have the "StdOrder" number 2. The team runs the experiment treatments in the "RunOrder" and records the yield and cost as shown in Figure 14.7. For example, the first treatment is "50, 150, Tapori" for "Time, Temp, Catalyst", and the yield is 45.1531 grams and the cost is $33.0854 for that treatment. Label two empty columns as "Yield" and "Cost" and enter the data as shown in Figure 14.7. Ensure that what you enter in your worksheet for each "StdOrder" number matches what is in Figure 14.7. For example, for StdOrder # 7, "Yield" should be "44.7077" and "Cost" should be "34.6241".

Minitab® randomizes block numbers as well. The team considers Block 2 the morning shift and Block 1 the afternoon shift, because Block 2 runs appear before Block 1 runs in this example (see Figure 14.7).

14.3 Analyze Phase

Now that the design is ready, the team analyzes the experiment. The following is the approach to analyzing an experimental design.

Click on "Analyze Factorial Design" as shown in Figure 14.8. As a result, the dialog box shown in Figure 14.9 opens. For the responses, select Yield and Cost as shown in Figure 14.9. Click on "Terms…" and the dialog box shown in Figure 14.10 opens. Notice that in the initial analysis, the team wishes to consider the potential effects of all of the possible "terms" (the factors and their interactions). For example, in Figure 14.10, "AB" is the term for the interaction between factor A and B, and "ABC" is the term for the interaction among A, B, and C. (As shown in Figure 14.5, "A" stands for Time, "B" stands for Temp, and "C" stands for Catalyst.) Check the box for "Include blocks in the model" if it is not already checked. Then, click on "OK". This takes you back to the dialog box shown in Figure 14.9. Click on "Graphs" and the dialog box shown in Figure 14.11 opens. For the "Effects Plots", check the boxes for "Normal" and "Pareto" as shown in Figure 14.11. Then, click on "OK" and it takes you back to the dialog box shown in Figure 14.9. Click on "OK" and the graphs shown in Figures 14.12–14.15 appear.

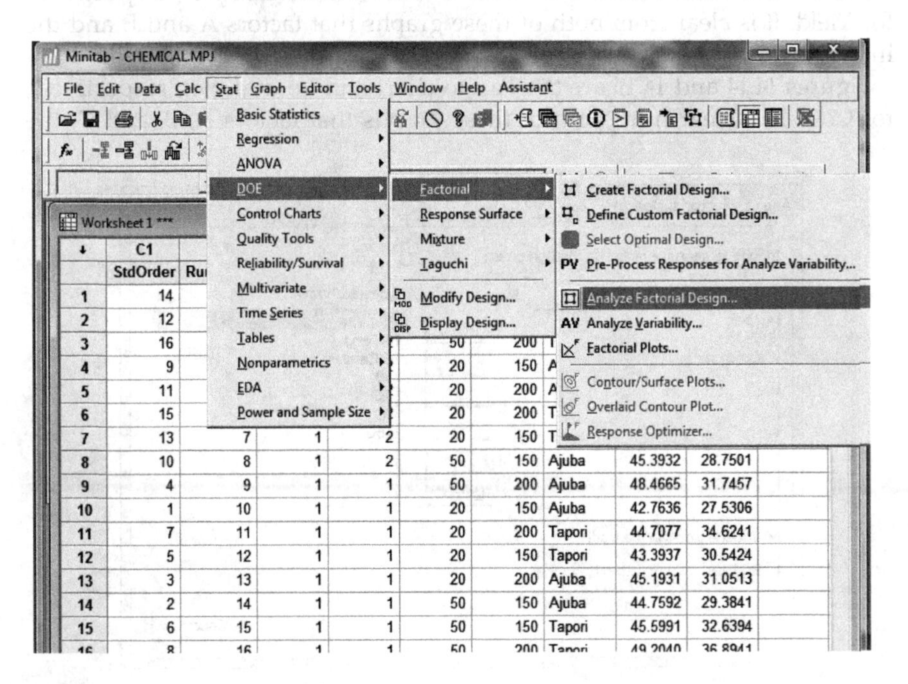

FIGURE 14.8
Selection of "Analyze Factorial Design".

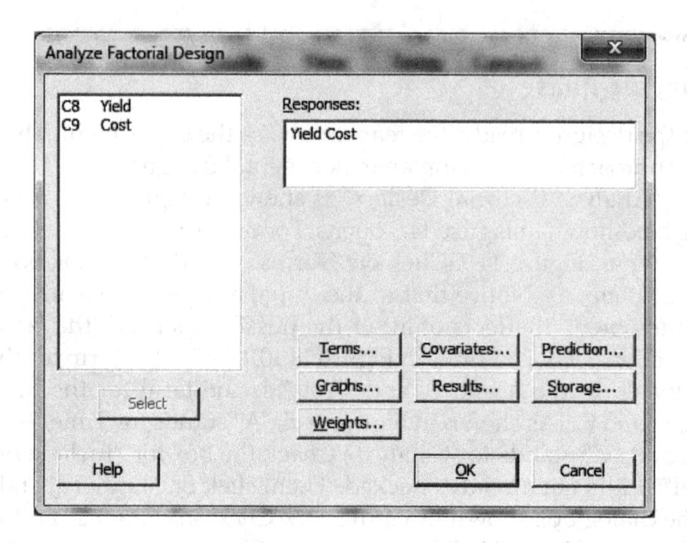

FIGURE 14.9
Selection of responses.

Figures 14.12 and 14.13 are the Pareto chart and normal plot, respectively, for Yield. It is clear from both of these graphs that factors A and B and the interaction between factors A and B have a significant effect on Yield.

Figures 14.14 and 14.15 are the Pareto chart and normal plot, respectively, for Cost. It is clear from both of these graphs that factors A, B, and C, the

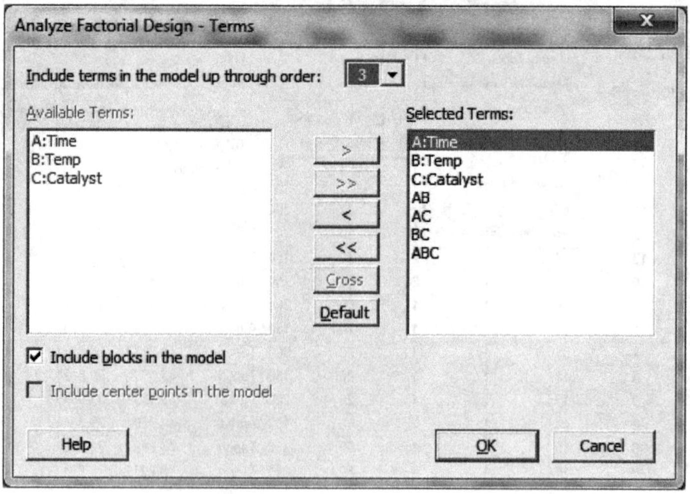

FIGURE 14.10
Selection of terms.

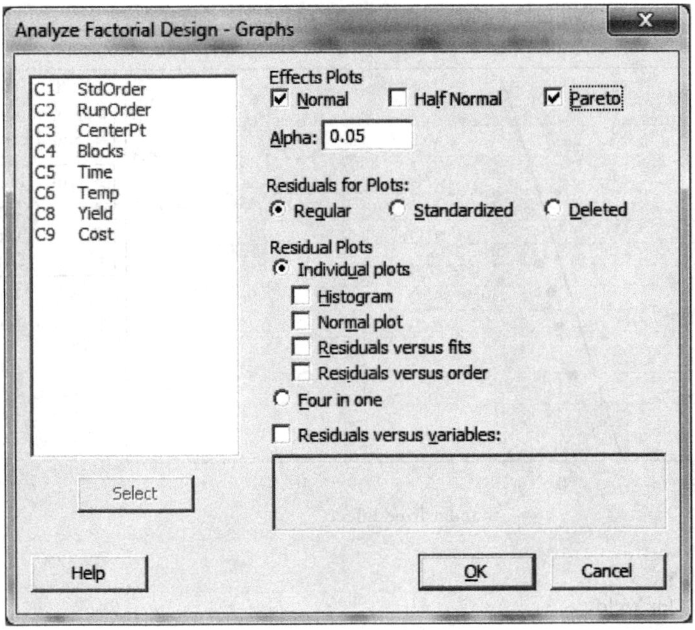

Analyze Factorial Design - Graphs

C1 StdOrder
C2 RunOrder
C3 CenterPt
C4 Blocks
C5 Time
C6 Temp
C8 Yield
C9 Cost

Effects Plots
Normal Half Normal Pareto
Alpha: 0.05

Residuals for Plots:
Regular Standardized Deleted

Residual Plots
Individual plots
Histogram
Normal plot
Residuals versus fits
Residuals versus order
Four in one
Residuals versus variables:

Select

Help OK Cancel

FIGURE 14.11
Selection of plots.

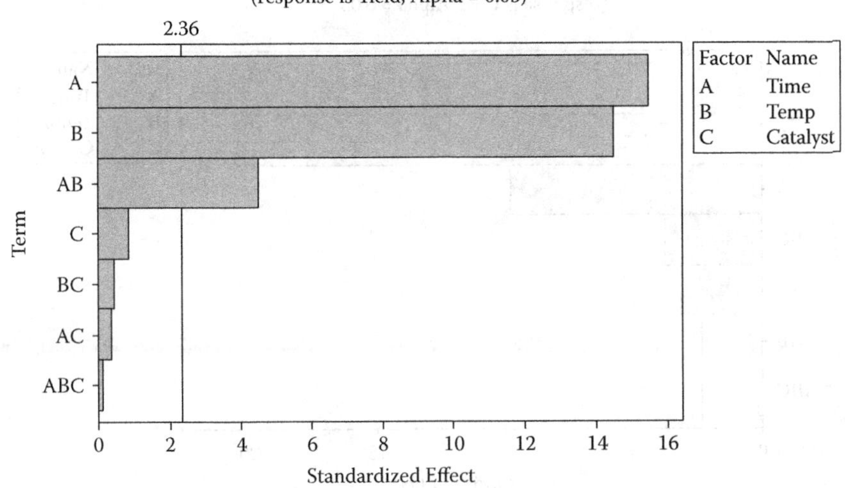

FIGURE 14.12
Pareto chart for Yield.

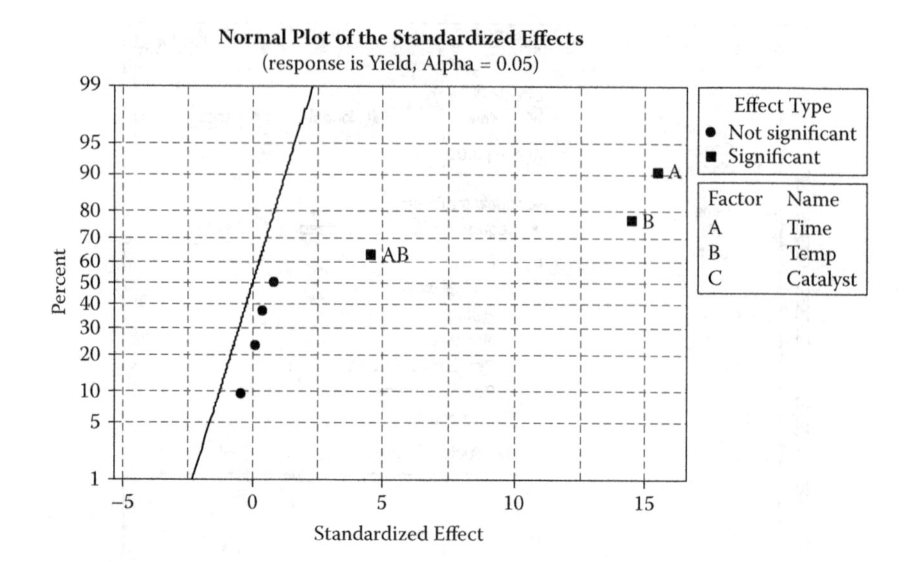

FIGURE 14.13
Normal plot for Yield.

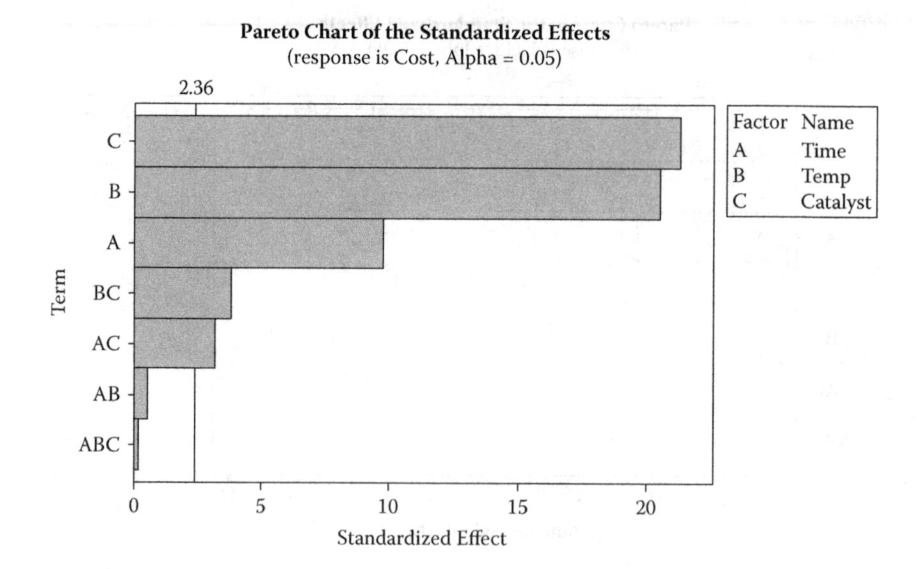

FIGURE 14.14
Pareto chart for Cost.

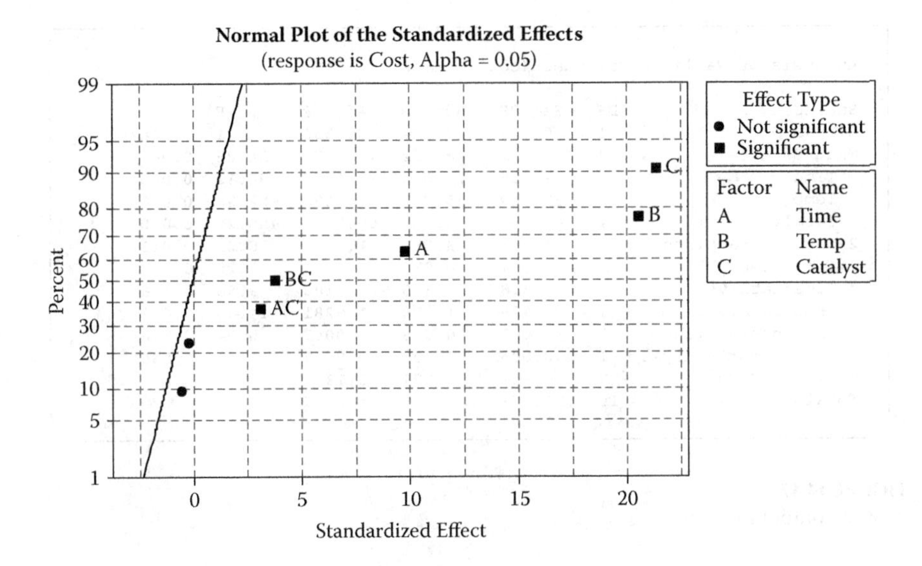

FIGURE 14.15
Normal plot for Cost.

interaction between factors B and C, and the interaction between factors A and C have a significant effect on Cost.

The session window file also produces the analysis of variance outputs shown in Figures 14.16 and 14.17 for both Yield and Cost, respectively. As in the Pareto chart and normal plot, the terms that have a significant effect on a response have *P*-values lower than 0.05. For example, for Yield, factors A and B and the interaction between factors A (Time) and B (Temp) have *P*-values lower than 0.05.

Analysis of Variance for Yield (coded units)

Source	DF	Seq SS	Adj SS	Adj MS	F	P
Blocks	1	0.0374	0.0374	0.0374	0.26	0.628
Main Effects	3	65.6780	65.6780	21.8927	150.15	0.000
Time	1	35.0328	35.0328	35.0328	240.27	0.000
Temp	1	30.5405	30.5405	30.5405	209.46	0.000
Catalyst	1	0.1047	0.1047	0.1047	0.72	0.425
2-Way Interactions	3	3.0273	3.0273	1.0091	6.92	0.017
Time*Temp	1	2.9751	2.9751	2.9751	20.40	0.003
Time*Catalyst	1	0.0222	0.0222	0.0222	0.15	0.708
Temp*Catalyst	1	0.0301	0.0301	0.0301	0.21	0.663
3-Way Interactions	1	0.0021	0.0021	0.0021	0.01	0.907
Time*Temp*Catalyst	1	0.0021	0.0021	0.0021	0.01	0.907
Residual Error	7	1.0206	1.0206	0.1458		
Total	15	69.7656				

FIGURE 14.16
ANOVA output for Yield.

Analysis of Variance for Cost (coded units)

Source	DF	Seq SS	Adj SS	Adj MS	F	P
Blocks	1	0.134	0.134	0.1336	1.01	0.348
Main Effects	3	128.722	128.722	42.9074	324.88	0.000
Time	1	12.695	12.695	12.6946	96.12	0.000
Temp	1	55.769	55.769	55.7688	422.26	0.000
Catalyst	1	60.259	60.259	60.2587	456.26	0.000
2-Way Interactions	3	3.283	3.283	1.0944	8.29	0.011
Time*Temp	1	0.037	0.037	0.0371	0.28	0.613
Time*Catalyst	1	1.318	1.318	1.3180	9.98	0.016
Temp*Catalyst	1	1.928	1.928	1.9281	14.60	0.007
3-Way Interactions	1	0.005	0.005	0.0047	0.04	0.856
Time*Temp*Catalyst	1	0.005	0.005	0.0047	0.04	0.856
Residual Error	7	0.924	0.924	0.1321		
Total	15	133.068				

FIGURE 14.17
ANOVA output for Cost.

The next step is to repeat the analysis for each response by eliminating one term at a time from the experiment, starting with the term with the highest *P*-value, and check for additional terms that may become significant or for the significant terms that become insignificant. For example, for Yield, eliminate ABC first because it has the highest *P*-value, and then repeat the analysis.

Figure 14.18 shows the dialog box that opens by following the steps shown in Figure 14.8. Notice that the analysis is repeated for one response at a time and that is why Cost is removed from the "Responses" box. The dialog box shown in Figure 14.19 opens when you click on "Terms". Figure 14.19 shows

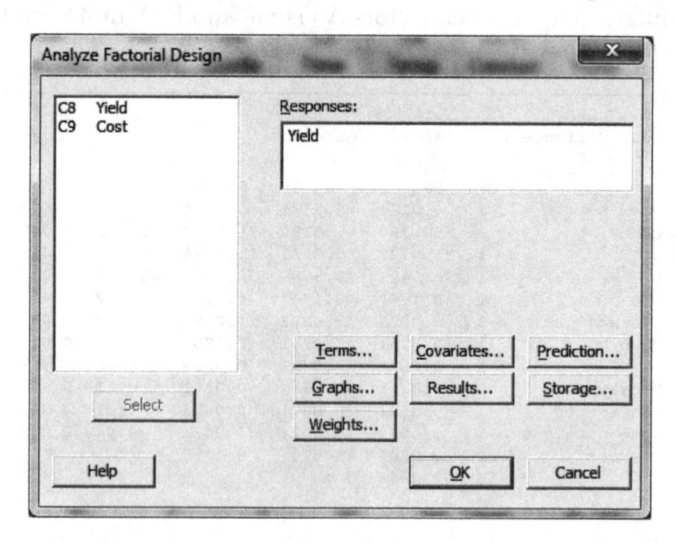

FIGURE 14.18
Selection of Yield response for reanalysis.

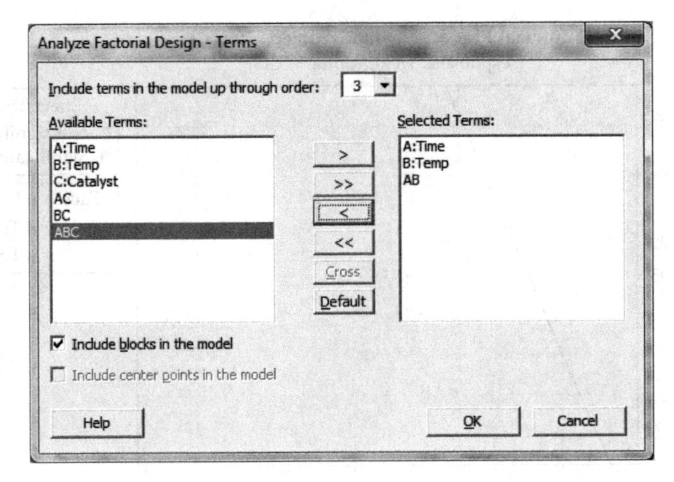

FIGURE 14.19
Selection of terms for Yield for reanalysis.

the "Selected Terms" as A, B, and AB, but note that this figure is obtained only after eliminating one term at a time and then repeating the analysis, as explained above. As shown in Figure 14.19, keep the box checked for "Include blocks in the model" despite the high *P*-value for blocks in the analysis of variance output, because the experiment is in fact performed in two blocks.

Figures 14.20 and 14.21 are the Pareto chart and normal plot, respectively, for Yield after all insignificant terms are removed (one at a time) from the analysis. It is clear from both of these graphs that factors A and B and the interaction between factors A and B have a significant effect on Yield.

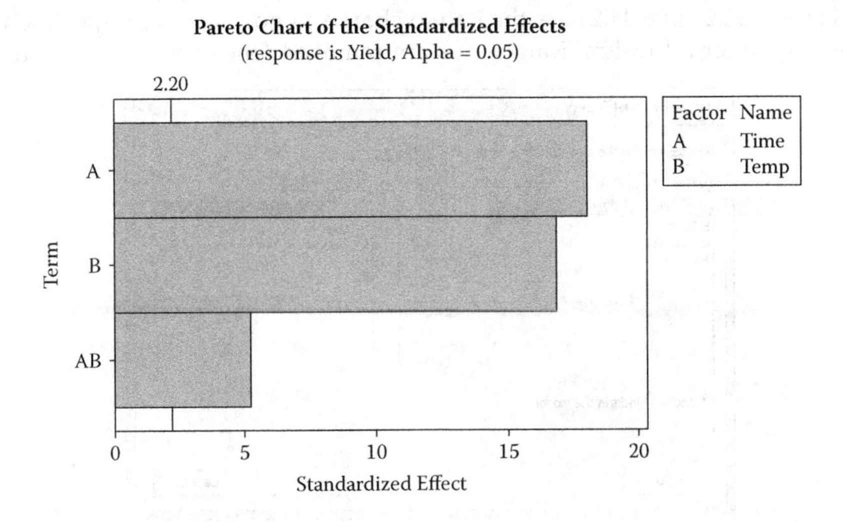

FIGURE 14.20
Pareto chart for Yield after reanalysis.

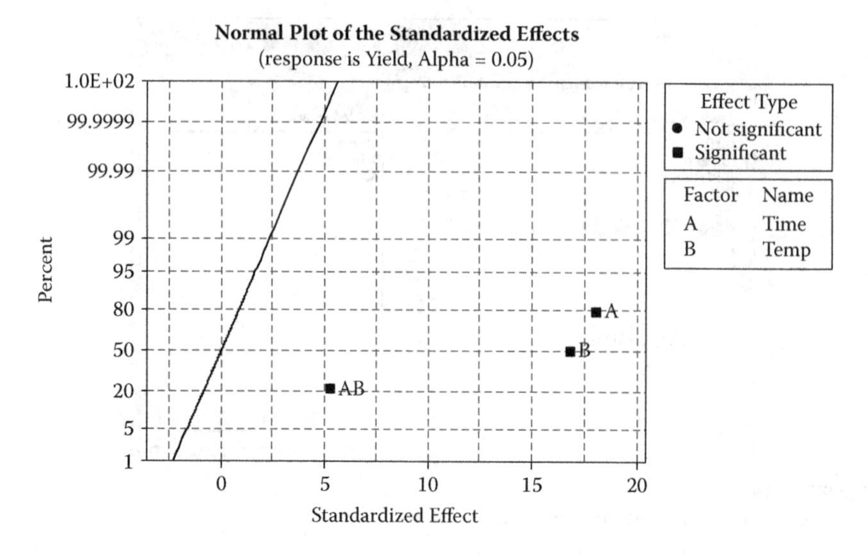

FIGURE 14.21
Normal plot for Yield after reanalysis.

The next step is to repeat the analysis for Cost by eliminating one term at a time from the experiment, starting with the term with the highest *P*-value, and check for additional terms that may become significant or for the significant terms that become insignificant. Figure 14.22 shows the "Selected Terms" for Cost, as A, B, C, AC, and BC, but note that this figure is obtained only after eliminating one term at a time and then repeating the analysis, as explained earlier.

Figures 14.23 and 14.24 are the Pareto chart and normal plot, respectively, for Cost, after all insignificant terms are removed (one at a time) from the

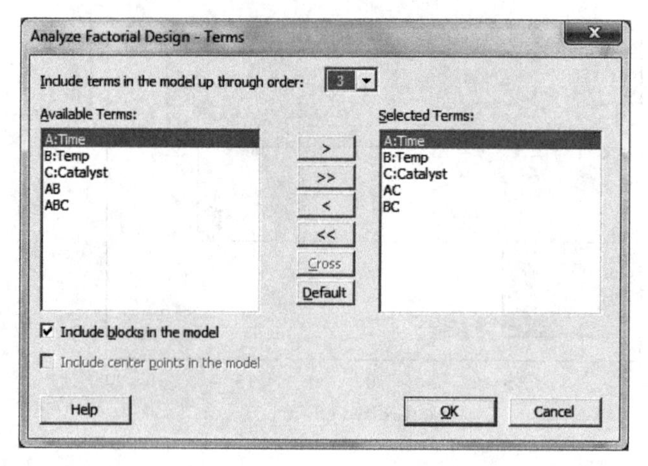

FIGURE 14.22
Selection of terms for Cost for reanalysis.

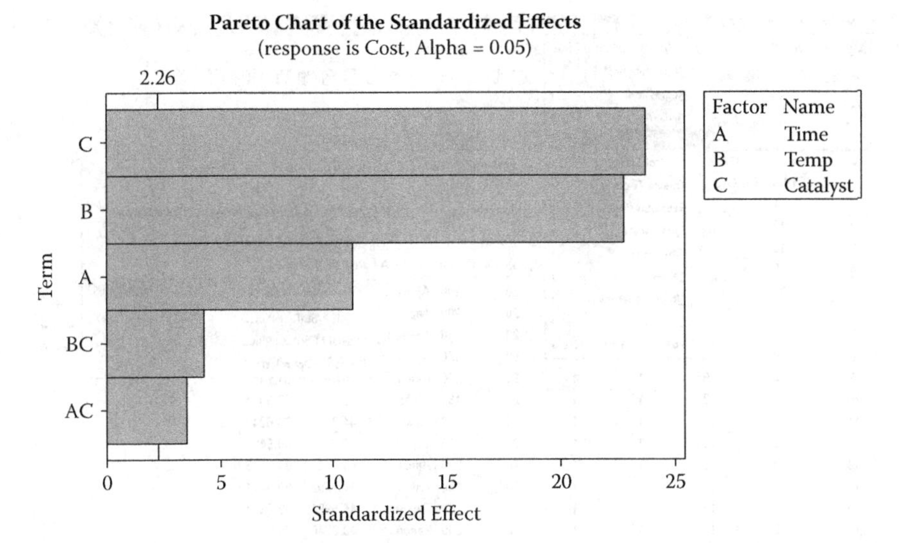

FIGURE 14.23
Pareto chart for Cost after reanalysis.

analysis. It is clear from both of these graphs that factors A, B, and C, the interaction between factors B and C, and the interaction between factors A and C have a significant effect on Cost.

Now that the team knows what terms have a significant effect on the responses (Yield and Cost), the team decides to perform further analysis of

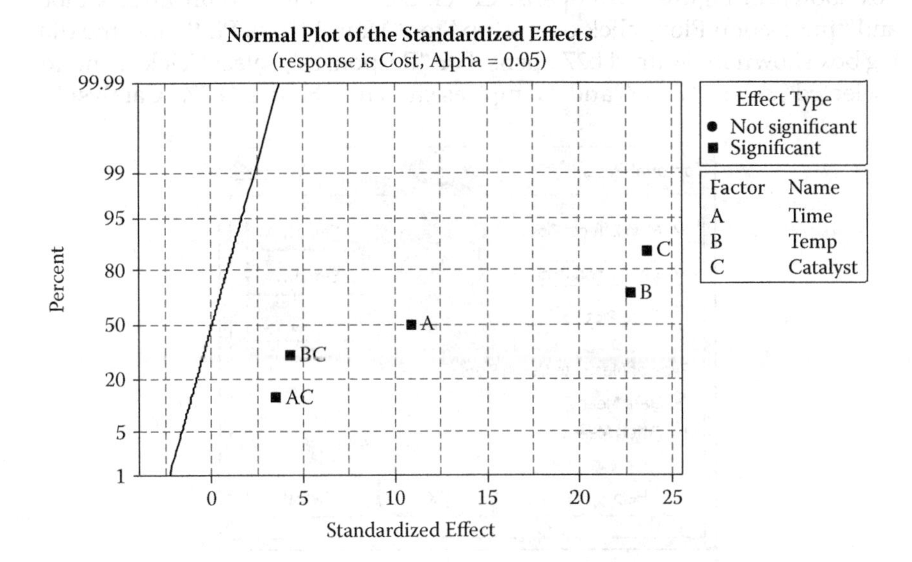

FIGURE 14.24
Normal plot for Cost after reanalysis.

	C1	C2							
	StdOrder	RunOrd							
1	14								
2	12								
3	16								
4	9								
5	11				20	200	Ajuba		
6	15				20	200	Tapori		
7	13				20	150	Tapori		
8	10				50	150	Ajuba		
9	4	9	1	1	50	200	Ajuba	[illegible]	[illegible]
10	1	10	1	1	20	150	Ajuba	42.7636	27.5306
11	7	11	1	1	20	200	Tapori	44.7077	34.6241
12	5	12	1	1	20	150	Tapori	43.3937	30.5424
13	3	13	1	1	20	200	Ajuba	45.1931	31.0513
14	2	14	1	1	50	150	Ajuba	44.7592	29.3841
15	6	15	1	1	50	150	Tapori	45.5991	32.6394
16	8	16	1	1	50	200	Tapori	49.2040	36.8941
17									

FIGURE 14.25
Selection of "Factorial Plots" for Yield.

that effect. To this end, main effects plots and interaction plots are produced. The following is the approach to doing so.

Click on "Factorial Plots" as shown in Figure 14.25. As a result, the dialog box shown in Figure 14.26 opens. Check the boxes for "Main Effects Plot" and "Interaction Plot", click on "Setup" for "Main Effects Plot", and the dialog box shown in Figure 14.27 opens. For "Responses", select "Yield", and for "Selected", select "Time" and "Temp" as shown in Figure 14.27. "Catalyst" is

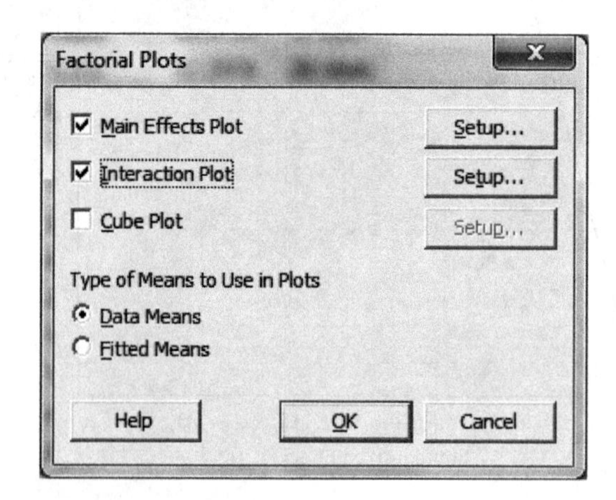

FIGURE 14.26
Selection of "Main Effects Plot" and "Interaction Plot" for Yield.

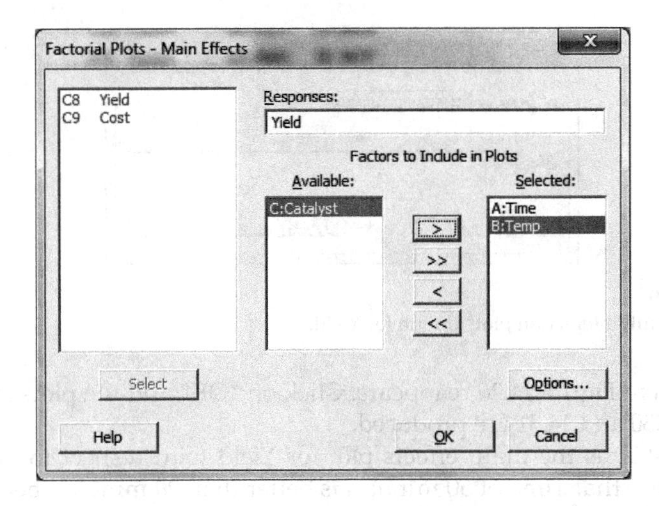

FIGURE 14.27
Setup for "Main Effects Plot" for Yield.

not selected because it is evident from Figures 14.12 and 14.13 that Catalyst does not have a significant effect on Yield. Click on "OK" and the dialog box shown in Figure 14.26 reappears. Click on "Setup" for "Interaction Plot" and the dialog box shown in Figure 14.28 opens. For "Responses", select "Yield", and for "Selected", select "Time" and "Temp" as shown in Figure 14.28. "Catalyst" is not selected because it is evident from Figures 14.12 and 14.13 that AB is the only interaction term that has a significant effect on Yield. Click on "Options" and the dialog box shown in Figure 14.29 opens. Check the box for "Draw full interaction plot matrix". Click on "OK" and the dialog

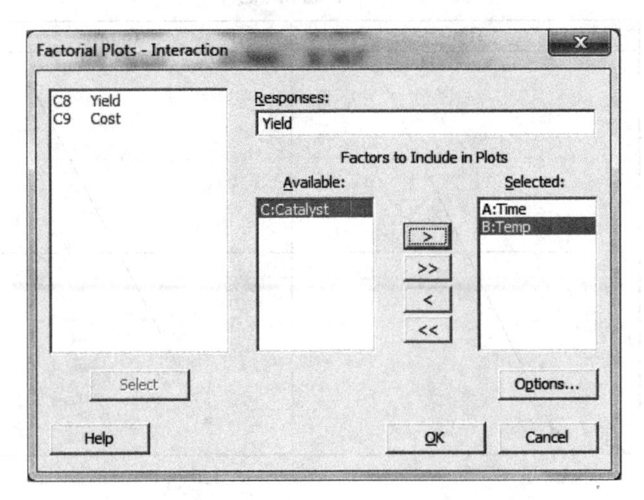

FIGURE 14.28
Setup for "Interaction Plot" for Yield.

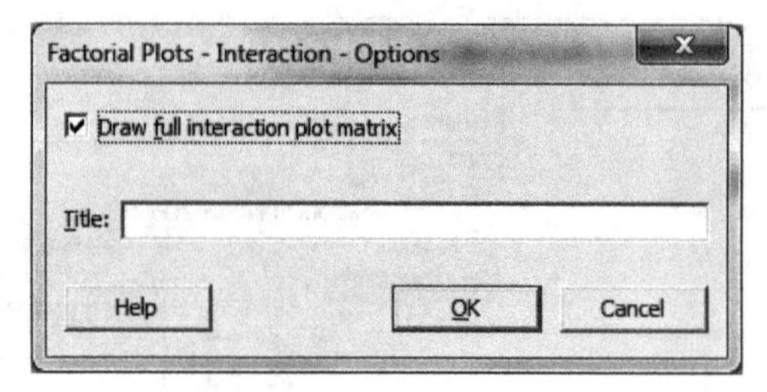

FIGURE 14.29
Selection of "full interaction plot" option for Yield.

box shown in Figure 14.26 reappears. Click on "OK" and the plots shown in Figures 14.30 and 14.31 are produced.

Figure 14.30 is the main effects plot for Yield with respect to Time and Temp. Notice that Time of 50 minutes is better than 20 minutes, because the average Yield is higher for the former. Also, Temp of 200 Fahrenheit is better than 150 Fahrenheit, because the average Yield is higher for the former.

Figure 14.31 is the interaction plot for Yield with respect to Time and Temp. Notice that when Temp is changed from 150 Fahrenheit to 200 Fahrenheit, the increase in the average Yield is higher for Time of 50 minutes than it is for Time of 20 minutes. Also, when Time is changed from 20 minutes to 50 minutes, the increase in average Yield is higher for Temp of 200 Fahrenheit than it is for Temp of 150 Fahrenheit.

Click on "Factorial Plots" as shown in Figure 14.32. As a result, the dialog box shown in Figure 14.33 opens. Check the boxes for "Main Effects Plot"

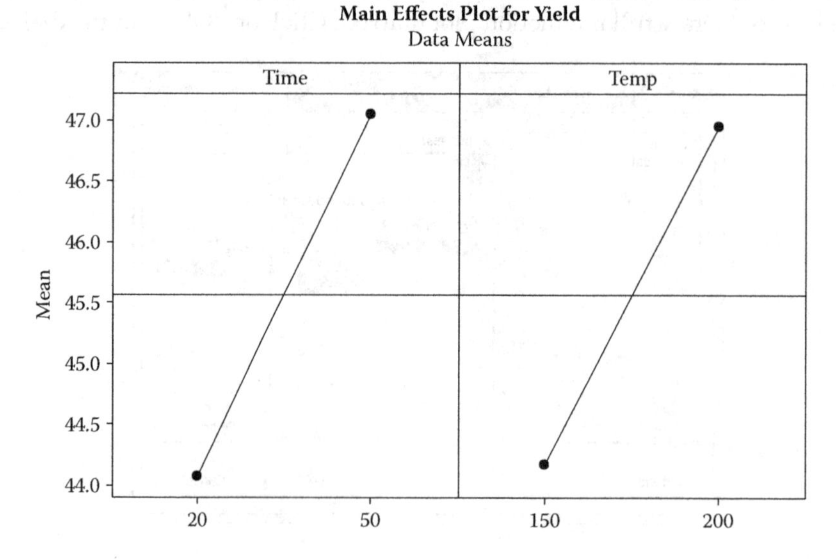

FIGURE 14.30
Main effects plot for Yield.

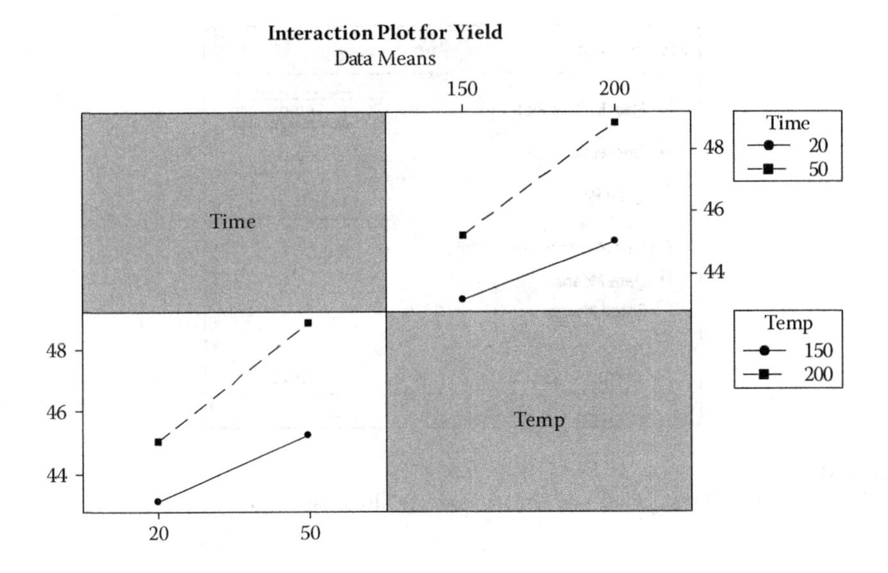

FIGURE 14.31
Interaction plot for Yield.

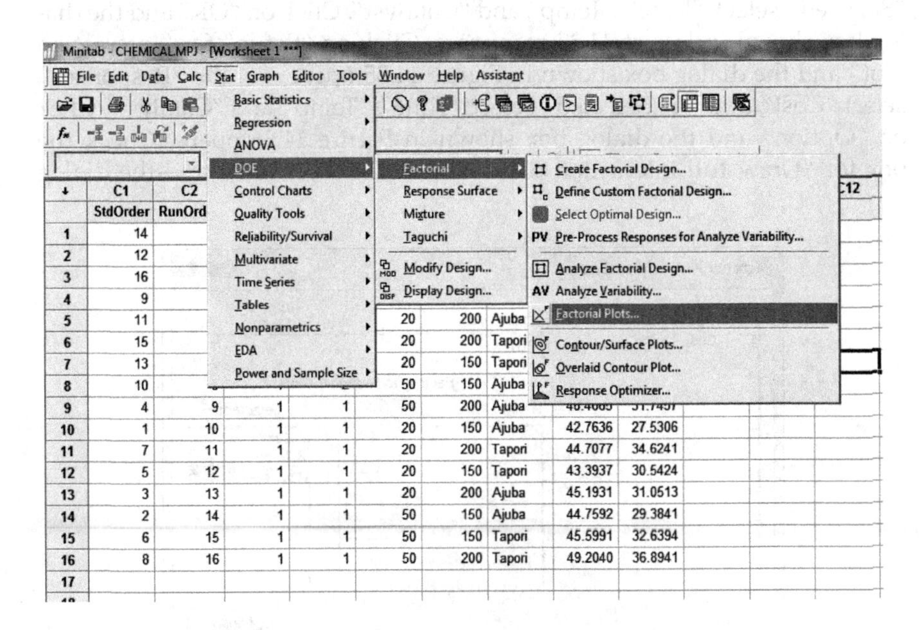

	StdOrder	RunOrd							
1	14								
2	12								
3	16								
4	9								
5	11				20	200	Ajuba		
6	15				20	200	Tapori		
7	13				20	150	Tapori		
8	10				50	150	Ajuba		
9	4	9	1	1	50	200	Ajuba		
10	1	10	1	1	20	150	Ajuba	42.7636	27.5306
11	7	11	1	1	20	200	Tapori	44.7077	34.6241
12	5	12	1	1	20	150	Tapori	43.3937	30.5424
13	3	13	1	1	20	200	Ajuba	45.1931	31.0513
14	2	14	1	1	50	150	Ajuba	44.7592	29.3841
15	6	15	1	1	50	150	Tapori	45.5991	32.6394
16	8	16	1	1	50	200	Tapori	49.2040	36.8941
17									

FIGURE 14.32
Selection of "Factorial Plots" for Cost.

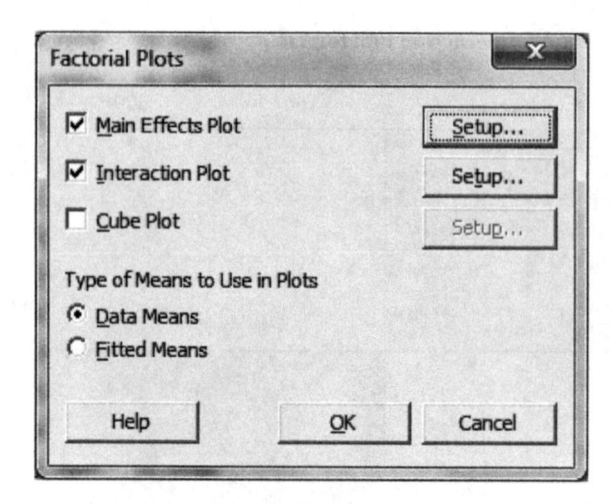

FIGURE 14.33
Selection of "Main Effects Plot" and "Interaction Plot" for Cost.

and "Interaction Plot". Click on "Setup" for "Main Effects Plot" and the dialog box shown in Figure 14.34 opens. For "Responses", select "Cost", and for "Selected", select "Time", "Temp", and "Catalyst". Click on "OK" and the dialog box shown in Figure 14.33 reappears. Click on "Setup" for "Interaction Plot" and the dialog box shown in Figure 14.35 opens. For the "Responses", select "Cost", and for "Selected", select "Time", "Temp", and "Catalyst". Click on "Option" and the dialog box shown in Figure 14.36 opens. Check the box for "Draw full interaction plot matrix". Click on "OK" and the dialog

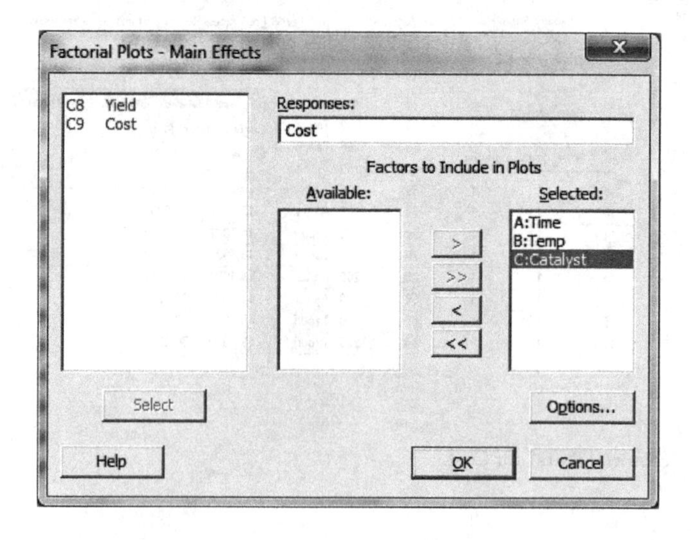

FIGURE 14.34
Setup for "Main Effects Plot" for Cost.

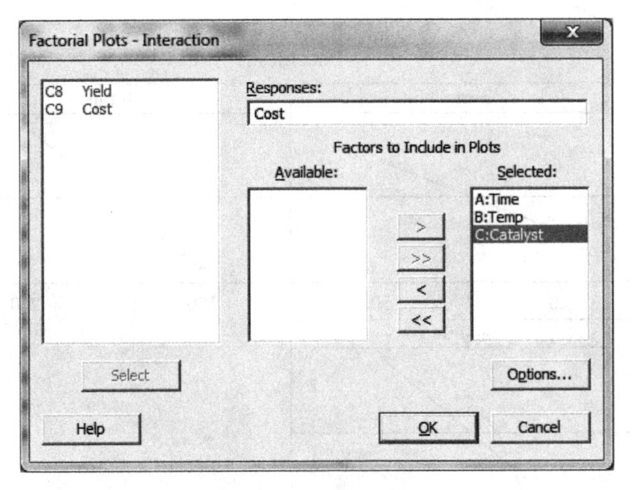

FIGURE 14.35
Setup for "Interaction Plot" for Cost.

box shown in Figure 14.33 reappears. Click on "OK" and the plots shown in Figures 14.37 and 14.38 are produced.

Figure 14.37 is the main effects plot for Cost with respect to Time, Temp, and Catalyst. Notice that Time of 20 minutes is better than 50 minutes, because the average Cost is lower for the former. Also, Temp of 150 Fahrenheit is better than 200 Fahrenheit, because the average Cost is lower for the former. Finally, Catalyst Ajuba is better than Catalyst Tapori, because the average Cost is lower for the former.

Figure 14.38 is the interaction plot for Cost with respect to Time, Temp, and Catalyst. Recall from Figures 14.14 and 14.15 that the term AB is not significant for Cost. Hence, as expected, in Figure 14.38 the lines for Time and Temp are parallel (which means interaction between Time and Temp have no significant effect on Cost). Terms BC and AC have a significant effect on

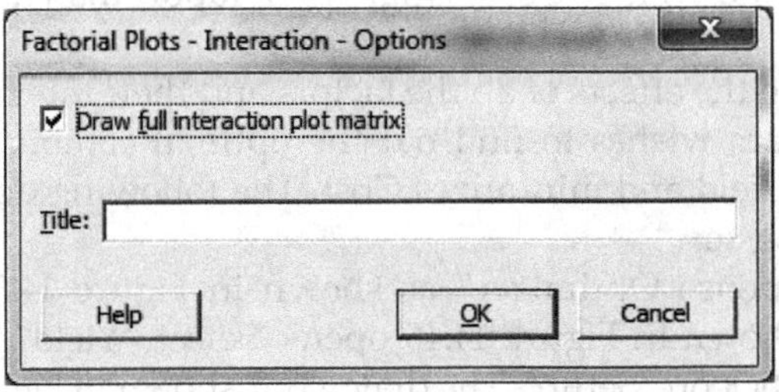

FIGURE 14.36
Selection of "full interaction plot" option for Yield.

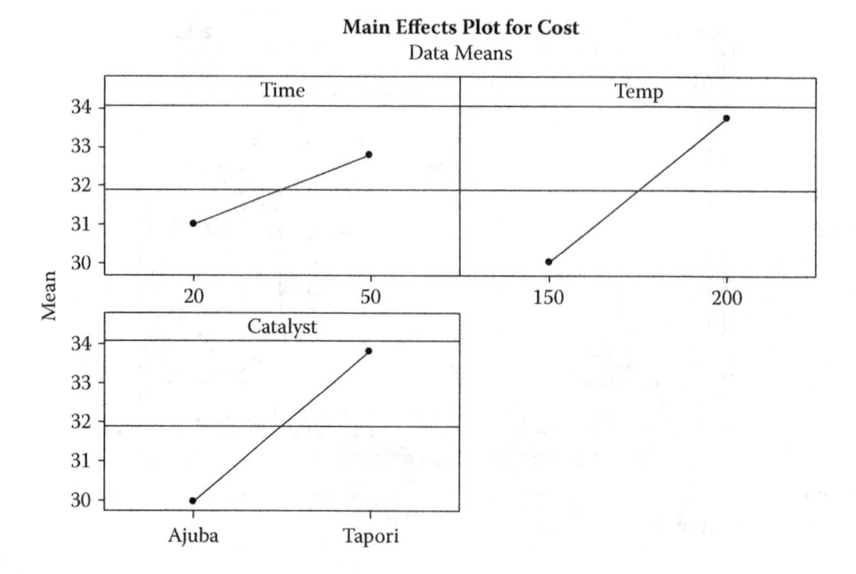

FIGURE 14.37
Main effects plot for Cost.

Cost, and the following interpretations are made from the interaction plot in Figure 14.38:

- When the Catalyst is changed from Ajuba to Tapori, the increase in the average Cost is higher for Time of 50 minutes than it is for Time of 20 minutes.
- When the Catalyst is changed from Ajuba to Tapori, the increase in the average Cost is higher for Temp of 200 Fahrenheit than it is for Temp of 150 Fahrenheit.
- When the Temp is changed from 150 Fahrenheit to 200 Fahrenheit, the increase in the average Cost is higher for Tapori than it is for Ajuba.
- When the Time is changed from 20 minutes to 50 minutes, the increase in the average Cost is higher for Tapori than it is for Ajuba.

After analyzing the effects of all the factors and their interactions on Yield and Cost, the team wishes to find out the optimal combination of factors that maximizes Yield and minimizes Cost. The following demonstrates the approach to doing so.

Click on "Response Optimizer" as shown in Figure 14.39. As a result, the dialog box shown in Figure 14.40 opens. Select "Yield" and "Cost" for "Selected", and click on "Setup". The dialog box shown in Figure 14.41 opens. For the "Goal" column, choose "Maximize" for "Yield" and "Minimize" for "Cost". As shown in Figure 14.41, the team believes that a target of 51 grams

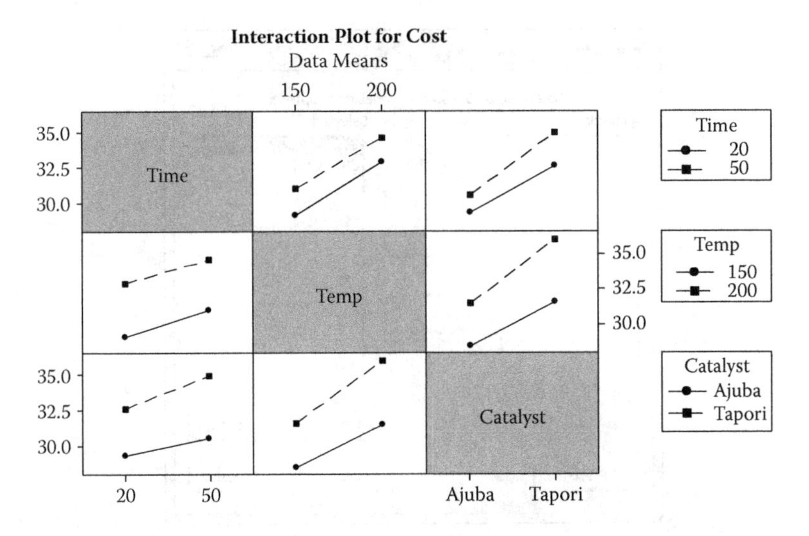

FIGURE 14.38
Interaction plot for Cost.

is appropriate for Yield, with no upper limit and a lower limit of 41 grams. The team also opines that a target of $31 is appropriate for Cost, with no lower limit and an upper limit of $41. Click on "OK" and the dialog box shown in Figure 14.40 reappears. Click on "OK" and the optimal output shown in Figure 14.42 is produced. As is evident from Figure 14.42, the

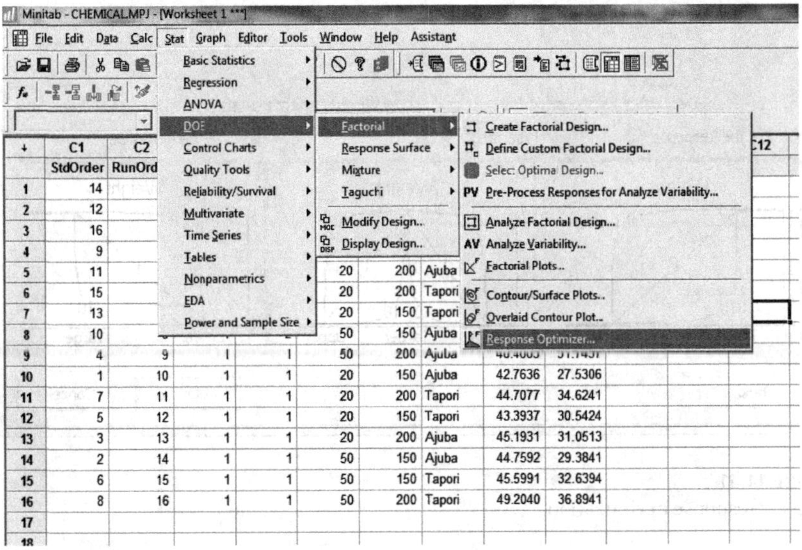

FIGURE 14.39
Selection of "Response Optimizer".

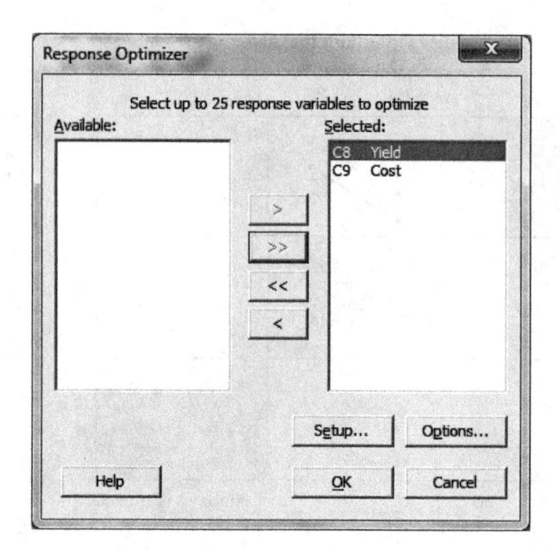

FIGURE 14.40
Selection of responses for optimization.

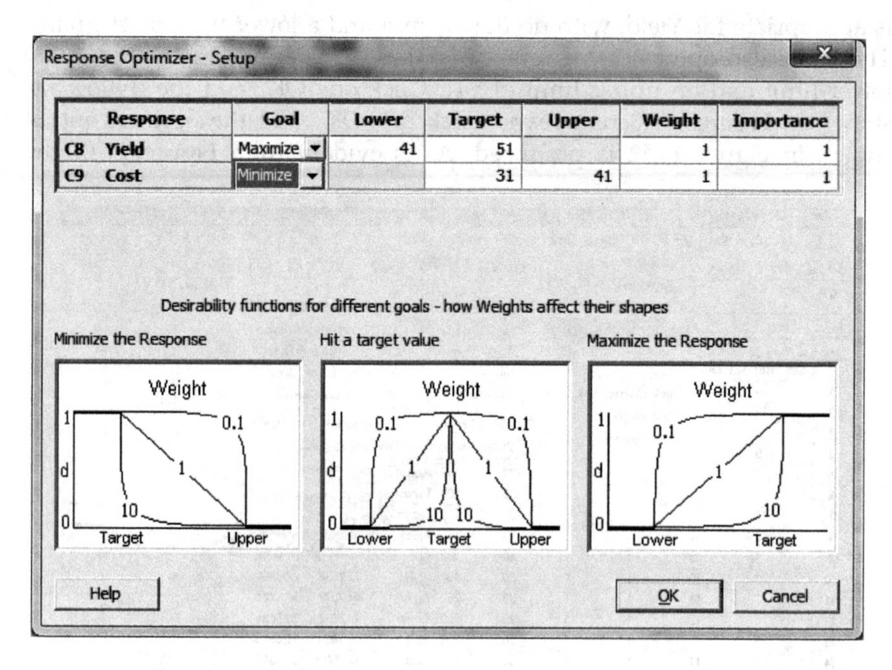

FIGURE 14.41
Setup for response optimization.

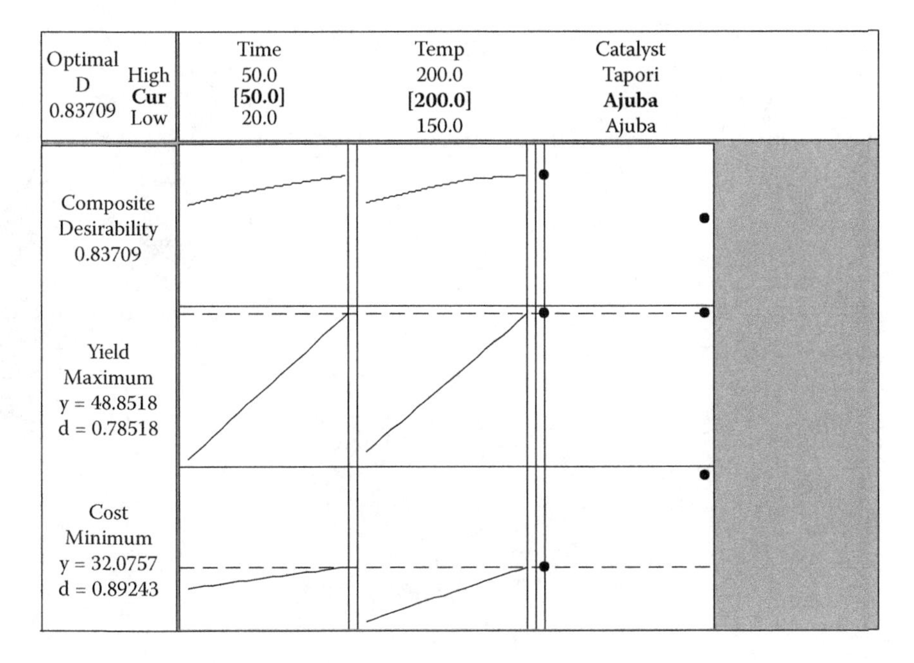

FIGURE 14.42
Optimal solution.

optimal treatment with the highest composite desirability (0.83709) is Time = 50 minutes, Temp = 200 Fahrenheit, and Catalyst = Ajuba.

14.4 Improve and Control Phases

The process is improved by setting Time at 50 minutes, Temp at 200 Fahrenheit, and Catalyst as Ajuba. Fail-safes are implemented to control any deviations from these settings.

15

Chi-Square Test to Verify Source Association with Parts Purchased and Products Produced

This case study is about a Six Sigma project at a company that wishes to reduce the number of defective products produced.

Sections 15.1 and 15.2 give brief descriptions of the define phase and the measure phase, respectively. Section 15.3 illustrates the analyze phase with detailed instructions for using Minitab®. Finally, Section 15.4 gives a brief description of the improve and control phases.

15.1 Define Phase

Lately, the company has been receiving complaints from customers that its products are defective. A critical part in the product, which is purchased from either of two suppliers, A and B, has been discovered to be defective more often than the other parts. The objective is to reduce the number of defective products.

15.2 Measure Phase

The company wants to check whether the quality of the critical part mentioned in the define phase depends on the supplier from whom it was purchased. To this end, a sample of 80 parts is taken from Suppliers A and B, and inspected as to whether each of them is conforming or nonconforming (defective). These data are stored in the CHAPTER_15_1.MTW worksheet (the worksheet is available at the publisher's website; the data from the worksheet are also provided in the Appendix). Figure 15.1 is a screenshot of the partial worksheet.

↓	C1-T	C2-T
	Supplier	Conforming?
1	A	Yes
2	B	No
3	A	No
4	A	No
5	B	No
6	B	Yes
7	B	Yes
8	B	No
9	A	No
10	B	Yes
11	B	Yes
12	A	No
13	A	No
14	B	No
15	A	No
16	B	No
17	A	No
18	A	No
19	A	No

FIGURE 15.1
Parts from suppliers A and B.

15.3 Analyze Phase

In order to check whether a part being defective depends on the supplier it is purchased from, the following set of hypotheses is considered for testing:

Null Hypothesis: There is no association between part conformance and supplier.

Alternative Hypothesis: There is an association between part conformance and supplier.

Open the CHAPTER_15_1.MTW worksheet and select "Cross Tabulation and Chi-Square" as shown in Figure 15.2. Doing so opens the dialog box shown in Figure 15.3. Enter "Supplier" for "For rows" and "Conforming?" for "For Columns". Then, check the box for "Counts" and click on "Chi-Square". It opens the dialog box shown in Figure 15.4. Check the box for "Chi-Square analysis" and click on "OK". It takes you back to the dialog box shown in Figure 15.3. Click on "Other Stats" and it opens the dialog box shown in

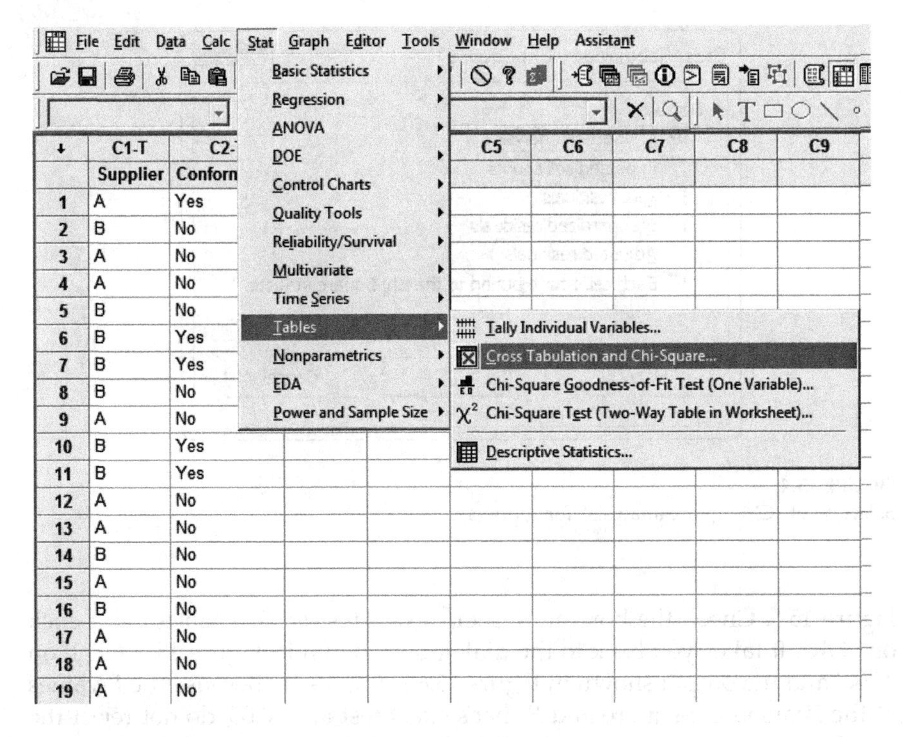

FIGURE 15.2
Selection of "Cross Tabulation and Chi-Square" for suppliers.

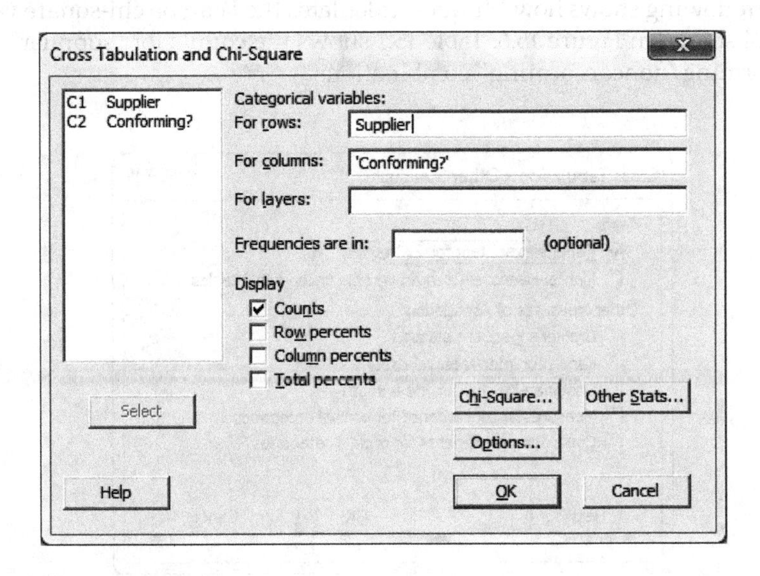

FIGURE 15.3
Selection of rows and columns for chi-square table for suppliers.

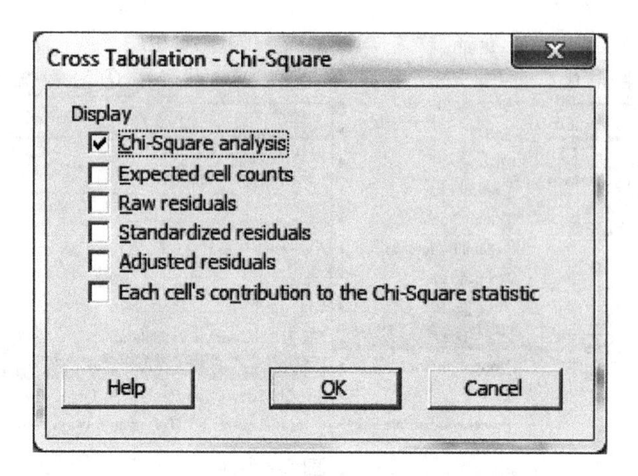

FIGURE 15.4
Selection of "Chi-Square analysis" for suppliers.

Figure 15.5. Check the box for "Fisher's exact test for 2x2 tables" and click on "OK". It takes you back to the dialog box shown in Figure 15.3. Click on "OK" and the output shown in Figure 15.6 is the result. Because the *P*-values of the Pearson chi-square and Fisher's exact test are >0.05, do not reject the null hypothesis. In other words, there is no significant association between part conformance and supplier.

The following shows how Minitab® calculates the Pearson chi-square value of 0.924 shown in Figure 15.6. Table 15.1 shows the counts for "supplier" and "conforming/nonconforming" combinations.

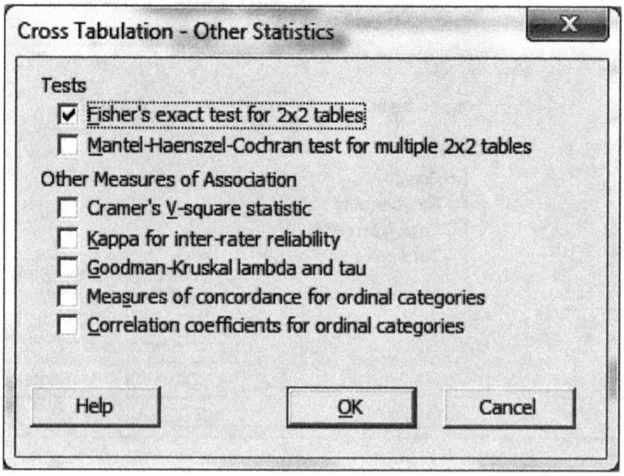

FIGURE 15.5
Selection of "Fisher's exact test for 2 × 2 tables" for suppliers.

```
Tabulated statistics: Supplier, Conforming?

Rows: Supplier   Columns: Conforming?

       No  Yes  All

A      26    4   30
B      39   11   50
All    65   15   80

Cell Contents:      Count

Pearson Chi-Square = 0.924, DF = 1, P-Value = 0.336
Likelihood Ratio Chi-Square = 0.961, DF = 1, P-Value = 0.327

Fisher's exact test: P-Value =  0.390509
```

FIGURE 15.6
Chi-square analysis output for suppliers.

The expected frequencies for chi-square analysis are calculated as follows:

Expected Frequency of Supplier A and Nonconforming = (30/80) * 65 = 24.375.

Expected Frequency of Supplier A and Conforming = (30/80) * (15) = 5.625.

Expected Frequency of Supplier B and Nonconforming = (50/80) * (65) = 40.625.

Expected Frequency of Supplier B and Conforming = (50/80) * (15) = 9.375.

Because each of the expected frequencies is at least 5, the condition to use chi-square distribution is satisfied.

Table 15.2 shows the counts and expected frequencies for "supplier" and "conforming/nonconforming" combinations.

$$\chi^2 = \frac{(26-24.375)^2}{24.375} + \frac{(4-5.625)^2}{5.625} + \frac{(39-40.625)^2}{40.625} + \frac{(11-9.375)^2}{9.375} = 0.92444$$

with 1 degree of freedom

TABLE 15.1
Chi-Square Table (Counts) for Suppliers

	Nonconforming	Conforming	
Supplier A	26	4	Total = 30
Supplier B	39	11	Total = 50
	Total = 65	Total = 15	Grand Total = 80

TABLE 15.2

Chi-Square Table (Counts and Expected Values) for Suppliers

	Nonconforming	Conforming
Supplier A	26 (24.375)	4 (5.625)
Supplier B	39 (40.625)	11 (9.375)

Notice above that, degrees of freedom = (rows – 1) * (columns – 1) = (2 – 1) * (2 – 1) = 1.

In order to view the above on a chi-square probability distribution plot, select "Probability Distribution Plot" as shown in Figure 15.7. Doing so opens the dialog box shown in Figure 15.8. Select "View Probability" as shown and click on "OK". It opens the dialog box shown in Figure 15.9. Select "Chi-Square" from the drop-down menu for "Distribution" and enter "1" for "Degrees of freedom". Click on "OK" and it opens the dialog box shown in

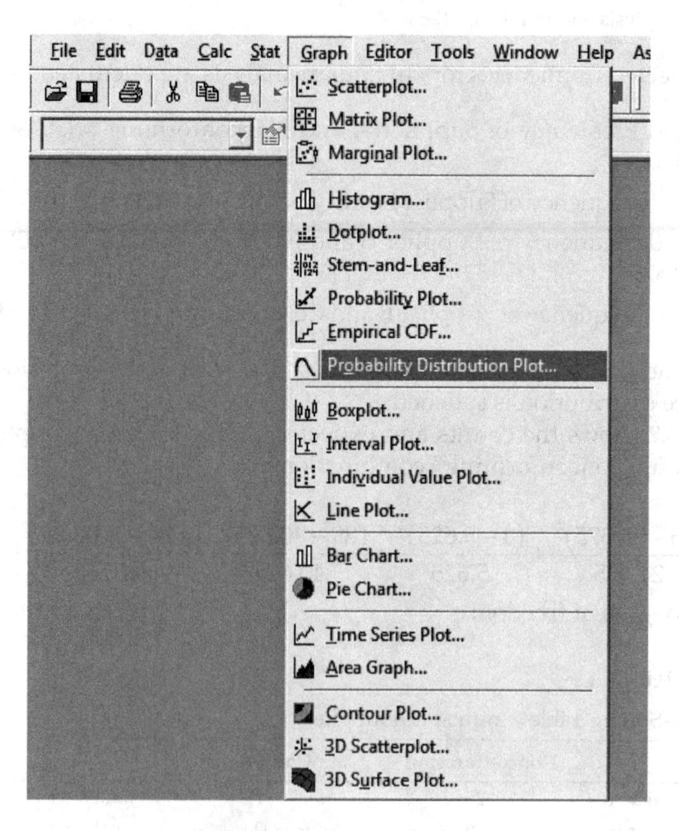

FIGURE 15.7
Selection of "Probability Distribution Plot" for suppliers.

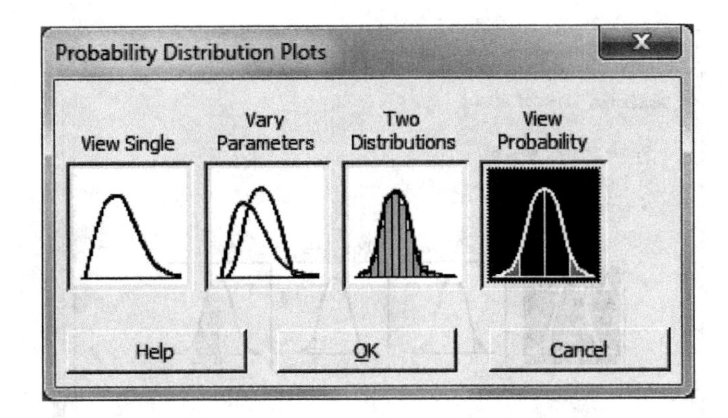

FIGURE 15.8
Selection of "View Probability" for suppliers.

Figure 15.10. Select "X" as shown and enter "0.92444" for "X value". Click on "OK" and the distribution plot shown in Figure 15.11 is the result.

Inasmuch as it is evident from the above analysis that the quality of the critical part does not depend on which supplier it is purchased from, the company now wonders whether the facility that processes the part to manufacture the product has an effect on the product failure. To this end, a sample of 80 products is taken from Facilities P and Q, and inspected as to whether

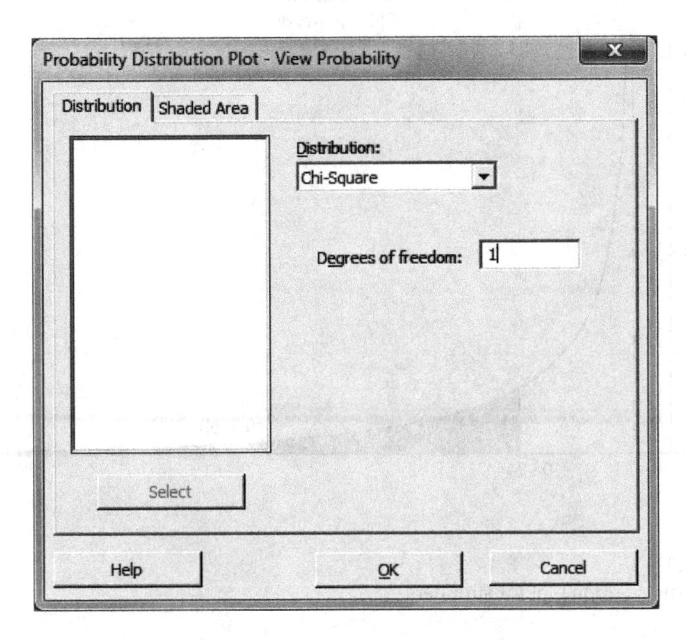

FIGURE 15.9
Entry of degrees of freedom for suppliers.

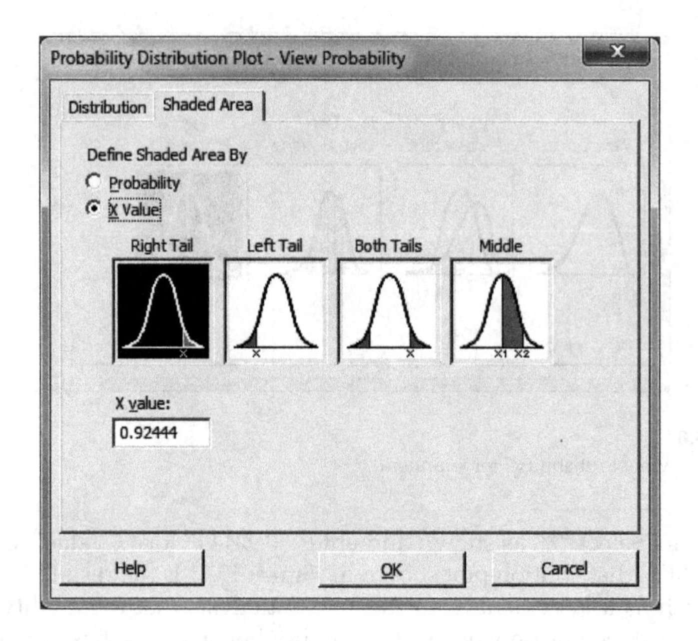

FIGURE 15.10
Entry of chi-square value for suppliers.

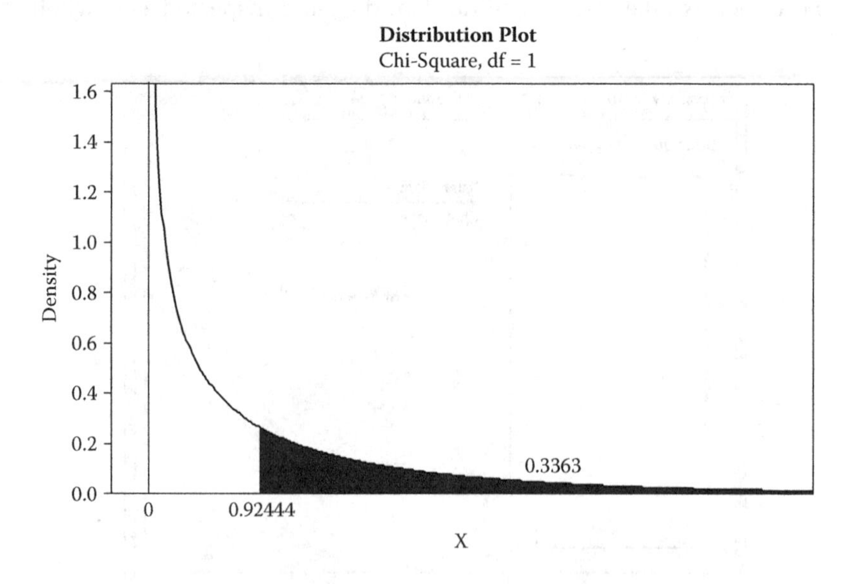

FIGURE 15.11
Chi-square distribution plot for suppliers.

↓	C1-T Facility	C2-T Conforming?
1	P	Yes
2	Q	No
3	Q	No
4	Q	No
5	Q	No
6	Q	Yes
7	P	Yes
8	P	No
9	P	No
10	P	Yes
11	P	Yes
12	Q	No
13	Q	No
14	Q	No
15	P	No
16	Q	No
17	P	No
18	Q	No
19	P	No

FIGURE 15.12
Products from facilities P and Q.

each of them is conforming or nonconforming (defective). These data are stored in the CHAPTER_15_2.MTW worksheet (the worksheet is available at the publisher's website; the data from the worksheet are also provided in the Appendix). Figure 15.12 is a screenshot of the partial worksheet.

In order to check whether a product being defective depends on which facility it is manufactured in, the following set of hypotheses is considered for testing:

Null Hypothesis: There is no association between product conformance and facility.

Alternative Hypothesis: There is an association between product conformance and facility.

Open the CHAPTER_15_2.MTW worksheet and select "Cross Tabulation and Chi-Square" as shown in Figure 15.13. Doing so opens the dialog box shown in Figure 15.14. Enter "Facility" for "For rows" and "Conforming?" for "For Columns". Then, check the box for "Counts" and click on "Chi-Square". It opens the dialog box shown in Figure 15.15. Check the box for "Chi-Square analysis" and click on "OK". It takes you back to the dialog box shown in

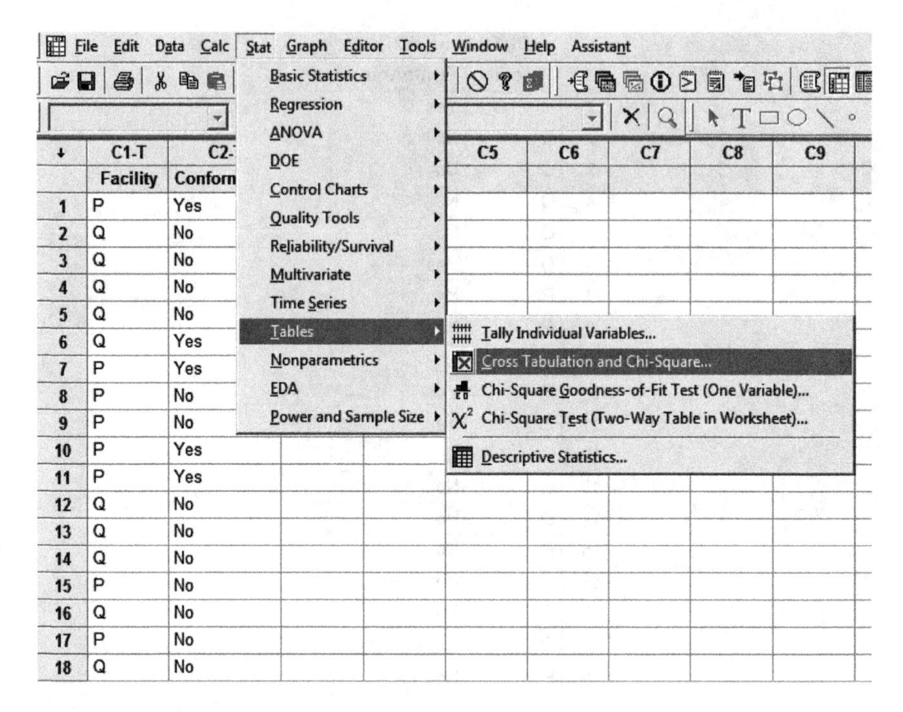

FIGURE 15.13
Selection of "Cross Tabulation and Chi-Square" for facilities.

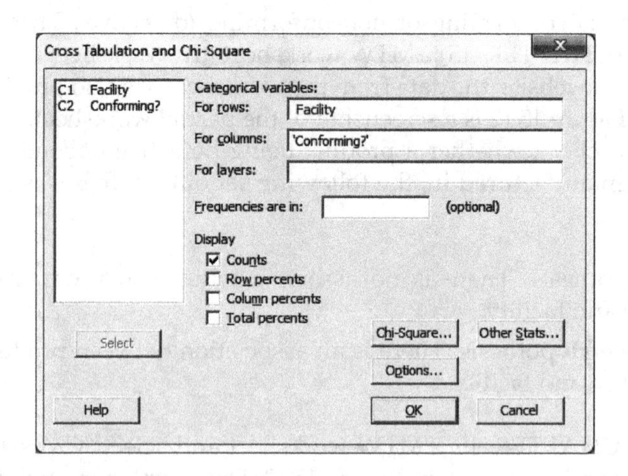

FIGURE 15.14
Selection of rows and columns for chi-square table for facilities.

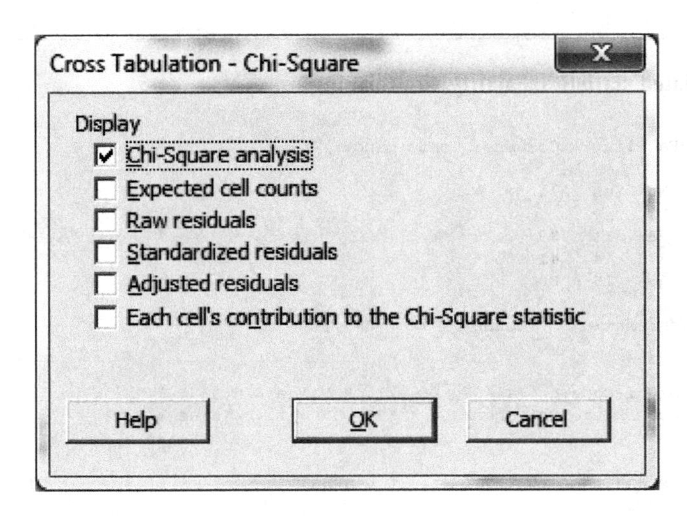

FIGURE 15.15
Selection of "Chi-Square analysis" for facilities.

Figure 15.14. Click on "Other Stats" and it opens the dialog box shown in Figure 15.16. Check the box for "Fisher's exact test for 2x2 tables" and click on "OK". It takes you back to the dialog box shown in Figure 15.14. Click on "OK" and the output shown in Figure 15.17 is the result. Because the *P*-values of the Pearson chi-square and Fisher's exact test are <0.05, reject the null hypothesis. In other words, there is a significant association between product conformance and facility.

The following shows how Minitab® calculates the Pearson chi-square value of 4.466 shown in Figure 15.17. Table 15.3 shows the counts for "facility" and "conforming/nonconforming" combinations.

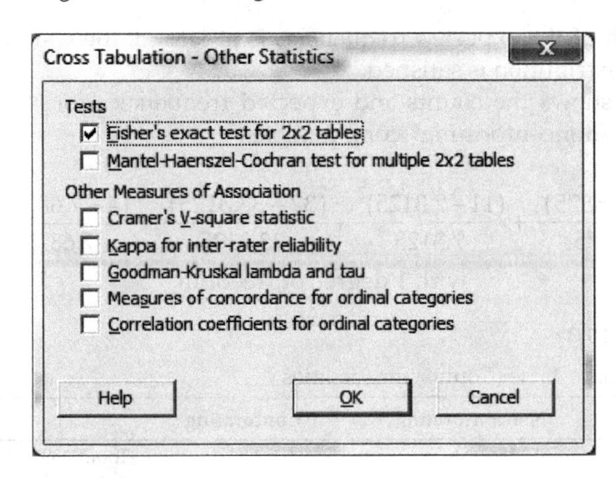

FIGURE 15.16
Selection of "Fisher's exact test for 2 × 2 tables" for facilities.

```
Tabulated statistics: Facility, Conforming?

Rows: Facility    Columns: Conforming?

       No  Yes  All

P      28   11   39
Q      37    4   41
All    65   15   80

Cell Contents:      Count


Pearson Chi-Square = 4.466, DF = 1, P-Value = 0.035
Likelihood Ratio Chi-Square = 4.597, DF = 1, P-Value = 0.032


Fisher's exact test: P-Value =  0.0460237
```

FIGURE 15.17
Chi-square analysis output for facilities.

The expected frequencies for Chi-Square analysis are calculated as follows:

Expected Frequency of Facility P and Nonconforming = (39/80) * (65) = 31.6875.

Expected Frequency of Facility P and Conforming = (39/80) * (15) = 7.3125.

Expected Frequency of Facility Q and Nonconforming = (41/80) * (65) = 33.3125.

Expected Frequency of Facility Q and Conforming = (41/80) * (15) = 7.6875.

Because each of the expected frequencies is at least 5, the condition to use chi-square distribution is satisfied.

Table 15.4 shows the counts and expected frequencies for "facility" and "conforming/nonconforming" combinations.

$$\chi^2 = \frac{(28-31.6875)^2}{31.6875} + \frac{(11-7.3125)^2}{7.3125} + \frac{(37-33.3125)^2}{33.3125} + \frac{(4-7.6875)^2}{7.6875} = 4.4656$$

with 1 degree of freedom

TABLE 15.3

Chi-Square Table (Counts) for Facilities

	Nonconforming	Conforming	
Facility P	28	11	Total = 39
Facility Q	37	4	Total = 41
	Total = 65	Total = 15	Grand Total = 80

TABLE 15.4

Chi-Square Table (Counts and Expected Values) for Facilities

	Nonconforming	Conforming
Facility P	28 (31.6875)	11 (7.3125)
Facility Q	37 (33.3125)	4 (7.6875)

Notice above that, degrees of freedom = (rows – 1) * (columns – 1) = (2 – 1) * (2 – 1) = 1.

In order to view the above on a chi-square probability distribution plot, select "Probability Distribution Plot" as shown in Figure 15.18. Doing so opens the dialog box shown in Figure 15.19. Select "View Probability" as shown and click on "OK". It opens the dialog box shown in Figure 15.20. Select "Chi-Square" from the drop-down menu for "Distribution" and enter "1" for "Degrees of freedom". Click on "OK" and it opens the dialog box shown in Figure 15.21. Select "χ" as shown and enter "4.4656" for "χ value". Click on "OK" and the distribution plot shown in Figure 15.22 is the result.

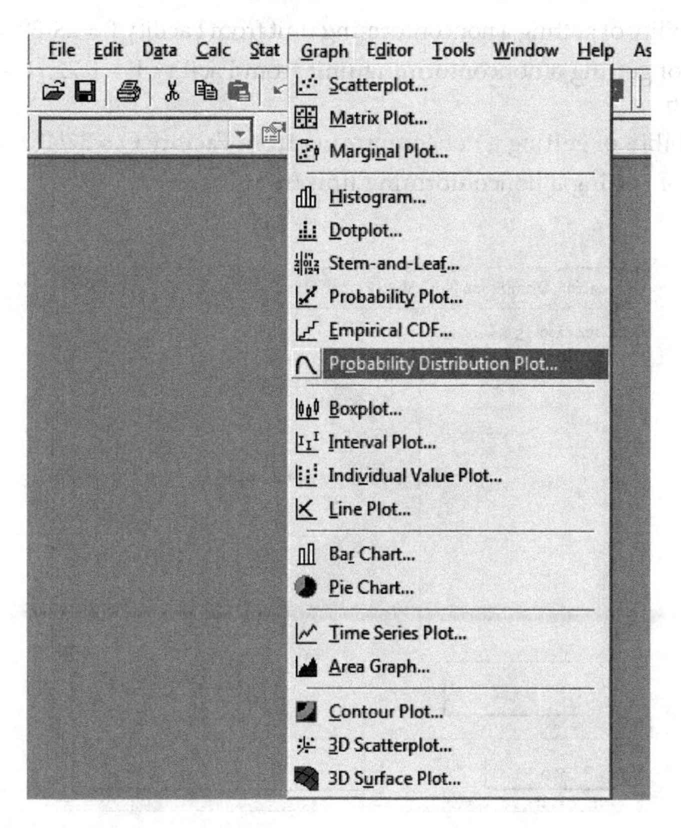

FIGURE 15.18
Selection of "Probability Distribution Plot" for facilities.

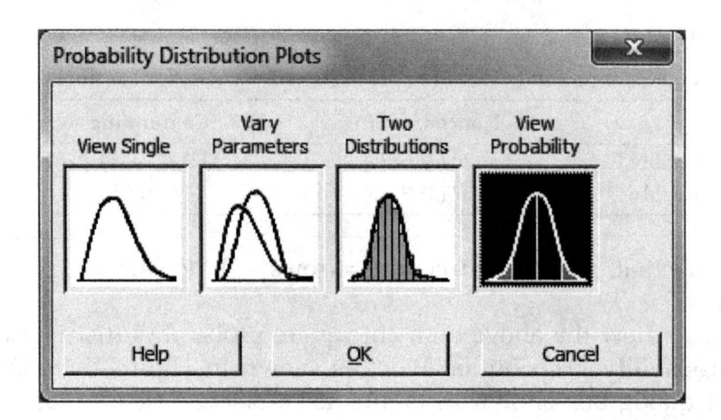

FIGURE 15.19
Selection of "View Probability" for facilities.

Based on Table 15.3, the following are calculated:

Probability of getting a nonconforming unit from Facility P = 28/39 = 0.72.

Odds of getting a nonconforming unit from Facility P = 0.72/(1 – 0.72) = 2.6.

Probability of getting a nonconforming from Facility Q = 37/41 = 0.90.

Odds of getting a nonconforming unit from Facility Q = 0.90/(1 – 0.90) = 9.

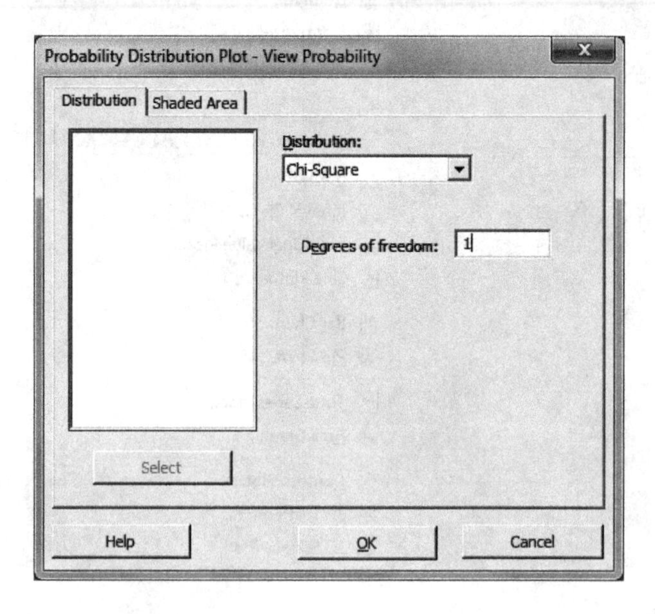

FIGURE 15.20
Entry of degrees of freedom for facilities.

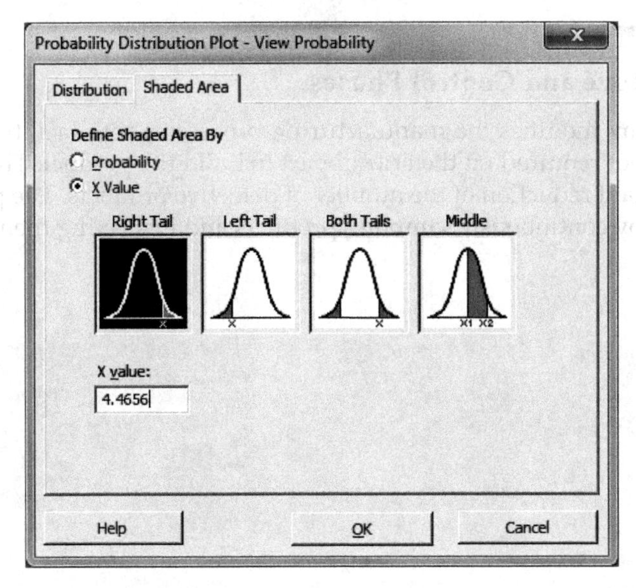

FIGURE 15.21
Entry of chi-square value for facilities.

Because the probability and odds of nonconformance for both facilities are quite high, the company decides to analyze the manufacturing processes at each facility. Upon analysis, it is discovered that the critical part that goes into each product undergoes high-performance drilling that reduces the part's strength and hence fails by the time the product is assembled.

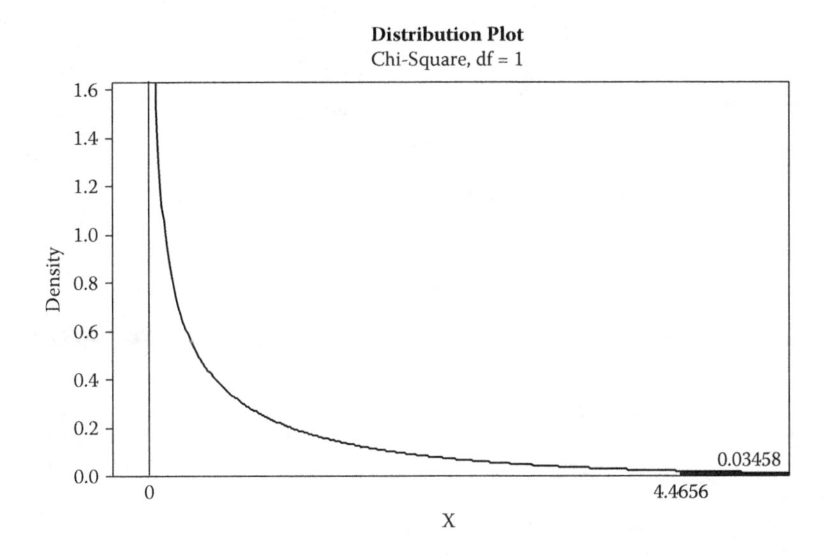

FIGURE 15.22
Chi-square distribution plot for facilities.

15.4 Improve and Control Phases

The company modifies the manufacturing process at both facilities so that drilling is not required on the critical part to build the product. This results in a significant reduction of the number of defective products. The process is controlled by continuously sampling products and inspecting them.

Appendix

Contents of Minitab® Worksheets

CHAPTER_3_BEFORE.MTW

Fat
15.5
12.3
15.4
16.5
15.9
17.1
16.9
14.3
19.1
18.2
18.5
16.3
20
19
15.6
13.5
14
16.5
19
18.6

CHAPTER_3_AFTER.MTW

Fat
14.9
15
15.4
15.3
15.2
15.1
14.9
14.8
15.6
14.5
15.3
15.8
15
15
14.3
15.3
15.2
14.7
15.1
14.7

CHAPTER_4_TEST_2.MTW

Machine Q
8.9844
7.4928
10.3180
8.7651
9.8638
8.6747
8.2423
9.8338
9.0193
9.9748
9.3649
8.8023
8.2759
8.0240
8.0074
9.5139
7.2129
7.4988
10.7370
9.1634

CHAPTER_4_TEST_3.MTW

Shift 1	Shift 2	Shift 3
10.1452	9.0988	9.8427
11.5773	11.1488	10.6511
10.9541	9.1446	10.8478
12.2203	9.4165	11.8373
9.7991	8.3927	9.8428
10.9832	8.6399	10.6842
10.3570	9.6037	10.6122
10.5090	10.0562	8.7841
8.1670	10.3295	9.4795
10.1447	10.2919	8.8236
9.1694	9.7747	9.7696
9.5420	10.5837	11.9735
10.1836	11.1162	10.8750
8.8772	9.2894	10.3372
10.3946	10.4884	10.1846
10.8078	10.6587	10.4060
9.3260	11.5430	9.4930
10.3787	9.6336	10.6049
9.9756	9.9233	9.9231
10.9430	10.1370	9.4900

CHAPTER_5_1.MTW

Importance of Candy Color	City	Observed
7	Boston	197
7	Cleveland	274
7	New York City	642
7	San Francisco	210
7	Chicago	197
6	Boston	257
6	Cleveland	405
6	New York City	304
6	San Francisco	252
6	Chicago	203
5	Boston	315
5	Cleveland	364
5	New York City	196
5	San Francisco	348
5	Chicago	250
4	Boston	480
4	Cleveland	326
4	New York City	263
4	San Francisco	486
4	Chicago	478
3	Boston	98
3	Cleveland	82
3	New York City	41
3	San Francisco	125
3	Chicago	100
2	Boston	63
2	Cleveland	46
2	New York City	36
2	San Francisco	70
2	Chicago	58
1	Boston	92
1	Cleveland	38
1	New York City	53
1	San Francisco	62
1	Chicago	29

CHAPTER_5_2.MTW

Color	Observed	Percent Expected
Yellow	29	0.14
Red	23	0.13
Orange	12	0.2
Blue	14	0.24
Green	8	0.16
Purple	20	0.13

CHAPTER_6_1.MTW[a]

52.9	60.7	53.8	57.2
55	53.2	55.7	55.6
55.5	56.2	51.7	56
54.1	54.2	52.1	52
55.9	54.9	56.8	57.8
54.3	57.7	54.4	54.3
55.7	52.6	55	54.4
55.9	54.6	56.2	54.1
58.1	55.7	54	52.6
55.1	54.8	53.7	51.8
49.3	54.6	55.8	52.1
53.4	56	54.8	53.2
52.7	53	53.2	52.4
51.1	56.4	57	54.4
56.5	51.4	56.7	52.3
55.9	52.7	56	55
51.9	54.5	53.3	54.4
56.2	51.3	53.8	54.5
55.1	55.7	56.7	57.1
53	52.5	55.7	52.7
54.5	55.7	54.1	53.6
58.8	55.9	53.4	55.4
54.4	52.9	54.4	56.9
56.1	53.9	53.7	53.1
57.3	59.1	57.4	53.4

[a] Data from the "Before" column are shown in four columns here in order to fit on the page.

CHAPTER_6_2.MTW

1	2	3	4	5
56	55	54	55.9	54.5
56	51	53.4	51.9	58.8
55.5	55.9	52.7	56.2	54.4
54.1	58.1	51.1	55.1	56.1
55.9	55.1	56.5	53	57.3
53.2	55.7	56	57.7	57.7
56.2	55.9	53.3	52.6	52.6
54.2	52.9	53.8	54.6	54.6
54.9	53.9	56.7	55.7	55.7
56	59.1	55.7	54.8	54.8
56	52.4	53.1	55	58
53	54.4	53.4	54.4	57

CHAPTER_7.MTW

Variety of Drinks	Number of Waiters	Free Refill	Satisfied	Sample Size
6	20	Yes	4	10
3	10	No	1	10
6	20	No	5	10
3	20	No	2	10
5	15	No	3	10
5	15	Yes	2	10
3	20	Yes	2	10
3	10	Yes	0	10
6	10	No	1	10
6	10	Yes	2	8

CHAPTER_8_1.MTW

Employee ID	Q1	Q2	Q3	Q4	Q5	Q6	Q7
21341	5	1	4	5	5	1	5
21392	3	2	3	5	4	3	4
21342	4	1	1	5	5	5	5
21391	4	2	4	4	2	4	2
21343	4	2	2	4	5	3	3
21390	5	2	3	4	4	2	4
21344	5	2	3	4	4	2	4
21389	5	1	3	4	4	1	5
21345	5	1	1	5	5	1	5
21388	5	1	3	4	4	2	3
21346	5	1	3	4	4	2	4
21387	4	2	4	4	4	2	3
21347	4	2	2	4	4	1	5
21386	5	1	2	4	4	3	4
21348	5	1	2	5	5	1	5
21385	5	2	4	5	4	3	3
21349	5	1	1	5	5	1	5
21384	4	1	3	5	5	1	5
21350	2	2	3	2	4	1	3
21383	4	2	4	4	5	4	5
21351	4	2	5	4	4	4	4
21382	4	2	3	5	5	1	4
21352	4	4	4	5	5	4	2
21381	4	2	2	5	5	1	5
21353	4	2	3	5	5	1	4
21380	4	2	4	3	4	2	4
21354	4	2	3	5	5	4	4
21379	4	2	5	4	4	1	2
21355	5	1	4	5	4	3	5
21378	4	1	3	5	5	2	4
21356	4	4	3	4	4	2	3
21377	4	2	4	5	5	1	5
21357	3	4	5	4	4	1	5
21376	5	1	3	5	5	2	5
21358	5	2	4	5	5	1	4
21375	4	2	5	2	4	1	2
21359	4	2	3	5	5	1	4
21374	5	3	3	5	5	1	5
21360	4	3	4	4	4	2	2
21373	4	2	2	4	4	2	4
21361	5	2	2	4	4	1	5
21372	4	4	4	4	4	1	4
21362	4	1	2	5	5	1	4
21372	4	2	3	4	4	2	3
21363	4	2	2	4	4	2	4
21371	4	1	3	5	5	2	4
21364	4	2	2	5	4	1	5

CHAPTER_8_1.MTW (Continued)

Employee ID	Q1	Q2	Q3	Q4	Q5	Q6	Q7
21370	4	2	5	4	4	1	5
21365	3	2	4	3	5	1	3
21366	4	2	4	5	5	2	4
31245	4	4	3	4	4	2	4
31246	4	4	4	3	4	2	4
31247	4	2	4	4	4	1	4
31248	4	2	2	4	5	2	4
31249	5	1	3	5	5	2	4
31250	4	2	2	4	4	2	4
31251	4	3	4	4	4	4	2
31252	4	4	5	4	4	2	4
31253	4	2	4	4	4	1	4
31254	4	2	2	4	4	2	3
31255	4	3	4	4	4	2	3
31256	5	1	4	5	5	1	3
31257	5	2	2	5	5	2	4

CHAPTER_8_2.MTW

Empowerment	Ease of Commute	Challenge
4.50000	5	1
3.83333	5	1
4.83333	4	3
3.00000	1	2
4.00000	1	4
4.00000	3	3
4.00000	4	3
4.33333	3	5
5.00000	3	3
4.00000	2	1
4.16667	4	5
3.50000	4	4
4.16667	1	3
4.33333	5	4
4.83333	4	1
3.83333	3	3
5.00000	2	4
4.50000	1	4
3.00000	5	1
4.00000	4	1
3.50000	2	1
4.16667	5	3
3.33333	5	2

CHAPTER_8_2.MTW (Continued)

Empowerment	Ease of Commute	Challenge
4.50000	3	5
4.16667	2	4
3.50000	3	3
4.16667	4	2
3.16667	4	1
4.33333	5	3
4.33333	1	4
3.33333	2	1
4.16667	1	3
3.16667	3	3
4.66667	4	2
4.16667	3	5
2.83333	3	5
4.16667	2	2
4.33333	1	3
3.16667	4	4
4.00000	5	1
4.33333	5	5
3.33333	5	4
4.50000	5	1
3.66667	2	3
4.00000	2	1
4.33333	2	5
4.33333	3	4
3.66667	4	3
3.33333	4	5
4.00000	5	1
3.50000	1	5
3.16667	2	5
3.66667	3	5
4.16667	4	5
4.50000	4	2
4.00000	3	2
3.16667	2	2
3.16667	1	3
3.66667	5	4
3.83333	4	4
3.33333	3	5
4.16667	4	5
4.50000	5	5

CHAPTER_9_NEO.MTW

StdOrder	RunOrder	PtType	Blocks	P	Q	R	Sulfation
1	1	1	1	100.000	0.000	0.000	3
2	2	1	1	0.000	100.000	0.000	2
3	3	1	1	0.000	0.000	100.000	3
4	4	2	1	50.000	50.000	0.000	9
5	5	2	1	50.000	0.000	50.000	6
6	6	2	1	0.000	50.000	50.000	1
7	7	0	1	33.333	33.333	33.333	4
8	8	1	1	100.000	0.000	0.000	35
9	9	1	1	0.000	100.000	0.000	11
10	10	1	1	0.000	0.000	100.000	10
11	11	2	1	50.000	50.000	0.000	10
12	12	2	1	50.000	0.000	50.000	12
13	13	2	1	0.000	50.000	50.000	12
14	14	0	1	33.333	33.333	33.333	16
15	15	1	1	100.000	0.000	0.000	17
16	16	1	1	0.000	100.000	0.000	12
17	17	1	1	0.000	0.000	100.000	1
18	18	2	1	50.000	50.000	0.000	13
19	19	2	1	50.000	0.000	50.000	11
20	20	2	1	0.000	50.000	50.000	2
21	21	0	1	33.333	33.333	33.333	3

CHAPTER_9_ZEO.MTW

StdOrder	RunOrder	PtType	Blocks	X	Y	Z	Temperature
1	1	1	1	60.0000	20.0000	20.0000	398
2	2	2	1	40.0000	40.0000	20.0000	484
3	3	2	1	40.0000	20.0000	40.0000	424
4	4	1	1	20.0000	60.0000	20.0000	400
5	5	2	1	20.0000	40.0000	40.0000	409
6	6	1	1	20.0000	20.0000	60.0000	342
7	7	0	1	33.3333	33.3333	33.3333	456
8	8	-1	1	46.6667	26.6667	26.6667	460
9	9	-1	1	26.6667	46.6667	26.6667	450
10	10	-1	1	26.6667	26.6667	46.6667	411

CHAPTER_10_1.MTW

Location	Service	Waiting Time
1	1	35
1	1	24
1	1	42
1	1	21
1	2	12
1	2	54
1	2	10
1	2	9
1	3	25
1	3	32
1	3	34
1	3	45
2	1	42
2	1	35
2	1	21
2	1	34
2	2	19
2	2	18
2	2	17
2	2	23
2	3	10
2	3	9
2	3	4
2	3	11

CHAPTER_10_2.MTW

Arrival	Reg Desk	Procedure	Waiting Time
M	1	Acne	22
M	1	Acne	19
M	1	Scars	34
M	1	Scars	23
M	1	Laser	34
M	1	Laser	45
M	2	Acne	45
M	2	Acne	59
M	2	Scars	32
M	2	Scars	45
M	2	Laser	33
M	2	Laser	63
A	1	Acne	10
A	1	Acne	13
A	1	Scars	11
A	1	Scars	12
A	1	Laser	34
A	1	Laser	23
A	2	Acne	10
A	2	Acne	9
A	2	Scars	15
A	2	Scars	9
A	2	Laser	11
A	2	Laser	10

CHAPTER_11_1.MTW

Recycled Trash	Number of Tons (in Hundreds)
Glass	200
Plastic	50
Paper	600
Aluminum	260
Yard waste	200
Iron	210

CHAPTER_11_2.MTW

Office	Kitchen	Other	Milk Cartons
Office paper	Milk cartons	Newspapers	Paint containers
Magazines	Paper towels	Telephone books	Floor protectors
		Junk mail	Bird feeders

Office paper	Magazines	Junk Mail
Print on both sides	Scrapbook	Envelopes
Use email instead	Room borders	Wrapping paper
Fit more text per page	Use pictures to teach kids	
Use one-sided printouts for drafts		

Telephone books	Newspapers	Paper Towels
Use pages as "kitchen roll"	Cover work tables	Buy reusable paper towels
Opt-out of receiving them	Wrapping paper	Use for bathroom cleaning
	Clean car windows	

CHAPTER_12_VARIABLE.MTW

Part	Operator	Diameter
8	1	9.013
2	1	9.012
1	1	9.014
4	1	9.013
3	1	9.012
9	1	9.012
5	1	9.013
6	1	9.010
7	1	9.013
9	2	9.012
4	2	9.013
7	2	9.013
1	2	9.014
6	2	9.010
5	2	9.012
3	2	9.012
2	2	9.011
8	2	9.013
6	1	9.010
1	1	9.014
2	1	9.011
5	1	9.012
7	1	9.013

CHAPTER_12_VARIABLE.MTW
(Continued)

Part	Operator	Diameter
8	1	9.013
4	1	9.013
3	1	9.012
9	1	9.012
4	2	9.013
6	2	9.010
7	2	9.013
9	2	9.012
2	2	9.012
1	2	9.014
3	2	9.012
8	2	9.013
5	2	9.013

CHAPTER_12_ATTRIBUTE.MTW

Response	Part	Appraiser	Standard
Fail	6	John	Fail
Pass	7	John	Pass
Pass	10	John	Pass
Pass	8	John	Pass
Pass	2	John	Fail
Pass	4	John	Pass
Fail	1	John	Fail
Fail	9	John	Fail
Pass	3	John	Pass
Pass	5	John	Pass
Fail	9	Mary	Fail
Pass	5	Mary	Pass
Pass	8	Mary	Pass
Pass	10	Mary	Pass
Pass	7	Mary	Pass
Pass	3	Mary	Pass
Fail	2	Mary	Fail
Pass	4	Mary	Pass
Pass	1	Mary	Fail
Fail	6	Mary	Fail
Pass	3	Buddy	Pass
Pass	4	Buddy	Pass
Fail	2	Buddy	Fail

CHAPTER_12_ATTRIBUTE.MTW (Continued)

Response	Part	Appraiser	Standard
Fail	6	Buddy	Fail
Pass	5	Buddy	Pass
Fail	9	Buddy	Fail
Pass	8	Buddy	Pass
Pass	10	Buddy	Pass
Pass	7	Buddy	Pass
Fail	1	Buddy	Fail
Pass	3	John	Pass
Fail	6	John	Fail
Pass	5	John	Pass
Pass	8	John	Pass
Pass	4	John	Pass
Pass	7	John	Pass
Fail	1	John	Fail
Pass	10	John	Pass
Fail	9	John	Fail
Pass	2	John	Fail
Fail	6	Mary	Fail
Pass	3	Mary	Pass
Pass	4	Mary	Pass
Pass	10	Mary	Pass
Pass	5	Mary	Pass
Pass	7	Mary	Pass
Fail	2	Mary	Fail
Fail	1	Mary	Fail
Fail	9	Mary	Fail
Pass	8	Mary	Pass
Fail	1	Buddy	Fail
Fail	9	Buddy	Fail
Pass	10	Buddy	Pass
Fail	6	Buddy	Fail
Pass	4	Buddy	Pass
Pass	5	Buddy	Pass
Pass	8	Buddy	Pass
Pass	7	Buddy	Pass
Pass	3	Buddy	Pass
Fail	2	Buddy	Fail

CHAPTER_15_1.MTW

Supplier	Conforming?
A	Yes
B	No
A	No
A	No
B	No
B	Yes
B	Yes
B	No
A	No
B	Yes
B	Yes
A	No
A	No
B	No
A	No
B	No
A	No
A	No
A	No
A	No
A	No
B	Yes
B	Yes
B	No
B	No
A	Yes
B	No
A	No
B	No
B	No
B	No
B	No
B	No
B	Yes
B	No
B	No
A	No
B	No
B	No
B	No
B	No

CHAPTER_15_1.MTW (Continued)

Supplier	Conforming?
B	Yes
B	No
B	Yes
A	No
A	No
A	No
A	Yes
B	No
A	No
A	No
B	No
B	No
B	Yes
A	Yes
A	No
B	No
B	No
B	No
A	No
B	No
A	No
B	No
A	No
A	No
A	No
B	No
B	No
B	No
B	No
B	No
B	No
B	No
A	No
B	Yes
A	No
B	No
B	No
B	No
B	No

CHAPTER_15_2.MTW

Facility	Conforming?
P	Yes
Q	No
Q	No
Q	No
Q	No
Q	Yes
P	Yes
P	No
P	No
P	Yes
P	Yes
Q	No
Q	No
Q	No
P	No
Q	No
P	No
Q	No
P	No
P	No
P	No
P	Yes
P	Yes
Q	No
Q	No
Q	Yes
Q	No
Q	No
Q	No
Q	No
P	No
P	No
P	No
P	Yes
P	No
Q	No
Q	No
Q	No
Q	No
Q	No
P	No
P	Yes

CHAPTER_15_2.MTW (Continued)

Facility	Conforming?
P	No
P	Yes
Q	No
Q	No
Q	No
Q	Yes
P	No
P	No
P	No
P	No
Q	No
P	Yes
Q	Yes
P	No
P	No
P	No
Q	No
Q	No
Q	No
Q	No
Q	No
Q	No
Q	No
Q	No
Q	No
Q	No
P	No
P	No
P	No
P	No
P	No
P	No
P	Yes
P	No
P	No
Q	No
Q	No
Q	No

Index

Note: Page numbers ending in "e" refer to equations. Page numbers ending in "f" refer to figures. Page numbers ending in "t" refer to tables.

N

P

Q

9781482205572